ADVANCED SAFETY MANAGEMENT FOCUSING ON Z10 AND SERIOUS INJURY PREVENTION

ADVANCED SAFETY MANAGEMENT FOCUSING ON Z10 AND SERIOUS INJURY PREVENTION

Second Edition

FRED A. MANUELE, CSP, PE
President
Hazards Limited

Published by John Wiley & Sons, Inc., Hoboken, New Jersey
Published simultaneously in Canada

For general information on our other products and services or for technical support, please contact
our Customer Care Department within the United States at 877-762-2974, outside the United States
at 317-572-3993 or fax 317-572-4002.

Wiley also publishes its books in a variety of electronic formats. Some content that appears in print
may not be available in electronic formats. For more information about Wiley products, visit our web
site at www.wiley.com.

Library of Congress Cataloging-in-Publication Data:

Manuele, Fred A., author.
 Advanced safety management focusing on Z10 and serious injury prevention /
Fred A. Manuele. – Second edition.
  p. ; cm.
 Includes bibliographical references and index.
 ISBN 978-1-118-64568-0 (cloth)
I. American National Standards Institute. II. American Industrial Hygiene Association. III. Title.
[DNLM: 1. Safety Management–standards–United States. 2. Occupational Health–United States.
3. Risk Management–standards–United States. 4. Wounds and Injuries–prevention &
control–United States. WA 485]
 RA645.T73
 363.11–dc23

                                                                                    2013038105

Printed in the United States of America

10 9 8 7 6

To Irene

CONTENTS

FOREWORD

My association with Fred Manuele began over 20 years ago. During that time we have frequently communicated, shared the podium at major conferences, critiqued one another's writing, and collaborated on and shared research findings. Fred is one of the most highly regarded experts in the field of operational risk management. He has received many of the most prestigious awards in his field, including recognition as a Fellow by the American Society of Safety Engineers, being given the Lifetime Achievement Award from the Board of Certified Safety Professionals, and others.

He is one of the most recognized writers in the field of occupational safety. His articles have received several professional paper awards. The first edition of *Advanced Safety Management* is used as a textbook in graduate and undergraduate programs in several universities.

As a technical resource to the ANSI/AIHA Z10 committee during the drafting of the 2005 edition and the 2012 revision, he attended meetings, participated in discussions, and submitted substantial text related to requirements and appendices. Fred wrote first drafts of some of the new appendices for the 2012 revision. ANS/AIHA Z10, in its present form, establishes the overarching framework necessary for management to mobilize the organization to proactively address system, process, and operational performance improvement.

As chair of ANSI/AIHA Z10 committee, I frequently conduct seminars, webinars, and workshops and deliver keynote speeches on management system improvement to members of management, union representatives, and safety and health professionals. Those who have or are currently implementing a Z10 management system very often cite the first edition of *Advanced Safety Management* as their most important and helpful resource.

This second edition will prove to be even more valuable because it has been updated based on the 2012 revision of Z10. By their very nature, ANSI standards are not always easy to understand and interpret. To improve usability, the 2012 revision of Z10 clarified and enhanced critical requirements, added explanatory language, significantly improved several appendices, and added five new appendices. The appendices provide implementation examples and offer advice as to how conformance to a requirement can be achieved.

This edition goes a step further in providing extensive background and context for Z10, detailing its relationship to other management system standards. Separate chapters on key standard sections and subsections provide detailed explanations of requirements and implementation recommendations. There are also chapters on key related subjects such as serious injury and fatality prevention, human error prevention, macro thinking, and the significance of an organization's culture.

Audience response data that I have collected during presentations and training sessions from participants using anonymous audience response devices reveals that:

- Most organizations' safety activities are largely "program based" rather than "system based."
- A high percentage recognize the need to move to a system-based approach.
- Participant knowledge of management systems is quite low—safety and health professionals score lower than higher-level managers.
- Non-systems thinking beliefs and assumptions are widespread, such as the assumption that reducing minor incidents will also prevent serious and fatal incidents.
- A high percentage of participants have great difficulty developing effective alternative safety process measures, metrics, and feedback loops.

These data dramatically underscores the timeliness and importance of this new edition of *Advanced Safety Management*. There is a growing understanding that real sustained safety performance improvement can only come from refining the relative management systems and from macro (systems) thinking.

Even though OSHA recordable rates have been reduced significantly in many companies, a large number of these organizations have not experienced a corresponding reduction in serious and fatal incidents. For decades, safety and health professionals have been misled into relying on the Heinrich premise—that reduction in minor incidents will result in equivalent reduction in serious and fatal incidents.

The chapter "Innovations in Serious Injury and Fatality Prevention" demonstrates how specific management system improvements will lead to a reduction in serious injuries. Organizations implementing Z10 management systems tend to be very successful in this crucial area because of enhanced management leadership, improved employee engagement, and a supportive safety culture.

Interest in occupational safety and health management systems is expanding, as evidenced by the beginning of activity to develop a global health and safety management system standard, a project known as ISO PC283. ANS/AIHA Z10 will be used as a primary resource by members of the U.S. Technical Advisory Group in the international standard writing process.

Fred Manuele's comprehensive and timely new edition will accelerate the growth and implementation of safety and health management systems worldwide.

JAMES HOWE, CSP
Chair, ANSI/AIHA Z10 Standard Committee

PREFACE TO THE SECOND EDITION

This book focuses on Z10, which is the national standard for *Occupational Health and Safety Management Systems*, and on serious injury and fatality prevention. The impetus for an updated version of the first edition derived from two developments.

- A revised version of Z10 was approved by the American National Standards Institute in June 2012. Any comments and guidance for the revisions and extensions would be beneficial.
- Although the rates for serious occupational injuries and fatalities have continued to drop significantly in past decades, it is now recognized that the rates have plateaued in recent years. I propose that major and somewhat shocking innovations in the content and focus of occupational risk management systems will be necessary to achieve additional progress.

The principal purpose of this book continues to be providing guidance to managements, safety professionals, educators, and students on having operational risk management systems that meet the requirements of Z10 and to be informative on reducing the occurrence of accidents that result in serious injuries and fatalities.

This book is used in several university-level safety degree programs. Input was sought from professors and experienced safety professionals who use the book as a reference on subjects that should be expanded or addressed.

- Additional emphasis is given to the most important section in Z10, Management Leadership and Employee Involvement, with particular reference to contributions that employees can make.
- A new provision, on risk assessment was added to Z10. Its importance is stressed. I propose that risk assessment be established as the core of an operational risk

management system as a separately identified element following closely after the first element. Comments on hazard identification and analysis and risk assessment techniques are made in several chapters.

- A significant departure from typical safety management systems is presented in the chapter "Innovations in Serious Injury and Fatality Prevention" in the form of a Socio-Technical Model for an Operational Risk Management System. This model stresses the significance of the organizational culture established at the board of director and senior management levels with respect to attaining and maintaining acceptable risk levels; the provision of adequate resources; risk assessment, prioritization, and management; prevention through design; maintenance for system integrity; and management of change. It is made clear that there is a notable connection between improving management systems to meet the provisions of Z10 and avoiding serious injuries and fatalities.
- Avoidance of human error is given expanded attention.
- New chapters have been researched and written. Their titles are:
  - Macro Thinking: The Socio-Technical Model
  - Safety Professionals as Culture Change Agents
  - Prevention through Design
  - A Primer on System Safety
- Chapters on "Management of Change" and "The Procurement Process" have been revised and expanded considerably.

Soon after approval for the revision of Z10 was given by the American National Standards Institute, the secretariat was transferred to the American Society of Safety Engineers (ASSE). Additional promotion has been given to Z10 by ASSE as the state-of-the-art occupational safety management system. Interest shown in the standard by safety professionals is impressive. Z10 has achieved recognition as a sound base from which to develop innovations in existing safety management systems.

For a huge percentage of organizations, adopting the provisions in Z10 will achieve major improvements in their occupational health and safety management systems and serve to reduce the potential for accidents that might result in a serious injury or fatality.

FRED A. MANUELE
President, Hazards Limited

PREFACE TO THE FIRST EDITION

The principal purpose of this book is to provide guidance to managements, safety professionals, educators, and students concerning two major, interrelated developments affecting the occupational safety and health discipline:

- Issuance, for the first time in the United States, of a national consensus standard for occupational safety and health management systems
- Emerging awareness that traditional systems to manage safety do not adequately address serious injury prevention

On July 25, 2005, the American National Standards Institute approved a new standard, entitled *Occupational Health and Safety Management Systems*, designated as ANSI/AIHA Z10-2005. This standard is a state-of-the-art, best practices guide. Over time, Z10 will revolutionize the practice of safety.

The chapter "An Overview of ANSI/AIHA Z10-2005" comments on all of the provisions in the standard. The chapter "Serious Injury Prevention" gives substance to the position that adopting a different mindset is necessary to reduce serious injury potential. Other chapters give implementation guidance with respect to the standard's principal provisions and to serious injury prevention.

Recognition of the significance of Z10 has been demonstrated. Its provisions are frequently cited as representing highly effective safety and health management practices. The sales record for Z10 is impressive. Safety professionals are quietly making gap analyses, comparing existing safety and health management systems to the provisions of Z10.

Even though the standard sets forth minimum requirements, very few organizations have safety and health management systems in place that meet all of the provisions of the standard. The provisions for which shortcomings will often exist, and for which emphasis is given in this book, pertain to:

- Risk assessment and prioritization
- Applying a prescribed hierarchy of controls to achieve acceptable risk levels
- Safety design reviews
- Including safety requirements in procurement and contracting papers
- Management of change systems

As ANSI standards are applied, they acquire a quasi-official status as minimum requirements for the subjects to which they pertain. As Z10 attains that stature, it will become the benchmark, the minimum, against which the adequacy of safety and health management systems will be measured.

The chapter "Serious Injury Prevention" clearly demonstrates that while the occupational injury and illness incident frequency rate is down considerably, incidents resulting in serious injuries are not down proportionally. The case is made that typical safety and health management systems do not adequately address serious injury prevention. Thus, major conceptual changes are necessary in the practice of safety to reduce serious injury potential. That premise permeates every chapter in this book.

Safety and health professionals are advised to examine and reorient the principles on which their practices are based to achieve the significant changes necessary in the advice they give. Guidance to achieve those changes is provided.

Why use the word "advanced" in the title of this book? If managements adopt the provisions in Z10 and give proper emphasis to the prevention of serious injuries, they will have occupational health and safety management systems as they should be rather than as they are. There is a strong relationship between improving management systems to meet the provisions of Z10, a state-of-the-art standard, and minimizing serious injuries.

ACKNOWLEDGMENTS

It is not easy to write a brief "thank you" paragraph that recognizes all the people who have contributed to my education, and thus to the content of this book. There are so many. I have had the good fortune throughout several decades of having highly competent associates who gave of their time to debate with me and to offer criticisms of what I had written. Their input has been exceptionally valuable. For all who have influenced my thinking and my work, I sincerely express my thanks and gratitude.

INTRODUCTION

This book provides guidance on applying the provisions of ANSI/AIHA Z10-2012, the *Occupational Health and Safety Management Systems* standard, and on serious injury and fatality prevention as interrelated subjects. An abstract is provided here for each chapter. A Professor who uses my books in his classes suggested that each chapter be a stand-alone essay. Partial success with respect to that suggestion has been achieved. Although doing so requires a little repetition, the reader benefits by not having to refer to other chapters while perusing the subject at hand.

1. Overview of ANSI/AIHA Z10-2012

In this chapter I comment on all of the sections in Z10. Encouragement is given to safety and health professionals to acquire a copy of the standard and to move forward in applying its provisions. Some of the subjects that are emphasized are the most important section in the standard, which is Management Leadership and Employee Participation; significance of this state-of-the-art consensus standard, societal implications, and management review provisions. Also, specific provisions in the standard that are not included in typical safety management systems are highlighted, such as the safety through design processes, risk assessments and prioritization, management of change, a prescribed hierarchy of controls, and including safety specifications in purchasing and procurement forms and documents. The case is made that bringing safety and health management systems up to the Z10 level will reduce the probability of incidents occurring that result in serious injury and illness.

Advanced Safety Management: Focusing on Z10 and Serious Injury Prevention, Second Edition. Fred A. Manuele.

2. Achieving Acceptable Risk Levels: The Operational Goal

Reference is made in several parts of Z10 to the need to achieve and maintain acceptable risk levels. Other recently modified standards do the same. In this chapter I provide a worldwide history of the adoption of the concept of acceptable risk, establish a base for determining risk levels, discuss risk assessment matrices, and outline the logic for arriving at the definition of acceptable risk given.

3. Innovations in Serious Injury and Fatality Prevention

My recent research requires the conclusion that major and somewhat revolutionary innovations in safety management systems will be necessary to further reduce serious injuries and fatalities. Statistics show that such injury rates have plateaued. In this chapter I propose several innovations in A Socio-Technical Model for an Operational Risk Management System. The case is made that if organizations did a better job with respect to certain aspects of Z10, such as Management Leadership and Employee Participation, Risk Assessment and Prioritization, Design Reviews, Management of Change, application of the Hierarchy of Controls, and in the Procurement processes, the potential for serious injuries and fatalities would be reduced.

4. Human Error Avoidance and Reduction

In the chapter on Innovations in Serious Injury and Fatality Prevention, it is stated that reducing human errors as causal factors is necessary as attempts are made to reduce serious injuries. We focus on human errors occurring above the worker level that derive from deficiencies in an organization's safety culture, safety management systems, design and engineering decision making, and the organization of work.

5. Macro Thinking: The Socio-Technical Model

Impetus for this chapter arises out of a transition taking place in the human error prevention field which diminishes the emphasis on changing worker behavior and expands on the idea of improving the design of the systems in which people work. Mention of system thinking and taking a holistic, socio-technical approach in hazard and risk control appears more often in safety-related literature. This is powerful stuff—and needed. In this chapter the term *macro thinking* is used principally rather that *system thinking* because *macro* implies taking a much larger view. The socio-technical model is described and promoted.

6. Safety Professionals as Culture Change Agents

This chapter promotes the idea that the overarching role of a safety professional is that of a culture change agent. The case is made that proposals made by safety professionals for improvement in operational risk management systems pertain to deficiencies in systems or processes and that the deficiency can be corrected only if there is a modification in an organization's culture—a modification in the way things get done, a modification in the *system of expected performance*. Thus, the primary role for a safety professional is that of a culture change agent. The definition given is: "A change agent is a person who serves as a catalyst to bring about organizational change. Change agents assess the present, are controllably dissatisfied with it, contemplate a future that should be, and take action to achieve the culture changes necessary to achieve the desired future."

7. The Plan-Do-Check-Act Concept (PDCA)

Writers of Z10 made it clear that continual improvement for occupational health and safety management systems is to be achieved through applying the recognized quality concept of Plan-Do-Check-Act. However, information is not provided on the PDCA concept and methodology. In this chapter we discuss the origin and substance of the PDCA concept, relate the PDCA concept to basic problem-solving techniques, and provide guidance on initiating a PDCA process.

8. Management Leadership and Employee Participation: Section 3.0 of Z10

This is the most important section in Z10. Why? Because safety is culture driven, and management creates the culture. As top management makes decisions directing an organization, the outcomes of those decisions establish its safety culture. In this chapter we comment on requirements of managements to attain superior results; policy statements; defining roles, assigning responsibilities and authority, providing resources, and establishing accountability; employee participation; relating management leadership to preventing serious injuries; and making a safety culture analysis. Employee participation particularly is emphasized.

9. Planning: Section 4.0 of Z10

The success of an occupational health and safety management system is largely contingent on the thoroughness of the planning processes. In Z10 the planning process goal is to identify and prioritize the "issues," which are defined as "hazards, risks, management system deficiencies, and opportunities for improvement." Reviews are to be made to identify those issues, priorities are to be set, objectives are to be established for the prioritized issues, and actions are to be outlined for continual improvement. In this chapter we discuss all of the provisions in the planning section. Special emphasis is given to the assessment and prioritization requirements in Section 4.2.

10. Implementation and Operation: Section 5.0 of Z10

All of the chapters described previously related to Z10 provisions pertain to the "plan" step in the Plan-Do-Check-Act process. The implementation and operation section moves into the "do" step. The standard says that elements in this section "provide the backbone of an occupational health and safety management system and the means to pursue the objectives from the planning process." This is a very brief chapter. Comments are made only on certain of its provisions: Contractors; Emergency preparedness; Education, Training, Awareness, and Competence; Communications; and Document and Record Controls.

Mention was made in the abstract of the planning section of the special emphasis given to the assessment and prioritization requirements in Section 4.2. Several provisions in implementation and operation (Section 5.0) relate directly to assessment and prioritization. Separate chapters are devoted to the assessment and prioritization provisions and the implementation and operation provisions:

11. Primer on Hazard Analysis and Risk Assessment
12. Provisions for Risk Assessments in Standards and Guidelines
13. Three- and Four-Dimensional Risk Scoring Systems
14. Hierarchy of Controls
15. Safety Design Reviews
19. Management of Change
20. The Procurement Process

Also, lean concepts as discussed in this book (Chapter 18) relate to the safety design review provisions in Z10.

An important addition was made in Z10's 2012 version that can be found in Section 5.1.1, "Risk Assessments." Several of the chapters enumerated previously pertain to risk assessment.

11. Primer on Hazard Analysis and Risk Assessment: Sections 4.2 and 5.1.1 of Z10

My intent in this chapter is to provide sufficient knowledge of hazard analysis and risk assessment methods to serve most of a safety and health professional's needs. Explored are what a hazard analysis is, how a hazard analysis is extended into a risk assessment, the steps to be followed in conducting a hazard analysis and a risk assessment, several commonly used risk assessment techniques, and risk assessment matrices.

12. Provisions for Risk Assessment in Standards and Guidelines: Sections 4.2 and 5.1.1 of Z10

Several safety standards and guidelines issued in recent years contain hazard analysis and risk assessment provisions. This is a significant, worldwide trend. Comments are made on the content of several of those standards and guidelines. Taken as a whole, they are convincing indicators, along with the hazard analysis and risk assessment provisions in Z10, that safety and health professionals will be expected to know, as a matter of career enhancement, how to make risk assessments.

13. Three- and Four-Dimensional Risk Scoring Systems: Sections 5.1.1 and 5.1.2 of Z10

For risk assessments, the practice—broadly—is to establish qualitative risk levels by considering only two dimensions: probability of event occurrence, and the severity of harm or damage that could result. Translating those assessments into numerical risk scores is not usually necessary. However, systems now in use may be three- or four-dimensional and require numerical risk scorings. In this chapter we review several numerical risk scoring systems. A three-dimensional numerical risk scoring system that I developed to serve the needs of those who prefer to have numbers in their risk assessment systems is presented.

14. Hierarchy of Controls: Section 5.1.2 of Z10

This section of Z10 states: "The organization shall implement and maintain a process for achieving feasible risk reduction based on the following order of controls." Achieving an understanding of this order of controls is a step forward in the practice of safety. In this chapter we review the evolution of hierarchies of control, discuss the Z10 hierarchy and provide guidelines on its application, comment on the logic of applying the hierarchy of controls, and place the hierarchy within good problem-solving techniques, as in the safety decision hierarchy.

15. Safety Design Reviews: Section 5.1.3 of Z10

Design Review and Management of Change requirements are addressed jointly in Section 5.1.3. Although the subjects are interrelated, each has its own importance and uniqueness. Guidance on the concept of management of change is provided in Chapter 19. In this chapter we discuss the design review processes in Z10 and include a review of safety through design concepts, comment on how some safety professionals are engaged in the design process, and provide a composite of safety through design procedures in place, a safety design review and operations requirements guide, and a general design safety checklist.

16. Prevention through Design: Sections 5.1.1 to 5.1.4 of Z10

The core of the concept of prevention through design is the identification and analysis of hazards and the making of risk assessments in the design and redesign processes. This material has a particular kinship to risk assessment in Section 5.1.1. That section requires establishing "a risk assessment process(es) appropriate to the nature of hazards and level of risks." The following principles are offered in support of the premise that prevention through design should have a prominent place within an operational risk management system.

1. Hazards and risks are avoided, eliminated, reduced, or controlled most effectively and economically in design and redesign processes.

2. Hazard analysis is the most important safety process in that if it fails, all other processes are likely to be ineffective.

3. Risk assessment should be the cornerstone of an operational risk management system.

4. If hazards are identified and analyzed and risks are assessed and dealt with properly in the design and redesign processes, the potential for serious injury is reduced significantly.

We comment on all of the elements in ANSI/ASSE Z590.3, the standard entitled *Prevention through Design: Guidelines for Addressing Occupational Hazards and Risks in Design and Redesign Processes*. A major section deals with relationships with suppliers. Guidelines are provided on how to avoid bringing hazards and their related risks into a workplace.

17. A Primer on System Safety: Sections 4.0, 4.2, 5.1.1, 5.1.2, and Appendix F of Z10

This chapter relates to several sections in Z10. There is a direct relationship between system safety concepts and the processes necessary to implement Z10 and to the practice of safety as a whole. System safety is hazards based and risk based and is employed in the design processes. I believe that generalists in the practice of safety will improve the quality of their performance by acquiring knowledge of applied system safety concepts and practices. In this chapter we relate the generalist's practice of safety to applied system safety concepts, provide a history of the origin, development, and application of system safety methods, outline the system safety idea in terms applicable to the generalist's practice of safety, and encourage safety generalists to acquire knowledge and skills in system safety.

18. Lean Concepts—Emphasizing the Design Process: Section 5.1.3 of Z10

Applied lean concepts are to eliminate waste, improve efficiency, and lower production costs. The direct and ancillary costs of accidents are among the elements of waste that should be addressed in the lean process. In this chapter we discuss the origin of lean concepts and how broadly they are being applied, comment on the opportunity for effective involvement in lean initiatives by safety professionals, and outline a unique design process in which lean safety and environmental needs are addressed.

19. Management of Change: Section 5.1.3 of Z10

The objective of a management of change system is to identify potential hazards and risks prior to beginning activities such as new construction, alteration of an existing facility, shutdowns and startups, and unplanned maintenance. In this chapter we discuss the management of change provisions in Z10, define the purpose and methodology of a management of change system and relate the system to the change analysis concept, establish the significance of these provisions in preventing serious injuries and fatalities and major property damage incidents, and outline management of change procedures. Four real-world examples of management of change procedures are included and an approach is provided for access to six others.

20. The Procurement Process: Section 5.1.4 of Z10

The purpose of the procurement process is to avoid bringing hazards and risks into the workplace. In this chapter we establish the importance of including safety specifications in purchasing orders and contracts, and provide resources and guidance on design specifications that become purchasing specifications to be met by vendors who supply machinery, equipment, and materials. Two addenda are examples of procedures that relate to the procurement process.

21. Evaluation and Corrective Action: Section 6.0 of Z10

Section 6.0 requires that processes be in place to determine whether the results intended for the occupational health and safety management system were achieved. In this chapter we comment on monitoring, measurement, and assessment requirements; provisions for taking corrective actions; and communications on the lessons learned being fed back into the planning and management review initiatives. Separate chapters of this book deal with two provisions considered vital in the evaluation and corrective action section: incident investigation (chapter 22) and audit requirements (chapter 23).

22. Incident Investigation: Section 6.2 of Z10

In this chapter we encourage that incident investigation be given a higher place within a safety management system; comment on the cultural difficulties that may be faced by those who try to have incident investigations improved; review the content of a good incident investigation form; provide materials to assist in crafting an investigation procedure; and promote the adoption of contributing factor identification, analysis, and resolution systems.

23. Audit Requirements: Section 6.3 of Z10

Provisions in Z10 require that audits be made to determine whether an organization has effectively implemented occupational health and safety management systems. In this chapter we establish that the principal purpose of an audit is to improve the culture, discuss the implications of hazardous situations observed, explore management expectations, comment on auditor qualifications, discuss the need to have safety and health management system audit guides tailored to the location being audited, and provide resources to develop suitable audit guides.

24. Management Review: Section 7.0 of Z10

The requirements for periodic top management reviews of the occupational health and safety management system and for continuous improvement are defined in Section 7.0. In this chapter we discuss progress made with respect to risk reduction, risk assessment and prioritization, employee participation, actions taken on recommendations made in audits, the extent to which objectives have been met, performance related to expectations, and taking action based on the findings in the management review.

25. Comparison: Z10 and Other Standards and Guidelines, and VPP Certification

Users of Z10 often ask how Z10 compares with other standards and guidelines. Such comparisons are included in this chapter. The desire that some companies have for their safety management systems to be certified is recognized. Organizations are encouraged to consider being certified as meeting OSHA Voluntary Protection Program (VPP) requirements, which are included in an addendum.

CHAPTER 1

OVERVIEW OF ANSI/AIHA Z10-2012

On July 25, 2005, the American National Standards Institute (ANSI) approved a new standard entitled *Occupational Health and Safety Management Systems*. Its designation was ANSI/AIHA Z10-2005. That was a major development. *For the first time in the United States, a national consensus standard for a safety and health management system applicable to organizations of all sizes and types was issued.* Z10 is an ANSI-approved standard. Other safety management system guidelines have been issued that do not have the approval of an accrediting organization.

In accord with ANSI requirements, standards must be reviewed at least every five years for revision or reaffirmation. As appropriate, the secretariat, the American Industrial Hygiene Association, formed a committee to review Z10. The outcome of its work is the revised standard approved on June 27, 2012 and designated ANSI/AIHA Z10-2012. Shortly after the approval, the secretariat was transferred to the American Society of Safety Engineers.

All persons who give counsel on occupational safety and health within an organization or who give counsel on occupational safety and health management systems to entities other than their own should have a copy of this revised standard and be thoroughly familiar with its content. Significant changes have been made in the revision, and valuable support information has been added in appendices. With its appendices, the standard is a brief safety and health management system manual.

This standard provides senior management with a well-conceived state-of-the-art concept and action outline to improve a safety and health management system. Drafters of Z10 adopted many of the best worldwide practices. As employers make improvements to meet the standard's requirements, it can be expected that the frequency and

Advanced Safety Management: Focusing on Z10 and Serious Injury Prevention,
Second Edition. Fred A. Manuele.
© 2014 John Wiley & Sons, Inc. Published 2014 by John Wiley & Sons, Inc.

severity of occupational injuries and illnesses will be reduced. The beneficial societal implications of Z10 are substantial.

Adoption of the Z10 standard or parts of it is believed to be quite broad, but a precise measure of its use and influence would be difficult to develop. Nevertheless, it is significant that:

- Over 7000 copies of the 2005 version of the standard were sold by the American Industrial Hygiene Association (AIHA) and the American Society of Safety Engineers (ASSE). That's a very large number for sales of a standard.
- ASSE, which is now the secretariat for Z10-2012, advised that sales of the latter version have been brisk.
- Several universities have used the first edition of *Advanced Safety Management: Focusing on Z10 and Serious Injury Prevention* in safety degree courses.
- Comments often appear in safety-related literature on Z10 provisions and their application.

This standard has had and will continue to have a significant and favorable impact on the content of the practice of safety and on the knowledge and skill requirements for safety and health professionals. Over time, Z10 will revolutionize the practice of safety.

Since Z10 represents the state of the art, it is not surprising that many organizations do not have management systems in place that meet all its provisions. To identify the shortcomings and to develop an improvement plan, a gap analysis should be made in which the safety and health management systems in place are compared with Z10 requirements.

To assist in developing an understanding of the content and impact of this standard, in this overview chapter we comment on:

- Each section of the standard
- Its history and development as the standard writing committee reached consensus
- A prominent and major theme within Z10
- How that major theme relates to serious injury prevention
- Z10 being a management system standard, not a specification standard
- International harmonization and compatibility
- Long-term influences and societal implications
- The continual improvement process: the Plan-Do-Check-Act concept

HISTORY, DEVELOPMENT, AND CONSENSUS

The American Industrial Hygiene Association obtained approval as the ANSI Accredited Standards Committee for this standard in March 1999. The first full meeting of the committee took place in February 2001. Over a six-year period, as many as 80 safety professionals were involved as committee members, alternates, resources, and interested commenters. They represented industry, labor, government, business associations, professional organizations, academe, and persons of general interest.

Thus, broad participation in the development of and acceptance of the standard was achieved, and the breadth of that participation is significant. One of the reasons for the Z10 committee's success was its strict adherence to the due-diligence requirements applicable to the development of an ANSI standard. There was a balance of stakeholders providing input and open discussion which resulted in their vetting each issue raised to an appropriate conclusion.

In the early stages of the committee's work, safety and health, quality, and environmental standards and guidelines were collected from throughout the world. They were examined and considered for their applicable content. In crafting Z10, the intent was not only to achieve significant safety and health benefits through its application, but also to have a favorable impact on productivity, financial performance, quality, and other business goals. The standard is built on the well-known Plan-Do-Check-Act process for continuous improvement. We address that subject in Chapter 7.

For the 2012 version, the committee applied the same due-diligence provisions as those required by ANSI. Well over 50 committee members represented industry, government, unions, and educational institutions. And consensus was reached on several revisions and additions representing the current state of the art.

Employers who have a sincere interest in reducing employee injuries and illnesses will welcome discussions on how their safety and health management systems can be improved. A significant number of companies have issued safety policy statements in which they affirm that they will comply with or exceed all relative laws and standards. Those employers, particularly, will want to implement provisions in the standard that are not a part of their safety and health management systems.

A MAJOR THEME

Throughout all the sections of Z10, starting with management leadership and employee participation through the management review provisions, the following theme is prominent. Processes for continual improvement are to be in place and implemented to assure that:

- Hazards are identified and evaluated.
- Risks are assessed and prioritized.
- Management system deficiencies and opportunities for improvement are identified.
- Risk elimination, reduction, or control measures are taken to assure that acceptable risk levels are attained.

In relation to the foregoing, the following definitions as given in the standard are particularly applicable.

Note: *Wherever the wording in this chapter appears in italic type, the material is a direct quote from the standard.*

- **Hazard**: *a condition, set of circumstances, or inherent property that can cause injury, illness, or death*
- **Exposure**: *contact with or proximity to a hazard, taking into account duration and intensity*
- **Risk**: *an estimate of the combination of the likelihood of an occurrence of a hazardous event or exposure(s) and the severity of injury or illness that may be caused by the event or exposures*
- **Probability**: *the likelihood of a hazard causing an incident or exposure that could result in harm or damage—for a selected unit of time, events, population, items or activity being considered*
- **Severity**: *the extent of harm or damage that could result from a hazard-related incident or exposure*
- **Risk assessment**: *process(es) used to evaluate the level of risk associated with hazards and system issues*

In Appendix F, which gives guidance on risk assessment, the definitions above are duplicated. Although *acceptable risk* is not a term included in the standard's definitions, it is made clear in several places in the standard that the goal is to achieve acceptable risk levels. For example, later in this chapter it is shown that Section 6.4, "Corrective and Prevention Actions," states clearly that an organization is to have processes in place to ensure that acceptable risk levels are achieved and maintained. Also, Appendix F states: The goal of the risk assessment process including the steps taken to reduce risk is to achieve safe working conditions with an acceptable level of risk. Chapter 2 deals with "Achieving Acceptable Risk Levels: The Operational Goal".

Understanding the standard's major theme and these definitions is necessary to apply this standard successfully.

RELATING THIS MAJOR THEME TO SERIOUS INJURY PREVENTION

A plea is made in Chapter 3, "Innovations in Serious Injury and Fatality Prevention," for organizations to improve their safety cultures so that a focus on the prevention of serious injuries is embedded into every aspect of their safety and health management systems. In our current economic world, staffs at all levels are expected to do more with less. Seldom will all the resources, money, and personnel be available to address all risks. To do the greatest good with the limited resources available, risks presenting the potential for the most serious harm must be given higher priority for management consideration and action.

Z10 IS A MANAGEMENT SYSTEM STANDARD

Z10 is not a specification standard—it is a management system standard. What's the difference between the two? In a management system standard, general process and system guidelines are given for a provision without specifying in detail how the provision is to be carried out. In a specification standard, such details are given. Section 5.2-B of Z10 is used here to illustrate the difference.

Section 5.2: Education, Training, Awareness, and Competence. The organization shall establish processes to:

B. Ensure through appropriate education, training, or other methods that employees and contractors are aware of applicable OHSMS requirements and their importance are competent to carry out their responsibilities as defined in the OHSMS.

That is the extent of the requirements for Section 5.2B. Comments are made in the advisory part of the standard on certain subjects for which training should be given, such as safety design, incident investigation, hazard identification, good safety practices, and the use of personal protective equipment, but those comments are not a part of the standard.

If Z10 were written as a specification standard, requirements comparable to the following might be extensions of Section 5.2B.

a. At least 12 hours of training shall be given initially to engineers and safety professionals in safety through design, to be followed annually with a minimum of 6 hours of refresher materials.
b. All employees shall be given a minimum of 3 hours of training annually in hazard identification.
c. All employees shall be given a minimum of 4 hours of training annually in the use of personal protective equipment.
d. All training activities conducted as a part of this provision shall be documented and the records shall be retained for a minimum of 5 years.

COMPATIBILITY, HARMONIZATION, AND POSSIBLE INTERNATIONAL IMPLICATIONS

One of the goals of the drafters of the standard was to assure that it could be integrated easily into whatever management systems an organization has in place. As to structure, the standard is compatible and harmonized with quality and environmental management system standards: the ISO 9000 and ISO 14000 series. Also, Z10 is written as a generic standard and patterned after the style of those standards. In this context, *generic* means that the standards can be applied to *all*:

- *Organizations of any size or type.*
- *Sectors of activity, whether a business enterprise, a non-profit service provider, or a government entity.*

ISO is the designation for the International Organization for Standardization, which is based in Geneva, Switzerland. It is the world's largest nongovernmental developer of standards, working with a network of the national standards institutes in 148 countries. The United States is represented at the ISO by the American National Standards Institute. On two occasions, in 1996 and 2000, votes were taken at the ISO on developing a standard for an occupational safety and health management system. In the latter case, the vote against carried by a narrow margin. The membership of ISO is worldwide, and a consensus among its members for such a standard had not yet emerged.

Of particular note is the recognition given in Z10's introduction to the International Labour Organization's *Guidelines on Occupational Health and Safety Management Systems* as a resource. The designation for the *Guidelines* is ILO-OSH 2001. It is a good, additional reference for safety and health management systems. The *Guidelines* can be downloaded at http://us.yhs4.search.yahoo.com/yhs/search?p=ilo+osh+2001+ management+systems&hspart=att&hsimp=yhs-att_001&type=att_lego_portal_home.

ILO is an international organization of considerable influence. Intentionally, Z10 adopts from and is in harmony with ILO-OSH 2001. Similarities between the *Guidelines* and Z10 are notable; But Z10 goes beyond the *Guidelines* in some respects.

Z10 was approved by a recognized standards-approving organization (i.e., ANSI) and represents current best practices. Since consideration will probably again be given to the development of an international safety and health management system standard at ISO, one can easily speculate on Z10 becoming the model for that standard. Continue the speculation: International requirements for accredited safety and health management system audits related to the provisions of Z10 can be envisioned.

LONG-TERM INFLUENCE: SOCIETAL IMPLICATIONS

As the provisions of this ANSI standard continue to be brought to the attention of employers as they strive to have safety management systems that are compatible with its provisions, its impact on what employers and society believe to be an effective safety management system will be extensive. Over time, Z10 will become the benchmark against which the adequacy of safety and health management systems will be measured. Societal expectations of employers with respect to their safety and health management systems will be defined by the standard's provisions.

Employment Implications

A recent and brief verbal survey of professors engaged in safety degree programs indicates that employers of safety professionals are seeking candidates who are equipped with the knowledge and skill to give counsel on meeting many of the provisions in the standard. In that respect, certain provisions of the standard are of particular note—provisions to which safety professionals should give particular attention. Those provisions are given in "Planning," Section 4.0; "Implementation and Operations," Section 5.0; and "Evaluation and Corrective Action," Section 6.0.

In summary, they state that employers "shall" establish and implement processes to:

- Identify and control hazards in the design process and when changes are made in operations. That requires that safety design reviews be made for new and altered facilities and equipment.
- Have an effective management of change system in place—through which hazards and risks are identified and evaluated in the change process.
- Assess the level of risk for identified hazards—for which knowledge of risk assessment methods will be necessary.
- Utilize a prescribed hierarchy of controls in dealing with hazards to achieve acceptable risk levels—for which the first step is to attempt to design out or otherwise eliminate the hazard.
- Avoid bringing hazards into the workplace—by incorporating design and material specifications in procurement contracts for facilities, equipment, and materials.

Educational Implications

Since one of the criteria for success of a technical degree program is the employment possibilities for graduates, prudent professors responsible for safety programs are assuring that core courses equip students properly to meet employer needs. In many cases that has necessitated substantive curricula modifications. This textbook has been adopted in several university safety degree programs at both the bachelors and masters, degree levels.

Certification Implications

Provisions in Z10 have a direct relationship to the content of examinations for the Certified Safety Professional (CSP) designation. Those examinations are reviewed about every five years to assure that they are current with respect to what safety professionals actually do. In the review process, safety professionals are asked to tell the surveyor about the reality of the content of their work at the time the survey is made.

For a review of current educational requirements to prepare a student to enter the practice of safety, I instituted a study to compare the content of the Comprehensive Practice Examination Guide issued by the Board of Certified Safety Professionals in 2011 with that issued in 2006. Substantial changes were made in the later edition, and many of them relate to the principal requirements of Z10. It is not said here the changes resulted from the issuance of Z10. But it is said that Z10 represents sound practice with respect to the actuality of an occupational health and safety management system.

OSHA Implications

A good reference on the possible implications of Z10 with respect to OSHA and to the legal liability potential is the March 2006 publication entitled *Legal Perspectives— ANSI Z10-2005 Standard: Occupational Health and Safety Management Systems.*

It was written by Adele Abrams, an attorney and American Society of Safety Engineers advocate in Washington, DC. I recommend that safety professionals read the full version of the paper, which is on the Internet. Access is achieved by entering the title of the paper into a search engine. Briefly, Abrams writes:

> Although it is unlikely that OSHA will resume regulatory activity to adopt a federal safety and health management systems standard at this time, if such activity was commenced in the future, OSHA would be obligated to consider adopting Z10 as that standard. Federal legislation and administrative rules direct agencies to use voluntary consensus standards in lieu of developing government-unique standards, except when such use would be inconsistent with the law or otherwise impractical. [*Author's note*: This was written in 2006. OSHA did begin activity on an injury and illness prevention program, which has stalled. The premise cited here is one of the many obstacles. As a matter of principle, it is not conceivable that OSHA could issue an injury and illness prevention program whose provisions required less than those of an accredited national consensus standard].

> Z10 could also have enforcement ramifications under OSHA's General Duty Clause (Section 5a), which requires that employers maintain a place of employment that is free from recognized hazards that are causing or are likely to cause death or serious injuries. Meeting the requirements of Z10 could be agreed upon during discussions between OSHA and employers as they developed consent orders to resolve citations made during inspections.

Legal Liability Implications

For safety consultants who give advice on safety management systems to employers other than their own employer, the issuance of this standard presents legal liability potentials for which they should be knowledgeable. These excerpts from Abrams' paper are pertinent.

> Safety and health professionals have an obligation to keep abreast of the latest knowledge and to include "best practices" in their safety programs and consultation activities, to the maximum extent feasible. Knowledge and comprehension of the ANSI Z10 standard may be imputed to safety professionals, in terms of determining what a "reasonable person" with similar training would be likely to know. Willful ignorance of the best practices set forth in Z10 and/or failure to incorporate such preventive measures in the workplace or programs under the safety and health professional's direction or oversight could lead to personal tort liability or professional liability.

Consider the following scenario. An employer receives a citation from OSHA. In the negotiations that follow, the employer agrees with OSHA that the safety management system must be improved. You, a safety consultant, receive a phone call from the obviously stressed employer asking that you provide counsel on the improvements to be made so that the safety management system meets good standards.

You call on the employer, agree on a course of action and a price, and the arrangements are confirmed through a letter contract. You decide that the framework you will use to help the employer is a typical safety management system, which does not contain the provisions in Z10 pertaining to safety design reviews, management of change, risk assessments and prioritization, a hierarchy of controls, and including safety specifications in purchasing agreements. Your counsel is well received and acted upon. Your contract is fulfilled and you have been paid.

Later, an incident occurs in the employer's operations and an employee is injured severely. Since workers' compensation laws govern, the employee cannot sue the employer. The employee's lawyer casts a large net to identify defendants. She discovers that you provided counsel on improvements to be made in the employer's safety management system.

You are on the witness stand. The employee's lawyer is ready. She studied the safety management system document on which you based consultation with your client and she has knowledge of the ANSI standard *Occupational Health and Safety Management Systems*, approved in 2012. You are led through the entirety of the substance of your advice to the employer. Then the lawyer establishes that you, a safety professional, have knowledge of ANSI standards. She gets you to agree that ANSI standards establish the minimum requirements for the subjects to which they apply and that, over time, they acquire a quasi-official status. She takes the position that Z10 represents the state of the art.

The lawyer works you through the elements in Z10 that were not addressed in the counsel you gave to your client and relates your omissions to the causal factors for the incident and injuries that occurred to her client. She establishes that you, as a safety professional, have an obligation to be familiar with and apply the state of the art in the counsel you give. She emphasizes, particularly, that your counsel was not based on the state of the art. Since you were negligent, you are liable.

Consultants who give advice to organizations to improve their safety and health management systems on a fee basis have reviewed the foregoing scenario and say that it is plausible.

TABLE OF CONTENTS IN Z10-2012

To provide a base for review and comparison with the safety management systems with which safety practitioners are familiar, the table of contents is duplicated here.

Table of Contents

Foreword

1.0   Scope, Purpose, and Application
    1.1   Scope
    1.2   Purpose

J. Contractor Safety and Health (Section 5.1.5)

K. Incident Investigation Guidelines (Section 6.2)

L. Audit (Section 6.3)

M. Management Review Process (Sections 7.1 and 7.2)

N. Management System Standard Comparison (Introduction)

O. Bibliography and References

New appendices added to the 2012 version of Z10 are: F, "Risk Assessment"; I, "Procurement"; J, "Contractor Safety and Health"; M, "Management of Change"; and N, "Management System Standard Comparison."

Some appendices provide extensive detail on the subjects covered. Others give explanatory comments, examples of forms and procedures, and reference sources for many of the sections to which they apply. Although the appendices are not part of the standard, they can be helpful to those who have implementation responsibility.

THE CONTINUAL IMPROVEMENT PROCESS: THE PDCA CONCEPT

Z10 is built on the well-known Plan-Do-Check-Act (PDCA) process for continual improvement. Understanding the PDCA concept is necessary to effectively implement the standard. A review of the concept is given in Chapter 7. "The Plan-Do-Check-Act Concept (PDCA)."

In Z10's Introduction there is a chart based on the PDCA concept. A slightly reduced form of the chart is presented at the beginning of each of the standard's major sections. That version is shown in Chapter 7. Similar continual improvement charts based on the PDCA concept are shown in the Quality Management Systems Standards—ANSI/ISO/ASQ Q9000-2000 series. The ISO 14000 series on environmental management was revised in 2004 to make it compatible with the ISO 9000 series. It is also based on the PDCA concept. In addition, the U.S. Environmental Protection Agency (EPA) suggests building an environmental management system on a PDCA model.

Throughout the standard, the words *process*, *processes*, *implemented*, and *continual improvement* are often repeated. That is also the case in the standards on quality and environmental management cited previously. Z10 is based on a continual improvement approach. The standard outlines the *processes* to be put in place, *not the specifics*, to have an effective safety and health management system.

Brief comments will be made here to provide an overview of the major sections of the standard. With respect to these remarks, keep in mind the intent of the terms *shall* and *should*. As is common in ANSI standards, requirements are identified by the word *shall*. An organization that chooses to conform to the standard is expected to fulfill the *shall* requirements. The word *should* is used to describe recommended practices or to give an explanation of the requirements. Recommended practices and advisory comments are not requirements of the standard.

SECTION 1.0: THE SCOPE, PURPOSE, AND APPLICATION OF Z10

Section 1.1: Scope

This section defines the *minimum requirements* (my emphasis) for occupational health and safety management systems (OHSMS). Even though the standard says that it sets forth minimum requirements, only a small segment of employment locations have safety management systems in place that include all of its elements, particularly those pertaining to safety through design and management of change concepts.

The emphasis in the advisory data is on a generic and systems approach for continual improvement in safety and health management and the avoidance of specifications. Further, the writers of the standard recognized the uniqueness of the culture and organizational structures of individual organizations and the need for each entity to "define its own specific measures of performance."

ANSI standards acquire a quasi-official status and may be viewed as containing only the minimum requirements—that is, the fewest requirements—which may not be sufficient in a particular situation. Repeating for emphasis: Safety consultants who give counsel on safety management to employers other than their own employer should recognize the status that ANSI standards acquire from a legal liability viewpoint.

Section 1.2: Purpose

This section states that the primary purpose of this standard is to provide a management tool to reduce the risk of occupational injuries, illnesses, and fatalities.

Section 1.3: Application

This section states that this standard is applicable to organizations of all sizes and types. As is the case in the ISO 9000 and ISO 14000 series of standards, there are no limitations or exclusions in Z10 by industry or business type or number of employees. Z10 applies to all employers. In the introduction and in comments in the advisory column opposite Section 1.3, it is made clear that the structure of the standard is to allow integration with quality and environmental management systems. Doing so is a good idea.

SECTION 2.0: DEFINITIONS

As is typical in ANSI standards, definitions are given of certain of the terms used in the standard. Safety professionals should become familiar with them. One addition was made to the definitions in the 2012 version: which is – *Risk Assessment: Process(es) used to evaluate the level of risk associated with hazards and system issues.*

SECTION 3.0: MANAGEMENT LEADERSHIP & EMPLOYEE PARTICIPATION

Section 3.0 is dealt with only briefly here. In Chapter 8 "Management Leadership and Employee Participation" we emphasize the significance of management leadership, the culture derived from management leadership, and managing change.

It should be understood that Section 3.0 is the standard's most important section. Safety professionals will surely agree that *Top management leadership and effective employee participation are crucial for the success of an Occupational Health and Safety Management System (OHSMS).* Top management leadership is vital because it sets an organization's safety culture and because continual improvement processes cannot be successful without sincere top management direction. Key statements in the "shall" column of the standard follow.

3.1.1 *Top management shall direct the organization to establish, implement and maintain an OHSMS.*
3.1.2 *The organization's top management shall establish a documented occupational health and safety policy.*
3.1.3 *Top management shall provide leadership and assume overall responsibility*
3.2 *The organization shall establish a process to ensure effective participation in the OHSMS by its employees at all levels.*

As management provides direction and leadership, assumes responsibility for the OHSMS, and ensures effective employee participation, the purpose of the standard must be kept in mind—to reduce the risk of occupational injuries, illnesses, and fatalities. That will be done best if personnel in the organization understand that in the application of every safety and health management process, the outcome is to achieve acceptable risk levels, and that a special focus must be given to identifying the causal factors for incidents that result in serious injuries. In Chapter 2, "Achieving Acceptable Risk Levels: The Operational Goal," we provide guidance on achieving acceptable risk levels.

In some incident investigation reports on serious injuries and fatalities it is apparent that contributing causal factors derived from severe expense and staff reductions. Maintenance staffs were reduced significantly. Preventive maintenance schedules could not be maintained. Safety-related work orders were given lower priority. Section 3.1.3A requires that management provide appropriate resources.

Over the long term, not providing resources to replace equipment at the end of its expected life and severely reducing maintenance capability increase serious injury and fatality potential significantly. This section—providing adequate resources—has more significance in the economic times being experienced at present.

There is supporting data in Annexes A, B, and C on policy statements, roles and responsibilities, and employee participation. Another good reference on management leadership and employee involvement is the chapter "Superior Safety Performance: A Reflection of an Organization's Culture" in my book *On the Practice of Safety, 4th edition.*

SECTION 4.0: PLANNING

In the PDCA process, planning is the first step. As would be expected, this section sets forth the planning process to implement the standard and to establish plans for improvement. *The planning process goal is to identify and prioritize OHSMS issues (defined as hazards, risks, management system deficiencies and opportunities for improvement).* (Note the emphasis on hazards, risks, and management system deficiencies.)

In the continual improvement process, as elements in the standard are applied, information defining opportunities for further improvement in the safety and health management system, and thereby risk reduction, is to be fed back into the planning process for additional consideration.

4.1 Requires that a review be made to identify the differences between existing operational safety management systems and the requirements of the standard. *The review shall include information regarding*:
A. *Relevant business systems and operational processes;*
B. *Operational issues such as, hazards, risks, and controls;*
C. *Previously identified OHSMS issues;*
D. *Allocation of resources;*
E. *Applicable regulations, standards, and other health and safety requirements;*
F. *Risk assessments and evaluations;*
G. *Process and mechanisms for employee participation;*
H. *Results of audits; and*
I. *Other relevant activities.*

4.2 Sets forth the requirements for Assessment and Prioritization.
The organization shall establish a process to assess and prioritize OHSMS issues on an ongoing basis. The process shall:
A. *Assess the impact on health and safety of OHSMS issues and assess the level of risk for identified hazards;*
B. *Establish priorities based on factors such as the level of risk, potential for system improvement, standards, regulations, feasibility, and potential business consequences; and*
C. *Identify underlying causes and other contributing factors related to system deficiencies that lead to hazards and risks.*

For clauses 4.2A and 4.2B, the following are selected explanatory notes.

E4.2A: The assessment of risks should include factors such as: identification of potential hazards; exposure, measurement data; sources and frequency of exposure; human behavior, capabilities, and other human factors; types of measures used to control hazards, and potential severity of hazards.

E4.2B: Business consequences may include either increased or decreased productivity, sales or profit or public image.

So, employers are to have processes in place to identify and analyze hazards, assess the risks deriving from those hazards, and establish priorities for amelioration which, when acted upon, will attain acceptable risk levels. Appendix D provides guidance on assessment and prioritization.

Section 4.3: Objectives

The organization shall establish a process to set documented objectives, quantified where practicable, based on issues that offer the greatest opportunity for OSHMS improvement and risk reduction.

Section 4.4: Implementation Plans and Allocation of Resources

This section follows logically in accord with a sound problem-solving procedure. After hazards, risks, and shortcomings in safety management systems have been identified and objectives have been outlined, a plan should be established and implemented to achieve the objectives. Item B in Section 4.4 reads as follows: *"Assign resources to achieve the established objectives of the implementation plans"*. It is an absolute that if adequate resources are not provided, over time, acceptable risk levels cannot be maintained.

SECTION 5.0: IMPLEMENTATION AND OPERATION

This section *defines the operational elements that are required for implementation of an effective OHSMS. These elements provide the backbone of an OHSMS and the means to pursue the objectives from the planning system.*

Section 5.1: OHSMS Operational Elements

Six operational elements are to be integrated into the management system. A new and important addition to Z10 was made in this section. It follows.

Section 5.1.1 Risk Assessment

The organization shall establish and implement a risk assessment process(es) appropriate to the nature of hazards and level of risk.

Adding this "shall" provision reflects a worldwide trend emphasizing the importance of risk assessments. Appendix F provides a six-page overview of risk assessment and includes data on a few techniques. Chapter 11 in this book is titled "A Primer on Hazard Analysis and Risk Assessment". It provides guidance with respect to the standard's risk assessment provision. Chapter 16 is titled "Prevention through Design", which relates to ANSI/ASSE Z590.3, the Prevention through Design standard. It is made clear in that chapter that hazard analysis and risk assessment are at the core of prevention through design concepts and of Z10. Brief descriptions are given in Chapter 16 of several hazard analysis and risk assessment techniques.

Having knowledge of preliminary hazards analysis, what-if/check analysis, and failure mode and effects analysis and how they are applied will satisfy the needs of safety professionals as they give counsel on risk assessment. It may be that a risk situation is so complex that consulting skills must be engaged to apply quantitative risk assessment methods—but that will be the exception.

In the application of these hazard analysis and risk assessment techniques, qualitative rather than quantitative judgments will be sufficient. Mathematical calculations required will not be extensive. Appendix F provides an example of a risk assessment matrix. Risk assessment matrices set forth incident probability categories, severity of harm or damage ranges, and resulting risk levels. A risk assessment matrix can serve as a valuable instrument in working with decision makers on setting risk levels and prioritizing ameliorating actions. Variations in published risk assessment matrices are substantial. A safety professional should develop a matrix that is suitable to the organization to which counsel is given. Appendix F also includes a hazard analysis and risk assessment guide, which is comparable to the guide shown in Chapter in 16 this book titled "Prevention through Design".

Section 5.1.2 Hierarchy of Controls

Although we said earlier that Z10 is a management system standard and not a specification standard, the provisions pertaining to a hierarchy of controls are the exception. Provisions for the use of a specifically defined hierarchy of controls are outlined. The organization "shall" apply the methods of risk reduction in the order prescribed. This is how the standard and the explanatory comments read.

> *The organization shall establish a process for achieving feasible risk reduction based upon the following preferred order of controls:*
> *A. Elimination;*
> *B. Substitution of less hazardous materials, processes, operations, or equipment;*
> *C. Engineering controls;*
> *D. Warnings;*
> *E. Administrative controls; and,*
> *F. Personal protective equipment.*
>
> *Feasible application of this hierarchy of controls shall take into account:*
> *a. The nature and extent of the risks being controlled;*
> *b. The degree of risk reduction desired;*
> *c. The requirements of applicable local, federal, and state statutes, standards and regulations;*
> *d. Recognized best practices in industry;*
> *e. Available technology;*

f. Cost-effectiveness; and,

g. Internal organization standards.

E5.1.2: The hierarchy provides a systematic way to determine the most effective feasible method to reduce the risk associated with a hazard. When controlling a hazard, the organization should first consider methods to eliminate the hazard or substitute a less hazardous method or process. This is best accomplished in the concept and design phases of any project. Refer to Section 5.1.3. If this is not feasible, engineering controls such as machine guards and ventilation systems should be considered. This process continues down the hierarchy until the highest level feasible control is found. Often a combination of controls is most effective. In cases where the higher order controls (elimination, substitution and implementation of engineering controls) do not reduce risk to an acceptable level, lower order controls (e.g. warnings, administrative controls, or personal protective equipment) are used to complement engineering controls to reduce risks to an acceptable level.

For example, if an equipment modification or noise enclosure (engineering control) is insufficient to reduce noise levels, then limiting exposure through job rotation and using hearing protection would be an acceptable supplemental means of control.

Note that this standard prescribes a hierarchy of controls that contains six elements, the first of which, in priority order, is to design out or otherwise eliminate the hazard. If the hazard is eliminated, the risk is eliminated. Also, the substitution element is separate from the elimination element. That may not be so in other published hierarchies of controls. Some hierarchies have as few as three elements.

Annex G provides a pictorial and verbal display of the hierarchy of controls listed in Section 5.1.1 with application examples for each element. In an occupational setting, these outcomes are to be achieved through application of the hierarchy of controls.

1. Acceptable risk levels
2. Work methods and processes for which the probability is as low as reasonably practicable for:
   a. Errors being made by supervisors and workers because of design inadequacy
   b. Supervisors and workers defeating the system

Comparable outcomes should be expected through application of the hierarchy of controls for such as the design and use of industrial or consumer products, and environmental management systems.

The hierarchy of controls in Z10 is very close in substance to the model shown in Chapter 14, "Hierarchy of Controls" in this book. In that chapter, to move the state of the art forward, the hierarchy of controls is contained within a sound problem-solving technique. Also, the chapter includes a dissertation on the logic of taking action in an order of effectiveness, which relates directly to the hierarchy of controls in Z10.

Section 5.1.3 Design Review and Management of Change

The following excerpts indicate what the standard *requires* for design reviews and management of change and replicate the explanatory information given in its right-hand column. To repeat for emphasis: These are "shall" provisions.

The organization shall establish a process to identify, and take appropriate steps to prevent or otherwise control hazards at the design and redesign stages, and for situations requiring Management of Change to reduce potential risks to an acceptable level. The process for design and redesign and Management of Change shall include:

A. *Identification of tasks and related health and safety hazards;*

B. *Recognition of hazards associated with human factors including human errors caused by design deficiencies;*

C. *Review of applicable regulations, codes, standards, internal and external recognized guidelines;*

D. *Application of control measures (hierarchy of controls—Section 5.1.2);*

E. *A determination of the appropriate scope and degree of the design review and management of change; and*

F. *Employee participation.*

E5.1.3E: The process for conducting design reviews and managing changes is designed to prevent injuries and illnesses before new hazards and risks are introduced into the work environment. The design review should consider all aspects, including design, construction, operation, maintenance, and decommissioning. The following are examples of conditions that should trigger a design review or management of change process:

- *New or modified technology (including software), equipment, or facilities;*
- *New or revised procedures, work practices, design specifications;*
- *Different types and grades of raw materials;*
- *Significant changes to the site's organizational structure and staffing, including use of contractors;*
- *Modification of health and safety devices; and*
- *New health and safety standards or regulations.*

The Design Process   For quite some time, I and others have professed that the most effective and economical way to achieve acceptable risk levels is to have the hazards from which they derive addressed in the design process. That's what this standard requires. This is an exceptionally important element in this standard. Its impact can be immense.

To become qualified to give counsel on establishing a management system to apply the design review requirements in Z10, a large percentage of safety practitioners will have to acquire new knowledge and skills. An introduction to this subject can be found in Chapter 15 titled "Safety Design Reviews". The book *Safety Through Design* is also a substantive reference for the design process.

If a management system for design safety reviews is not in place in an organization, safety professionals should anticipate a long-term effort to achieve the culture change necessary to meet the requirements of Z10. This often means establishing a management system that mobilizes engineering, purchasing, quality control, and other departments that may not be accustomed to working collaboratively. To develop an understanding of the depth of what is to be undertaken, the chapter "Achieving the Necessary Culture Change" in *Safety Through Design* will help.

Management of Change   Employers are to have processes in place to identify and take appropriate steps to prevent or otherwise control hazards and reduce the potential risks associated with them when changes are made to existing operations, products, services, or suppliers. Getting effective management of change procedures in place is not easy.

With respect to drafting and implementing management of change procedures, generalists in the practice of safety can learn from the safety personnel in organizations that have met the management of change requirements of *OSHA's Rule for Process Safety Management of Highly Hazardous Chemicals*, 29 CFR 1910.119, issued in 1992. Briefly, 1910.119 requires that employers establish and implement written procedures to manage changes. Z10 and 1910.119 requirements have similar purposes.

My research shows that for all occupations, many incidents resulting in serious injury occur when out-of-the-ordinary situations arise, particularly when unusual and nonroutine work is being done and when there are sources of high energy present. In support of that premise, consider this excerpt from the historical and explanatory data published with respect to 1910.119:

> *Management of Change*: OSHA believes that one of the most important and necessary aspects of a process safety management program is appropriately managing changes to the process. This is because many of the incidents that the Agency has reviewed resulted from some type of the change to the process. While the Agency received some excellent suggestions concerning minor changes to improve this proposed provision, there was widespread support for including a provision concerning the management of change in the final rule.

Note that there was widespread support for the management of change provisions. About two years after 1910.119 became effective, Thomas Seymour, a director at OSHA who was in a leadership role as the standard was developed, stated that the feedback that OSHA received from chemical plant operators was that the management of change requirement in the standard was the most difficult to apply. Safety directors in chemical companies verified that statement. It is not surprising that specially focused courses have been developed to assist those who have the responsibility to meet the 1910.19 management of change requirements.

It is suggested that safety professionals study thoroughly the management of change requirements of Z10 to determine how they might assist in achieving the culture change necessary for their implementation. Applying change management methods will be necessary. Fortunately, the literature on change management is extensive. Chapter 19 addresses "Management of Change."

Note that Sections 5.1.3.1 and 5.1.3.2 are extensions of Section 5.1.3, "Design Review and Management of Change."

5.1.3.1 Applicable Life-Cycle Phases   *During the design and redesign processes, all applicable life cycle phases shall be taken into consideration.*

5.1.3.2 Process Verification   *The organization shall have processes in place to verify that changes in facilities, documentation, personnel and operations are evaluated and managed to ensure safety and health risks arising from these changes are controlled.*

Section 5.1.4 Procurement

Although the requirements for procurement are plainly stated and easily understood, they are brief in relation to the enormity of what will be required to implement them. An interpretation of the requirements could be: Safety practitioners, you are assigned the responsibility to convince management and purchasing agents that in the long term, it can be very expensive to buy cheap. This is how the standard and the explanatory data read.

The organization shall establish and implement processes to:

A. *Identify and evaluate the potential health and safety risks associated with purchased products, raw materials, and other goods and related services before introduction into the work environment;*

B. *Establish requirements for supplies, equipment, raw materials, and other goods and related services purchased by the organization to control potential health and safety risks; and*

C. *Ensure that purchased products, raw materials, and other goods and related services conform to the organization's health and safety requirements.*

E5.1.4: The procurement process should be documented. See section E5.4.
E5.1.4A: For example, organizations should evaluate SDSs (Safety Data Sheets) and other health and safety information of a new chemical, or examine the design specifications and operations manual for a new piece of equipment being considered for purchase.

Only a small percentage of employers have included specifications in their purchasing agreements and contracts that require suppliers to identify the hazards and assess the potential risks in the equipment and materials being purchased. As a safety director in a major company said recently, the only safety specification in their contracts is that OSHA standards and other legislative requirements be met. Chapter 20, "The Procurement Process" provides guidance on how to avoid bringing hazards into the workplace. So does Appendix I.

This Z10 standard implies that safety through design concepts are to be applied in an organization's purchasing system with respect to both physical hazards and work methods. Adding an element to safety management systems that is to avoid bringing hazards into the workplace could have startling good results in reducing the frequency and severity of hazardous incidents and exposures.

Although procedures encompassing the procurement requirements will not be put in place easily, recognition builds slowly that they should be an integral part of a safety management system. Getting these procurement provisions established presents a huge challenge for safety professionals, but the benefits can be immense.

Section 5.1.5 Contractors

This section requires that an organization have processes in place to avoid injury and illness to the organization's employees from activities of contractors and to the contractor's employees from the organization's operations. Many entities have such procedures in place. One of the "shall" provisions indicates that the process is to include "contractor health and safety performance criteria." That implies, among other things, vetting the contractor with respect to its previous safety performance before awarding a contract.

Section 5.1.6 Emergency Preparedness

To meet the requirements of this provision, an organization is to have management systems in place *to identify, prevent, prepare for, and/or respond to emergencies.* Also, periodic drills are to be conducted to test the emergency plans, and they are to be updated periodically.

Section 5.2: Education, Training, Awareness, and Competence

An organization is required to determine the knowledge needed to achieve competence, ensure that employees are aware of the OHSMS requirements, remove any barriers to participation in education, ensure that training is given in a language trainees understand and that training is ongoing, and ensure that trainers are competent. This section has six alphabetically designated provisions. In three of them, the words *competence* or *competent* appear. Thus, competence is emphasized. Employees and contractors are to be competent to fulfill their responsibilities. Trainers are to be competent to train.

These provisions, the standard says, are applicable to contractors also, which could be difficult to do. Comments in the advisory column, which are of some length on training, do not mention contractors. It is interesting that in the examples of the training that should be given, both safety design and procurement are mentioned. This is how item E5.2A reads:

> *E5.2A: Training in OHSMS responsibilities should include, for example, training for: Engineers in safety design (e.g. hazard recognition, risk assessment, mitigation, etc....); Those conducting incident investigations and audits for identifying underlying OHSMS non-conformances; Procurement personnel on impact of purchasing decisions; and others involved with the identification of OHSMS issues, methods of prioritization, and controls.*

Section 5.3: Communication

An organization is to institute processes to communicate information about the progress being made on its implementation plan; ensure prompt reporting of incidents, hazards, and risks; promote employee involvement so that they make recommendations on hazards and risks; inform contractors and *relevant external interested parties* of changes made that affect them; and remove barriers to all of the foregoing. With respect to contractors, item E5.3D gives guidance as follows:

> *E5.3D: The work activities of contractors can pose additional hazards for both employees and others in the workplace. Processes established for consultation with contractors should ensure risks will be appropriately addressed using good OHS practices. This consultation should include discussion and resolution of issues of mutual concern.*

Section 5.4: Document and Record Control Process

Documentation requirements for certain systems are specified in several places in Z10. As a performance standard would say, the document and record control processes are to fit the requirements of the safety and health management system in place. In the informational column, sound advice is given on the documentation process as follows.

> *E5.4: The type and amount of formal documentation necessary to effectively manage an OHSMS should commensurate with the size, complexity, and risks of an organization.*

An organization shall have document and record-keeping processes to *(1) implement an effective OHSMS, and (2) demonstrate or assess conformance with the requirements of this standard.* Documents shall be updated as needed, legible, adequately protected against damage or loss, and retained as necessary.

SECTION 6.0: EVALUATION AND CORRECTIVE ACTION

This section of the standard outlines the requirements for processes to evaluate the performance of the safety management system and to take corrective action when shortcomings are found. Communications on lessons learned are to be fed back into the planning process. The expectation is that new objectives and action plans will be written in relation to what has been experienced.

Section 6.1: Monitoring, Measurement, and Assessment

The organization shall establish and implement a process to monitor and evaluate hazards, risks, and their controls to assess OHSMS performance. These processes shall include some or all of the following methods, depending on the nature and extent of identified hazards and risks: workplace inspections and testing, assessments of exposures, incident tracking, measuring performance in relation to legal or other requirements; or other methods selected by the organization. Results of the monitoring processes shall be communicated as appropriate.

Section 6.2: Incident Investigation

The organization shall establish a process to report, investigate and analyze incidents in order to address OHSMS non-conformances and other factors that may be causing or contributing to the occurrence of incidents. The investigations shall be performed in a timely manner.

Because of studies this author has made, greater emphasis is now given to the importance of incident investigation because of its value in identifying cultural, operational, and technical causal factors for incidents that result in serious injuries and illnesses. However, the requirement for incident investigation in Z10 is contained in one brief paragraph, with no subsections. For a subject as important as incident investigation, it might seem that it is dealt with briefly. But as a "shall" requirement, nothing else need be said.

Advisory comments on incident investigation are more extensive. One of the significant advisory comments states that there is a value in feeding lessons learned from investigations into the planning and corrective-action processes. That fits well with the emphasis being given in this book to serious injury and illness prevention. This subject is explored in depth in Chapter 22, "Incident Investigation."

Section 6.3: Audits

An organization shall have audits made periodically with respect to application of the provisions in the OHSMS, ensure that audits are made by competent persons not attached to the location being audited, document and communicate the results, and have auditors communicate immediately on potentials for serious injuries or illnesses and fatalities so that swift corrective action can be taken.

Audits are to measure the organization's effectiveness in implementing the elements of the occupational health and safety management system. Thus, audits are to determine whether the management systems in place do or do not effectively identify hazards and control risks.

Although many safety professionals are familiar with safety audit processes, it is suggested that they review what the standard requires and determine whether it will be to their benefit to revise their audit systems. Annex L is helpful in this respect. It contains a sample audit protocol that matches all the sections of Z10.

An expanded treatise on safety audits is presented here in Chapter 23, "Audits". In the advisory column, it is made clear that audits are to be "system" oriented rather than "compliance" oriented.

Also, and importantly, E.6.3B clarifies the independence of auditors. While it says that *audits should be conducted by individuals independent of the activities being audited*, it also says that *this does not mean that audits must be conducted by individuals external to the organization.*

Section 6.4: Corrective and Preventive Actions

Revisions made in the 2012 version of Z10 are substantial in relation to the earlier version. It is duplicated here as written. It defines what the organization "shall" do to fulfill the provisions of this section.

> *The organization shall establish and implement corrective and preventive action processes to*:
>
> A. *Address non-conformances and hazards that are not being controlled to an acceptable level of risk.*
> B. *Identify and address new and residual hazards associated with corrective and preventive actions that are not being controlled to an acceptable level of risk.*
> C. *Expedite action high risk hazards (those that could result in fatality or serious injury/illness) that are not being controlled to an acceptable level of risk; and*
> D. *Review and ensure effectiveness of corrective and preventive actions taken.*

This is a major change in relation to the Section 6.4 provisions in the original version of Z10 in that there were no specifically stated requirements in the initial document to achieve acceptable risk levels. These new provisions recognize the general acceptance throughout the world that in an occupational setting, a goal is to achieve and maintain an acceptable risk level. It is now clearly stated in Z10 that organizations are to identify shortcomings in safety management systems and take preventive action as necessary to achieve acceptable risk levels—expediting the appropriate actions on high-risk hazards.

Section 6.5: Feedback to the Planning Process

As stated in the advisory column, this is a communication provision pertaining to all shortcomings in the safety management system. Its purpose is to provide a base for revision in the planning process. The standard says:

The organization shall establish and maintain process to ensure that the results of monitoring and measurement, audits, incident investigation, and corrective and preventive actions are included in the ongoing planning process (Section 4.2) and the management review (Section 7).

SECTION 7.0: MANAGEMENT REVIEW

This section requires that the OHSMS performance be reviewed periodically and that management take appropriate actions in accord with the results of the review. The management review section and extensive advisory comments pertaining to it are "must" reading. We stated earlier in this chapter that Management Leadership and Employee Participation is the most important section in Z10. This section on Management Review is a close second. Making periodic reviews of management systems effectiveness is an important part of the Plan-Do-Check-Act process.

Section 7.1: Management Review Process

The organization shall establish and implement a process for top management to review the OHSMS at least annually, and to recommend improvements to ensure its continued suitability, adequacy, and effectiveness.

E.7.1: Management reviews are a critical part of the continual improvement of the OHSMS.

These are some of the subjects to be reviewed, at least annually: progress in the reduction of risk; effectiveness of processes to identify, assess, and prioritize risk and system deficiencies; effectiveness in addressing underlying causes of risks and system deficiencies; the extent to which objectives have been met; and performance of the OHSMS in relation to expectations.

Section 7.2: Management Review Outcomes and Follow-Up

This section requires that management determine whether changes need to be made in *the organization's policy, priorities, objectives, resources, or other OHSMS elements* to establish the *future direction of the OHSMS.* In accord with good management procedures, senior management is expected to give direction to implementing the changes needed in safety and health management processes to continually reduce risks. The standard requires that *results and action items from the management reviews shall be documented and communicated to affected individuals, and tracked to completion.* This provision gives the necessary importance to the management review process. Action items are to be recorded, communicated to those affected, and followed through to a proper conclusion.

ADVISORY CONTENT AND APPENDICES

This standard provides a large amount of exceptionally valuable explanatory and supportive data in the advisory column—the E column—and in the Appendices. With respect to the quantity of text, information in the E column exceeds the "shall" requirements of the standard by about one-half.

Pages 1 through 29 pertain to the requirements of the standard and the material in the advisory column. Pages 30 through 88 are devoted to the appendices. That's about a 65 % increase over the 2005 version in space devoted to appendices. A safety professional must have a copy of the standard to appreciate the value of the guidance material in the E column and the appendices.

Particular mention is made of one of the appendices because I have had to respond to many questions pertaining to how Z10 compares to other standards. Appendix N (Informative) is entitled "Management System Standard Comparison (Introduction)." The Introduction reads as follows.

This table compares ANSI/AIHA® Z10 with international standards, and other guidelines. The matrix is intended to demonstrate the significant similarity in essential elements of management systems and assist organization in integrating management systems during implementation. Element by element comparison is difficult and check marks simply indicate that the element is present in a standard or document. This is particularly true in the case of ISO 9001-2008 because the purpose of this standard is significantly different than the other management system standards and guidelines listed in this matrix.

In Appendix N the content of ANSI Z10 is compared to:

ISO 14001:2004
BS OHSAS 18001:2007
ILO-OHSMS:2001
VPP 2008
ISO 9001:2008

In rather small print, the table fills four pages. This is a valuable document. Whoever put it together had to spend many hours to complete it.

CONCLUSION

This standard, from its first issue, represents an important step in the evolution of the practice of safety. Realistically, it can be expected that over time it will become the benchmark against which safety and health management systems will be measured. As the quality of such systems improves, it is logical to expect that the frequency and severity of occupational injuries and illnesses will be reduced.

It would be folly for safety professionals to ignore the long-range impact that Z10 will have on societal expectations concerning the quality of the safety management systems that employers have in place and on the expectations that employers will have concerning the knowledge and capabilities of their safety staffs.

Prudent safety professionals will study the requirements of the standard to determine whether additional skills and capabilities are needed and will move forward to acquire those skills. Having done so, they will be equipped to give guidance to managements on putting in place safety management system elements that may not exist in the organizations to which they give counsel.

REFERENCES

Abrams, Adele. *Legal Perspectives—ANSI Z10-2005 Standard: Occupational Health and Safety Management Systems*. Des Plaines, IL: American Society of Safety Engineers, March 2006. Also on the Internet.

ANSI/AIHA Z10-2012. *Occupational Health and Safety Management Systems*. Fairfax, VA: American Industrial Hygiene Association. ASSE is now the secretariat. Available at https://www.asse.org/cartpage.php?link=z10_2005.

ANSI/ASSE Z590.3-2011. *Prevention Through Design: Guidelines for Addressing Occupational Hazards and Risks in Design and Redesign Processes*. Des Plaines, IL: American Society of Safety Engineers, 2011.

ANSI/ISO/ASQ Q9001-2000. *American National Standard: Quality Management Systems—Requirements*. Milwaukee, WI: American Society for Quality, 2000.

BS OHSAS 18001:2007. *Occupational Health and Safety Management Systems—Requirements*. London: British Standards Institution, 2007.

Christensen, Wayne C. and Fred A. Manuele, Editors. *Safety Through Design*. Itasca, IL: National Safety Council, 1999.

ILO-OHS:2001. *Guidelines on Occupational Health and Safety Management Systems*. Geneva, Switzerland: International Labour Organization, 2001. http://us.yhs4.search.yahoo.com/yhs/search?p=ilo+osh+2001+management+systems&hspart=att&hsimp=yhs-att_001&type=att_lego_portal_home.

ISO 14001:2004. *Environmental management systems: Requirements with guidance for use*. Geneva, Switzerland: International Organization for Standardization, 2004.

Manuele, Fred A. *On the Practice of Safety,* 4th ed. Hoboken, NJ: Wiley, 2013.

OSHA's Rule for Process Safety Management of Highly Hazardous Chemicals. 29 CFR 1910.119. Washington, DC: U.S. Department of Labor, 1992.

The Comprehensive Practice Examination Guide, 6th ed. Champaign, IL: Board of Certified Safety Professionals Apr. 2011 (downloadable without charge at www.bcsp.org).

VPP—OSHA's Voluntary Protection Program, Section C in CSP 03-01-002–TED 8.4, Voluntary Protection Programs (VPP): Policies and Procedures, 2003. At http://www.osha.gov/pls/oshaweb/owadisp.show_document?p_table=DIRECTIVES&p_id=2976.

OTHER READING

Casada, M. L., A. C. Remson, D. A. Walker, and J. S. Arendt. *A Manager's Guide to Implementing and Improving Management of Change Systems*. Washington, DC: Chemical Manufacturers Association, 1993.

Manuele, Fred A. "Risk Assessment and Hierarchies of Control." *Professional Safety*, May 2005.

CHAPTER 2

ACHIEVING ACCEPTABLE RISK LEVELS: THE OPERATIONAL GOAL

Terminology in Z10 clearly establishes that processes are to be in place to assure that hazards are identified and analyzed and that actions are taken to assure that their attendant risks are at acceptable levels. Although the term *acceptable risk* appears in several places in the standard and the appendices, reference is made here only to provisions for corrective and prevention actions. They have greater import because they are "shall" requirements (italic indicates a direct quote from the standard):

6.4 Corrective and Preventive Actions
The organization shall establish and implement corrective and preventive action processes to:

A. *Address non-conformances and hazards that are not being controlled to an acceptable level of risk*
B. *Identify and address new and residual hazards associated with corrective and preventive actions that are not being controlled to an acceptable level of risk.*

In the E column, the explanatory column, these comments support and expand on the 6.4 "shall" requirements:

E6.4: An effective OHSMS should identify system deficiencies and control hazards in any part of the system to an acceptable level of risk. An organization needs

Advanced Safety Management: Focusing on Z10 and Serious Injury Prevention,
Second Edition. Fred A. Manuele.

to address all identified system deficiencies and inadequately controlled hazards to an acceptable level (et al.).

E6.4B: Risk cannot typically be eliminated entirely, though it can be substantially reduced through application of the hierarchy of controls. Residual risk is the remaining risk after controls have been implemented. It's the organization's responsibility to determine whether the residual risk is acceptable for each task and associated hazard. Where the residual risk is not acceptable, further actions must be taken to reduce risk.

Writers of the Z10 standard were aware that although the term *acceptable risk* is now used more frequently in standards and guidelines throughout the world, a substantial percentage of the personnel who have safety, health, and environmental responsibilities are reluctant to use the term. Evidence of that reluctance often arises in discussions on the development of new or revised standards or technical reports. Possibly, that aversion derives from:

- A lack of awareness of the factors that determine risk—probability and severity.
- Aversion to the premise that some risks are acceptable.
- Not being familiar with the literature that demonstrates convincingly that if a facility exists or an operation proceeds, attaining a zero risk level is impossible.
- Concern over the subjective judgments made in the risk assessment process and the uncertainties that almost always exist.
- The unavailability of in-depth statistical probability and severity data that allow precise and numerically accurate risk assessments.
- Insufficient knowledge and experience from which an understanding could be achieved that operational risks are necessarily accepted every day.

However, in recent years the concept of acceptable risk has been interwoven into internationally applied standards and guidelines for a very broad range of types of equipment, products, processes, and systems. That has occurred as it has become recognized by many participants in the development of those standards and guidelines that:

- Risk acceptance decisions are made constantly in real-world applications.
- Knowledge has emerged on how to manage with respect to those risks.
- Society benefits if those decisions achieve acceptable risk levels.

In this chapter I provide a primer from which an understanding of risk and the concept of acceptable risk can be attained. For that purpose:

- A far-reaching premise is presented that is fundamental in dealing with risk.
- Several examples are given of the use of the term *acceptable risk* as taken from the applicable literature.

- Discussions address the impossibility of achieving zero risk levels.
- The inadequacy of using *minimum risk* as a replacement for *acceptable risk* is explored.
- The shortcomings that may result from designing only to the requirements of a standard are presented.
- A risk assessment matrix for use in determining acceptable risk levels is offered as an example.
- The ALARP (as low as reasonably practicable) concept is discussed with an example of how the concept is applied in achieving an acceptable risk level.
- Comments are made on social responsibilities.

Note: For simplicity of expression, the terms *hazard, risk*, and *safety* apply to all hazard-related incidents or exposures that could result in injury or illness or damage to property or the environmental.

FUNDAMENTAL PREMISE

The following general, all-encompassing premise is basic to the work of all personnel who give counsel to avoid injury and illness, and property and environmental damage.

The entirety of purpose of those responsible for safety, regardless of their titles, is to manage their endeavors with respect to hazards, so that the risks deriving from those hazards are acceptable.

That premise is supported by this hypothesis: If there are no hazards, if there is no potential for harm, risks of injury or damage cannot arise.

The focus of the practice of safety is on hazards and the risks that derive from them to achieve acceptable risk levels.

PROGRESSION WITH RESPECT TO USE OF THE TERM *ACCEPTABLE RISK*

The more frequent use of the term *acceptable risk* in standards and guidelines is notable, as the following citations show. It is to the advantage of safety professionals who are reluctant to adopt the concept implied in the term *acceptable risk* to consider the breadth and implication of that evolution. The term *acceptable risk* is becoming the norm. A list of selected references, intentionally lengthy, is presented here to show how broadly the concept of acceptable risk has been adopted.

- In *Of Acceptable Risk: Science and the Determination of Safety* (1976), Lowrance wrote: "A thing is safe if it risks are judged to be acceptable" (p. 8).

- This next citation, from a court decision made in 1980, developed significant importance because it has given long-term guidance with respect to U.S. Department of Labor policy and to the work done at the National Institute for Occupational Safety and Health.

   The Supreme Court's benzene decision of 1980 states that "before he can promulgate any permanent health or safety standard, the Secretary [of Labor] is required to make a threshold finding that a place of employment is unsafe—in the sense that significant risks are present and can be eliminated or lessened by a change in practices" (IUD v. API, 448 U.S. at 642). The Court broadly describes the range of risks OSHA might determine to be significant: It is the Agency's responsibility to determine in the first instance what it considers to be a "significant" risk. *Some risks are plainly acceptable and others are plainly unacceptable.* [Emphasis in italics added.]

   If, for example, the odds are 1 in 1 billion that a person will die from cancer by taking a drink of chlorinated water, the risk clearly could not be considered significant. On the other hand, if the odds are 1 in 1000 that regular inhalation of gasoline vapors that are 2% benzene will be fatal, a reasonable person might well consider the risk significant and take appropriate steps to decrease or eliminate it. The Court stated further:

   The requirement that a "significant" risk be identified is not a mathematical straitjacket. Although the Agency has no duty to calculate the exact probability of harm, it does have an obligation to find that a significant risk is present before it can characterize a place of employment as "unsafe" and proceed to promulgate a regulation.

- ISO/IEC Guide 51:1999, *Safety aspects – Guidelines for their inclusion in standards*, was first published in 1990. This guide was issued by the International Organization for Standardization (ISO) and the International Electrotechnical Commission (IEC) to provide standardized terms and definitions to be used in standards for "any safety aspect related to people, property, or the environment." A second edition was issued in 1999, from which the following definitions are cited. (Work on a third edition is in progress.)

   Safety: freedom from unacceptable risk. (3.1)

   Tolerable risk: risk which is accepted in a given context based on the current values of society. (3.7)

- In the World Health Organization (WHO) publication *Water Quality: Guidelines, Standards and Health* (2001), Chapter 10 is entitled "Water Quality: Guidelines, Standards and Health." The following guidelines are given in determining acceptable risk. A risk is acceptable when:
  - It falls below an arbitrary defined probability.
  - It falls below some level that is already tolerated.

- It falls below an arbitrary defined attributable fraction of total disease burden in the community.
- The cost of reducing the risk would exceed the costs saved.
- The cost of reducing the risk would exceed the costs saved when the "costs of suffering" are also factored in.
- The opportunity costs would be better spent on other, more pressing, public health problems.
- Public health professionals say that it is acceptable.
- The general public say that it is acceptable (or more likely, do not say that it is not).
- Politicians say that it is acceptable.

- In a March 2003 entry to the Internet, OSHA set forth its requirements for organizations to obtain certification under its Voluntary Protection Programs (VPPs). An excerpt follows.

    Worksite analysis. A hazard identification and analysis system must be implemented to systematically identify basic and unforeseen safety and health hazards, evaluate their risks, and prioritize and recommend methods to eliminate or control hazards to an acceptable level of risk.

- The following appears in ANSI/ASSE Z244.1-2003, *American National Standard for Control of Hazardous Energy: Lockout/Tagout and Alternative Methods* (reaffirmed in January 2009).

    A.2 *Acceptable level of risk*: If the evaluation in A.1.6 determines the risk to be acceptable, then the process is completed. …

- This definition appears in the United Nations publication UN ISDR, *Terminology: Basic Terms of Disaster Risk Reduction*, 2004, p. 1:

    Acceptable risk: The level of loss a society or community considers acceptable given existing social, economic, political, cultural, technical and environmental conditions.

- In the "Sci-Tech Dictionary" ("Answers," on the Internet), a definition of acceptable risk as used in geology is given.

    Acceptable risk: (geophysics) In seismology, that level of earthquake effects which is judged to be of sufficiently low social and economic consequence, and which is useful for determining design requirements in structures or for taking certain actions.

- This is the definition of acceptable risk as shown in the Australian/New Zealand AS/NZS 4360:2004, *Risk Management Standard*:

    1.3.16 *Risk acceptance*: An informed decision to accept the consequences and the likelihood of a particular risk.

- The Office of Hazards Materials Safety in the Pipeline and Hazardous Materials Safety Administration (PHMSA) in the U.S. Department of Transportation has issued *Risk Management Definitions* (2005), from which the following definition is taken.

  > *Acceptable risk*: An acceptable level of risk for regulations and special permits is established by consideration of risk, cost/benefit and public comments. Relative or comparative risk analysis is most often used where quantitative risk analysis is not practical or justified. Public participation is important in a risk analysis process, not only for enhancing the public's understanding of the risks associated with hazardous materials transportation, but also for insuring that the point of view of all major segments of the population-at-risk is included in the analyses process.

  > Risk and cost/benefit analysis are important tools in informing the public about the actual risk and cost as opposed to the perceived risk and cost involved in an activity. Through such a public process PHMSA [Pipeline and Hazardous Materials Safety Administration] establishes hazard classification, hazard communication, packaging, and operational control standards.

- In the 2007 revision of the British Standards Institution publication BS OHSAS 18001:2007, *Occupational health and safety: Management systems—Requirements*, a significant change was made as follows:

  > The term "tolerable risk" has been replaced by the term "acceptable risk." (3.1)

- In the International Electrotechnical Commission document IEC 60601-1-9, International Standard for *Environmentally Conscious Design of Medical Equipment*, published in July 2007, it says in the introduction:

  > The standard includes the evaluation of whether risks are acceptable (risk evaluation).

- This reference is a highly recommended document issued in 2009 by the Institute for research for safety and security at work and the Commission for safety and security at work in Quebec, Canada. The title is "Machine Safety: Prevention of mechanical hazards." The introduction states:

  > When machine-related hazards ... cannot be eliminated through inherently safe design, they must then be reduced to an acceptable level.

- In ANSI B11.0-2010. American National Standard, *Safety of Machinery—General Requirements and Risk Assessment*, a broad-ranging standard on machinery and risk assessment, the following is written:

Acceptable risk: A risk level achieved after protective measures have been applied. It is a risk level that is accepted for a given task (hazardous situation) or hazard. For the purpose of this standard the terms "acceptable risk" and "tolerable risk" are considered to be synonymous.

Informative Note 1: The expression "acceptable risk" usually, but not always, refers to the level at which further technologically, functionally and financially feasible risk reduction measures or additional expenditure of resources will not result in significant reduction in risk. The decision to accept (tolerate) a risk is influenced by many factors including the culture, technological and economic feasibility of installing additional protective measures, the degree of protection achieved through the use of additional protective measures, and the regulatory requirements or best industry practice.

- The following definition is taken from ANSI/PMMI B155.1-2011, *American National Standard for Safety Requirements for Packaging Machinery and Packaging-Related Converting Machinery.*

  Acceptable risk: risk that is accepted for a given task or hazard. For the purpose of this standard the terms "acceptable risk" and "tolerable risk" are considered synonymous. (3.1)

  Informative Note 1: The decision to accept (tolerate) a risk is influenced by many factors including the culture, technological and economic feasibility of installing additional risk reduction measures, the degree of protection achieved through the use of additional risk reduction measures, and the regulatory requirements or best industry practice. The expression "acceptable risk" usually, but not always, refers to the level at which further reduction measures or additional expenditure of resources will not result in significant reduction in risk.

- In ANSI/ASSE Z590.3-2011, *Prevention through Design: Guidelines for Addressing Occupational Hazards and Risks in the Design and Redesign Processes*, the following appear:

Scope and Purpose

The goals of applying prevention through design concepts are to achieve acceptable risk levels.

Definitions

Acceptable risk: That risk for which the probability of an incident or exposure occurring and the severity of harm or damage that may result are as low as reasonably practicable (ALARP)in the setting being considered.

WITH RESPECT TO THE FOREGOING CITATIONS

Since it is almost always the case that resources are limited, a phrase that appears in the citation taken from World Health Organization material—"the opportunity costs would be better spent on other, more pressing problems"—has a significant bearing on risk acceptance decision making, resource allocation, and priority setting.

Several citations relate to the fact that residual risk cannot be eliminated entirely and that residual risk acceptance decisions are made commonly and frequently. Whenever a production machine is turned on, a residual risk level is being accepted: Every time a design decision is made or a product design is approved, those making the decision approve a residual and acceptable risk level.

Replacing the term *tolerable risk* with *acceptable risk* in BS OHSAS 18001 by an organization as influential as the British Standards Institution is noteworthy. In some parts of the world, because of requirements in contract bid situations, companies are required to show that their safety management systems are "certified." BS OHSAS 18001 is often the base of such certification. This modification made by the British Standards Institution indicates that the goal to be achieved is acceptable risk levels.

SUMMARY TO THIS POINT

As the references cited illustrate, the concept of acceptable risk has been adopted broadly internationally, and the term *acceptable risk* is becoming the norm. Those who assert that they are safety professionals and are reluctant to adopt the concept of acceptable risk would do well to recognize that they have an obligation to be current with respect to the state of the art and reconsider their views.

THE NATURE AND SOURCE OF RISK

Risk is expressed as an estimate of the probability of a hazard-related incident or exposure occurring and the severity of harm or damage that could result. All risks with which safety professionals deal derive from hazards. There are no exceptions. A *hazard* is defined as the potential for harm. Hazards include all aspects of technology and activity that produce risk. Hazards include the characteristics of things (i.e., equipment, dusts, chemicals, etc.) and the actions or inactions of people.

The *probability* aspect of risk is defined as the likelihood of an incident or exposure occurring that could result in harm or damage for a selected unit of time, events, population, items, or activity being considered. The *severity* aspect of risk is defined as the degree of harm or damage that could reasonably result from a hazard-related incident or exposure.

Comparable statements and definitions appear in much of the current literature on risk and acceptable risk. One resource has been chosen for citation here because of its broad implications. The following excerpts are taken from the *Framework for Environmental Health Risk Management* issued by The Presidential/Congressional Commission on Risk Assessment and Risk Management.

What Is "Risk"?

Risk is defined as the probability that a substance or situation will produce harm under specified conditions. Risk is a combination of two factors:

- The *probability* that an adverse event will occur
- The *consequences* of the adverse event

Risk encompasses impacts on public health and on the environment, and arises from *exposure* and *hazard*. Risk does not exist if exposure to a harmful substance or situation does not or will not occur. Hazard is determined by whether a particular substance or situation has the potential to cause harmful effects. *Risk* is the probability of a specific outcome, generally adverse, given a particular set of conditions.

Residual risk is the health risk remaining after risk reduction actions are implemented, such as risks associated with sources of air pollution that remain after implementation of maximum achievable control technology.

A ZERO RISK LEVEL IS NOT ATTAINABLE

It has long been recognized by researchers, authors, and practitioners that a zero risk level is not attainable. If a facility exists or an activity proceeds, it is not possible to conceive realistically of a situation in which there is no probability of an adverse incident or exposure occurring. William F. Lowrance was one of the most significant and influential authors on the concept of acceptable risk. In his book *Of Acceptable Risk: Science and the Determination of Safety*, Lowrance writes:

> Nothing can be absolutely free of risk. One can't think of anything that isn't, under some circumstances, able to cause harm. Because nothing can be absolutely free of risk, nothing can be said to be absolutely safe. There are degrees of risk, and consequently there are degrees of safety. (p. 8)

Similar comments are made in ISO/IEC Guide 51: *Safety Aspects—Guidelines for their inclusion in standards*. Under the heading "The Concept of Safety" (Section 5), this appears:

> There can be no absolute safety: some risk will remain, defined in this Guide as residual risk. Therefore a product, process or service can only be relatively safe. Safety is achieved by reducing risk to a tolerable level, defined in this Guide as tolerable risk.

In the real world, whether in the design or redesign processes or in facility operations, attaining a zero risk level is not possible. That said, after risk avoidance, elimination, reduction, or control measures are taken, the residual risk must be acceptable, as judged by the decision makers.

Also, it is necessary to recognize that inherent risks that are acceptable and tolerable in some occupations would not be tolerable in others. For example, some work conditions considered tolerable in deep-sea fishing (e.g., a pitching and rolling work floor, the ship's deck) would not be tolerable elsewhere. In other and opposite situations, such as for certain chemical or radiation exposures that are designed to function at higher than commonly accepted permissible exposure levels, the residual risk will be judged as unacceptable, and operations at those levels would not be permitted.

Nevertheless, society accepts a continuation of certain operations in which the occupational and environmental risks are high. That is demonstrated by the fatality rate data published annually by the U.S. Bureau of Labor Statistics in its National Census of Fatal Occupational Injuries.

The latest data, shown in Table 2.1, are for 2011. The fatality rate is the rate per 100,000 workers. The national average fatality rate for all private industries in 2011 was 3.5.

TABLE 2.1 Occupations with the Five Highest, Fatality Rates, 2011

Occupation	Fatality Rate
Fishers and related fishing workers	121.2
Logging workers	102.4
Aircraft pilots and flight engineers	57.0
Refuse and recyclable material collectors	41.2
Roofers	31.8

Although the fatality rates among all employment categories are highest for the occupations shown in Table 2.1, the public has not demanded discontinuation of the operations in which they occur. Inherent risks in the high hazard categories are considered "tolerable" in relation to the benefit attained. However, it should be recognized that there has been considerable research to make those occupations "safer".

OPPOSITION TO IMPOSED RISKS

The literature is abundant on the resistance that people have to being exposed to risks they believe are imposed on them. For some, their aversion to adopting the acceptable risk concept derives from their view that imposed risks are objectionable and are to be rebelled against. Conversely, they accept the significant risks of activities in which they choose to engage (e.g., skiing, bicycle riding, driving an automobile). This idea needs exploration, beginning with a statement that can withstand a test of good logic. In *System Safety for the 21st Century* Richard A. Stephans states:

> The safety of an operation is determined long before the people, procedures, and plant and hardware come together at the worksite to perform a given task. (p. 13)

Start from the beginning in the process of creating a new facility as the creditability of Stephans' statement is validated. Consider, first, a site survey for ecological considerations, soil testing, and then move into the design, construction, and fitting out of a facility.

Thousands of safety-related decisions are made in the design processes that result in an imposed level of risk. Usually, those decisions meet (or exceed) applicable safety-related codes and standards with respect to issues such as the contour of exterior grounds, sidewalks, and parking lots; building foundations; facility layout and configuration; floor materials; roof supports; process selection and design; determination of the work methods; aisle spacing; traffic flow; hardware; equipment; tooling; materials to be used; energy choices and controls; lighting, heating, and ventilation; fire protection; and environmental concerns.

Designers and engineers make decisions on the foregoing during the original design processes. Those decisions establish what the designers believe implicitly to be acceptable risk levels. Thus, the occupational and environmental risk levels have largely been imposed before a facility begins operations.

Indeed, if persons employed in such settings conclude that the risks imposed are not acceptable, communication systems should be in place to allow them to express their views and to have them resolved.

MINIMUM RISK AS AN INADEQUATE SUBSTITUTE FOR ACCEPTABLE RISK

Invariably, those who oppose use of the term *acceptable risk* offer substitutions. One frequent suggestion is that designers and operators should achieve minimum risk levels, or minimize the risks. That sounds good until the applications of the terms are explored. *Minimum* means the least amount or the lowest amount. *Minimization* means to reduce something to the lowest possible amount or degree. Note particularly, "least" and "lowest possible".

Assume that the *threshold limit value* (TLV) for a chemical is 4 parts per million (ppm). For $10 million, a system can be designed, built, and installed that will operate at 2 ppm. But for an additional $100 million, a 1 ppm exposure level can be achieved. Increase the investment to $200 million and the result is an exposure level of 0.1 ppm. At 2 ppm, the exposure level is acceptable but not the minimum possible; a lower exposure level can be achieved.

Requiring that systems be designed and operated to minimum risk levels—that risks be minimized—is impractical because the investments necessary to do so may be so high that the cost of the product required to recoup the investment and make a reasonable profit would not be competitive in the marketplace.

DESIGNING TO STANDARDS MAY NOT ACHIEVE AN ACCEPTABLE RISK LEVEL

Discussions regarding the process of developing consensus standards can be lively, positions taken by individual participants can be strongly held and debated for hours, and typically, many compromises are made. Some standards set forth only the minimum requirements on the subject covered. For example, in the scope section of Z10 it is stated: "This standard defines minimum requirements for occupational health and safety management systems (OHSMS)". Also, if a standard is obsolete, using the standard as a design base may result in designing to obsolescence and perhaps unacceptable risk levels.

A convincing argument can be made, as in the excerpt below, of the need sometimes to go beyond issued safety standards in the design process and to have decisions on acceptable risk levels be based on risk assessments. The excerpt is taken from *Related Information 1: Equipment/Product Safety Program*, which is an adjunct to SEMI S2-0706, *Environmental, Health, and Safety Guideline for Semiconductor Manufacturing Equipment*.

> Compliance with design-based safety standards does not necessarily ensure adequate safety in complex or state-of-the-art systems. It often is necessary to perform hazard analyses to identify hazards that are specific with the system, and develop hazard control measures that adequately control the associated risk beyond those that are covered in existing design-based standards. (p. 2)

Designing to a particular safety standard may or may not achieve an acceptable risk level. In any case, the results of risk assessments and subsequent amelioration actions, if necessary, should be dominant in deciding whether acceptable risk levels have been reached.

CONSIDERATIONS IN DEFINING ACCEPTABLE RISK

If the residual risk for a task or operation cannot be zero, for what risk level does one strive? It is the norm that resources are always limited. There is never enough money to address every hazard identified. Such being the case, safety professionals have a responsibility to give counsel so that the greatest good to society, to employees, to employers, and to product users is attained through applying available resources practicably and economically to obtain acceptable risk levels.

Determining whether a risk is acceptable requires consideration of many variables. An additional excerpt from ISO/IEC Guide 51, Section 5, helps in understanding the concept of designing and operating for risk levels as low as reasonably practicable.

> Tolerable risk [acceptable risk] is determined by the search for an optimal balance between the ideal of absolute safety and the demands to be met by a product, process or service, and factors such as benefit to the user, suitability for purpose, cost effectiveness, and conventions of the society concerned.

Understanding cost-effectiveness has become a more important element in risk acceptance decision making. That brings the discussion to ALARA and ALARP. ALARA and ALARP are commonly used acronyms in risk assessment and risk reduction literature. ALARA stands for "as low as reasonably achievable"; ALARP is short for "as low as reasonably practicable."

Use of the ALARA concept as a guideline originated in the atomic energy field. This is taken from the "Reference Library, Glossary of Terms" at www.nrc.gov.

ALARA: Acronym for "As Low As Reasonably Achievable," means making every reasonable effort to maintain exposures to ionizing radiation as far below the dose limits as practical, consistent with the purpose for which the licensed activity is undertaken, taking into account the state of technology, the economics of improvements in relation to benefits to the public health and safety, and other societal and socioeconomic considerations, and in relation to utilization of nuclear energy and licensed materials in the public interest (see 10/CFR 20.1003).

The implication that decision makers are to "[make] every reasonable effort to maintain exposures to ionizing radiation as far below the dose limits as practical" provides conceptual guidance in striving to achieve acceptable risk levels in all classes of operations.

ALARP seems to be an adaptation from *ALARA*. But *ALARP* has become the more frequently used term for operations other than atomic and appears more often in the literature. *ALARP* is defined as that level of risk which can be lowered further only by an increment in resource expenditure that is disproportionate in relation to the resulting decrement of risk. The concept embodied in the terms *ALARA* and *ALARP* applies to the design of products, facilities, equipment, work systems and methods, and environmental controls.

In the real world of decision making, benefits represented by the amount of risk reduction to be obtained and the costs to achieve those reductions become important factors. Trade-offs are frequent and necessary. An appropriate goal is to have the residual risk be as low as reasonably achievable. Paraphrasing the terms contained in the definition of *ALARA* given previously helps in explaining the process:

1. Reasonable efforts are to be made to identify, evaluate, and eliminate or control hazards so that the risks deriving from those hazards are acceptable.
2. In the design and redesign processes for physical systems and for the work methods, risk levels for injuries and illnesses to employees and the public, and property and environmental damage, are to be as far below what would be achieved by applying current standards and guidelines as is economically practicable.
3. For items 1 and 2, decision makers are to take into consideration:
   • The purpose of the undertaking.
   • The state of the technology.
   • The costs of improvements in relation to benefits to be obtained.
   • Whether the expenditures for risk reduction in a given situation could be applied elsewhere with greater benefit.

Since resources are always limited, spending an inordinate amount of money to reduce the risk only a little through costly engineering and redesign is inappropriate, particularly if that money could be better spent otherwise. That premise can be demonstrated through an example that uses a risk assessment matrix as a part of the decision making.

RISK ASSESSMENT MATRICES

A risk assessment matrix (figure 2.1) that assigns numbers to risk levels at the beginning serves well for the purposes of this chapter and for the following demonstration showing the application of the ALARP principle. This matrix is recommended because of experience with it. Many matrices use verbal categories to establish risk levels—such as low, moderate, serious, and high—as does this one. But in operations in which this matrix was used at the shop floor level, employees said that it was easier to relate risk categories to each other, such as a moderate risk level to a serious risk level, by first establishing in their minds the relationship between numbers—such as 6 to 12.

An example follows that illustrates how a team used the matrix within a risk assessment process and applied the ALARP concept to make a decision about acceptable risk. At the end, the team took the opportunity to suggest to management that rather than spend money on additional risk reduction through major physical process changes, they would prefer that the money be spent to build a wellness center.

1. A chemical operation was built 15 years ago. Although engineering modifications have been made in the system over the years, management is aware that its operations are not at the current state-of-the-art level.
2. A risk assessment team is convened to consider the chemically related risks in a particular process in the overall system.
3. In the deliberations, the group refers to its established hierarchy of controls:
   a. Risk avoidance.
   b. Eliminate or reduce risks in the redesign process.
   c. Reduce risks by substituting less hazardous methods or materials.
   d. Incorporate safety devices.
   e. Provide warning systems.
   f. Apply administrative controls (e.g., work methods, training, work scheduling).
   g. Provide personal protective equipment.
4. The group first considers and holds open the possibility of redesigning and replacing the process completely. Substitution of materials or methods is considered and it is determined that action on such opportunities has already been taken. Safety devices and warning systems have been updated and are considered state of the art. Maintenance is considered superior.
5. The occurrence probability for a chemically related illness is judged to be moderate (3) and the severity level is moderate (3). Thus, the risk score is 9, which is in category 3, so remedial action is to be given high priority.
6. The team recognizes that to reduce the risk further, improvements must be made so that appropriate training is given and repeated, and standard operating procedures and the use of personal protective equipment are rigidly enforced.
7. Management agrees to fund the necessary administrative improvements.

Severity Levels and Values	Occurrence Probabilities and Values				
	Unlikely (1)	Seldom (2)	Occasional (3)	Likely (4)	Frequent (5)
Catastrophic (5)	5	10	15	20	25
Critical (4)	4	8	12	16	20
Marginal (3)	3	6	9	12	15
Negligible (2)	2	4	6	8	10
Insignificant (1)	1	2	3	4	5

Numbers were intuitively derived. They are qualitative, not quantitative. They have meaning only in relation to each other.

Incident or Exposure Severity Descriptions

Catastrophic: One or more fatalities, total system loss and major business down time, environmental release with lasting impact on others with respect to health, property damage or business interruption.

Critical: Disabling injury or illness, major property damage and business down time, environmental release with temporary impact on others with respect to health, property damage or business interruption.

Marginal: Medical treatment or restricted work, minor subsystem loss or property damage, environmental release triggering external reporting requirements.

Negligible: First aid or minor medical treatment only, non-serious equipment or facility damage, environmental release requiring routine cleanup without reporting.

Insignificant: Inconsequential with respect to injuries or illnesses, system loss or down time, or environmental release.

Incident or Exposure Probability Descriptions

Unlikely: Improbable, unrealistically perceivable.
Seldom: Could occur but hardly ever.
Occasional: Could occur intermittently.
Likely: Probably will occur several times.
Frequent: Likely to occur repeatedly.

Risk Levels

Combining the Severity and Occurrence Probability values yields a risk score in the matrix. The risks and the action levels are categorized below.

Risk Categories, Scoring, and Action Levels

Category	Risk Score	Action Level
Low risk	1 to 5	Remedial action discretionary.
Moderate risk	6 to 9	Remedial action to be taken at appropriate time.
Serious risk	10 to 14	Remedial action to be given high priority.
High risk	15 or greater	Immediate action necessary. Operation not permissible except in an unusual circumstance and as a closely monitored and limited exception with approval of the person having authority to accept the risk.

FIGURE 2.1   Risk assessment matrix.

8. Assuming that these administrative improvements are made, the risk assessment group decides that the probability of occurrence of an illness from a chemical exposure would be low (2) and that the severity of harm expected would also be low (2). Thus, the risk score is 4—in category 1.

9. Reengineering and replacing the process would reduce the probability level to very low (1) and the severity level to very low (1), thereby achieving a risk score of 1, which is also in category 1. The estimated cost of completely redesigning and replacing the process—$1,500,000—was considered disproportionate with respect to the amount of risk reduction to be obtained.

10. The risk assessment group took the opportunity to say to management that they would prefer having money spent on a wellness center.

THE ALARP PRINCIPLE

ALARP promotes a management review whose intent is to achieve acceptable risk levels. Practical, economic risk trade-offs are frequent and necessary in the benefit/cost deliberations that take place when determining whether the costs to reduce risks further can be justified "by the resulting decrement in risk."

Several depictions of the ALARP concept begin with an inverted triangle, the purpose being to indicate that the risk is greater at the top and much less at the bottom. Figure 2.2 shows the concept combined with the elements in the risk assessment matrix.

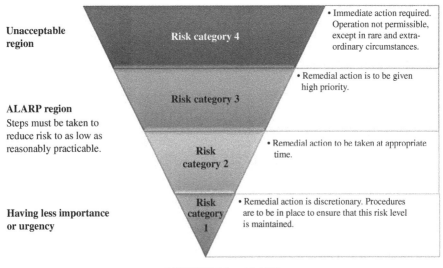

FIGURE 2.2   ALARP.

DEFINING ACCEPTABLE RISK

Risk acceptance is a function of many factors and varies considerably across industries (e.g., mining vs. medical devices vs. farming). Even at locations of a single global company, the acceptable risk levels can vary. The culture dominant in a company and the culture of a country in which a facility is domiciled play an important role in risk acceptability, as has been experienced by colleagues working in global companies. Training, experience, and resources can also influence acceptable risk levels. Risk acceptability is also time dependent, in that what is acceptable today may not be acceptable tomorrow, next year, or the next decade.

A sound and workable definition of acceptable risk must encompass hazards, risks, probability, severity, and economic considerations. I believe that the definition of acceptable risk included in ANSI/ASSE Z590.3-2011 represents the most practical use of the term developed in recent several years.

> *Acceptable risk*: That risk for which the probability of an incident or exposure occurring and the severity of harm or damage that may result are as low as reasonably practicable (ALARP) in the setting being considered.

All of the foregoing having been said, I believe that after applying all of the steps in the hierarchy of controls, the test for an acceptable risk level should be having the risk fall in the "low risk" category shown in Figure 2.1, but not higher than temporarily in the "moderate risk" category. There could be exceptions, such as for a rescue mission.

SOCIAL RESPONSIBILITY, AN EMERGING OPPORTUNITY

Formal consideration of the social responsibility concept by senior executives is a fairly recent development. What is social responsibility? An Internet search engine will reveal a large number of definitions. For the purposes of this chapter, two definitions have been chosen.

The World Business Council for Sustainable Development in its publication "Making Good Business Sense" by Lord Holme and Richard Watts, used the following definition:

> Corporate Social Responsibility is the continuing commitment by business to behave ethically and contribute to economic development while improving the quality of life of the workforce and their families as well as of the local community and society at large. (p. 1)

Gap, Inc. states in its Internet entry on social responsibility that the concept is fundamental to show how we do business.

> It means everything from ensuring that workers are treated fairly to addressing our environmental impact.

It is logical to suggest that if an organization initiates a social responsibility endeavor that is to include the well-being of workers, the environment, and the community at large, knowledge of and application of acceptable risk principles would be beneficial in its decision making.

The result would be efficient allocation of resources, fewer injuries and illnesses, property and environmental damage incidents, and serving the community well. That seems to present opportunities for safety professionals.

THE STATE OF THE ART IN RISK ASSESSMENT

Safety professionals must understand that risk assessment is as much an art as a science and that subjective (although educated) judgments are made on incident or exposure probability and the severity of outcome to arrive at a risk category. Also, it must be recognized that economically applicable risk assessment methodologies have not been developed to resolve all risk situations.

As an example, I was once asked: "How would you assess the cumulative risk in an operation in which there was an unacceptable noise level and toluene was used in the process?" It was hoped that search into relative resource material, such as EPA's *Framework for Cumulative Risk Assessment*, would provide an answer. That inquiry was not successful. EPA is cautionary about cumulative risk assessment methods. Their paper says:

> It should be acknowledged by all practitioners of cumulative risk assessment that in the current state of the science there will be limitations in methods and data available. (p. 31)

> Finding a common metric for dissimilar risks is not an analytical process, because some judgments should be made as to how to link two or more separate scales of risks. These judgments often involve subjective values, and because of this, it is a deliberative process. (p. 55)

> Calculating individual stressor risks and then combining them largely presents the same challenges as combination toxicology but also adds some statistical stumbling blocks. (p. 66)

Where multiple, diverse hazards exist, the practical approach is to treat each hazard independently, with the intent of achieving acceptable risk levels for all. In the noise and toluene example, the hazards are indeed independent. In complex situations, or when competing solutions to complex systems must be evaluated, the assistance of specialists with knowledge of more sophisticated risk assessment methodologies such as hazard and operability analysis (HAZOP) or fault tree analysis (FTA) may be required. However, for most applications, I do not recommend that diverse risks be summed through what could be a questionable methodology.

CONCLUSION

Risk acceptance is the deliberate decision to assume a risk. I postulate that an assumed risk is acceptable if the probability of a hazard-related incident or exposure occurring and the severity of harm or damage that may result are as low as reasonably practicable and tolerable in a given situation. In an ideal world, all personnel who are affected would be involved in or be informed of risk acceptance decisions.

Use of the term *acceptable risk* has arrived. It is becoming a norm. In organizations with advanced safety management systems, that idea—achieving practicable and acceptable risk levels throughout all operations—is a cultural value. It is suggested that safety professionals adopt the concept of attaining acceptable risk levels as a goal to be embedded in every risk avoidance, elimination, reduction, or control action proposed. To achieve that goal, safety professionals must educate others on the beneficial effects of applying the concept.

Also, safety professionals must have the ability to work through the greatly differing views that people can have about risk levels, incident and exposure probabilities, and severity potential. Workers views about risks should be considered for their value. With respect to environmental risks, community views must be taken into consideration as well. In arriving at acceptable risk levels where the hazard and risk scenarios are complex, it is best to gather a team of experienced personnel for their contributions and for their buy-in to the conclusions.

REFERENCES

ALARA, Reference Library, Glossary of Nuclear Terms, Nuclear Regulatory Commission, 2007. Enter "Glossary of Nuclear Terms" into a search engine.

ALARP "at a glance." London: Health and Safety Executive, (undated). At http://www.hse.gov.uk/risk/theory/alarpglance.htm.

ANSI/AIHA Z10-2012. American National Standard, *Occupational Health and Safety Management Systems*. Fairfax, VA: American Industrial Hygiene Association. ASSE is now the secretariat. Available at https://www.asse.org/cartpage.php?link=z10_2005.

ANSI/ASSE Z244.1-2003 (reaffirmed January 2009). American National Standard, for *Control of Hazardous Energy: Lockout/Tagout and Alternative Methods*. Des Plaines, IL: American Society of Safety Engineers.

ANSI/ASSE Z590.3-2011. *Prevention through Design: Guidelines for Addressing Occupational Hazards and Risks in Design and Redesign Processes*. Des Plaines, IL: American Society of Safety Engineers, 2011.

ANSI B11.0. *Safety of Machinery—General Safety Requirements and Risk Assessments*. Leesburg, VA: B11 Standards, Inc., 2010.

ANSI/PMMI B155.1-2006. American National Standard, *Safety Requirements for Packaging Machinery and Packaging-Related Converting Machinery*. Arlington, VA: Packaging Machinery Manufacturers Institute.

AS/NZS 4360:2004. *Risk Management Standard*. Strathfield, NSW, Australia: Standards Association of Australia.

BS OHSAS 18001:2007. *Occupational Health and Safety Management Systems—Requirements.* London: British Standards Institution.

Framework for Cumulative Risk Assessment. EPA/630/P-02/001 F, May 2003. Washington, DC: Risk Assessment Forum, U.S. Environmental Protection Agency.

Framework for Environmental Health Risk Management. Washington, DC: Presidential/ Congressional Commission on Risk Assessment and Risk Management, 1997.

Gap, Inc. Definition of Social Responsibility at http://www.gapinc.com/public/SocialResponsibility/ socialres.shtml.

IEC 60601-1-9. International Standard for *Environmentally Conscious Design of Medical Equipment.* Geneva, Switzerland: International Electrotechnical Commission, 2007.

ISO/IEC Guide 51:1999(E). *Safety aspects—Guidelines for their inclusion in standards.* Geneva, Switzerland: International Organization for Standardization, 1999 (IEC stands for the International Electrotechnical Commission.)

IUD v. API, 448 U.S. at 655. Supreme Court Decision, Benzene. 1980. osha.gov/pls/oshaweb/ owadisp.show_document?p_table=PREAMBLES&p_id=767.

Lowrance, William F. *Of Acceptable Risk: Science and The Determination Of Safety.* Los Altos, CA: William Kaufman, 1976.

Machine Safety: *Prevention of mechanical hazards.* (2009). Quebec, Canada: The Institute for research for safety and security at work and The Commission for safety and security at work in Quebec, 2009. Also at www.irsst.qe.ca.en/home.html.

National Census of Fatal Occupational Injuries in 2011. Bureau of Labor Statistics, U.S. Department of Labor, Washington, DC. USDL 08-1182 (released Aug. 20, 2008). Also at http://www.bls.gov.iif/oshcfoil.htm.

OSHA's Voluntary Protection Program (VPP). Section Cin CSP 03-01-002 – TED 8.4, Voluntary Protection Programs (VPP): Policies and Procedures, 2003. At http://www.osha.gov/pls/ oshaweb/owadisp.show_document?p_table=DIRECTIVES&p_id=2976.

Risk Management Definitions. Washington, DC: U.S. Department of Transportation, Pipeline and Hazardous Materials Safety Administration (PHMSA), 2005. (DOT, Office of Hazardous Materials Safety.) Also at www.phmsa.dot.gov/hazmat/risk.

"Sci-Tech Dictionary" 2009. www.answers.com.

SEMI S2-0706. *Environmental, Health, and Safety Guideline for Semiconductor Manufacturing Equipment,* 2006. San Jose, CA: SEMI (Semiconductor Equipment and Materials International). ("Related Information 1—Equipment/Product Safety Program" is an adjunct to these guidelines.) www.semi.org.

Social responsibility definition at http://www.mallenbaker.net/csr/definition.php.

Stephans, Richard A. *System Safety for the 21st Century.* Hoboken, NJ: Wiley, 2004.

UN ISDR. *Terminology: Basic Terms of Disaster Risk Reduction.* United Nations, New York, 2009. Also at http://www.unisdr.org/eng/library/lib-terminology-eng.htm.

Water Quality: Guidelines, Standards and Health, World Health Organization (WHO) 2001. Edited by Lorna Fewtrell and Jamie Bartram. Paul R. Hunter and Lorna Fewtrell wrote Chapter 10, "Water Quality: Guidelines, Standards and Health." London: IWA Publishing.

CHAPTER 3

INNOVATIONS IN SERIOUS INJURY AND FATALITY PREVENTION

Whereas the occupational fatality rate per 100,000 employees was reduced from 17.0 in 1971 to 4.0 in 2005, a 58% decline, the rate has been close to stationary from 2006 through 2011. In those six years, the rate has ranged from 3.9 to 3.5. Several attempts have been made to develop strategies to reduce the number of fatalities and the fatality rate without substantial progress.

In this chapter we propose that major and somewhat shocking innovations in the content and focus of occupational risk management systems will be necessary to achieve additional progress in fatality and serious injury prevention. Such innovations are discussed in this chapter as comments are made on:

- The history of activities focused on preventing serious injuries and fatalities
- Statistics displaying trends with respect to serious injuries and fatalities
- Developing specific industry and organizational statistics to identify opportunities
- Data on the types of activities out of which many serious injuries occur
- Methods to obtain interest in serious injury and fatality prevention
- Significant conceptual barriers to serious injury prevention
- Achieving a culture change
- Procedures to develop analytical data on serious incident experience
- Risk assessments as the core of an operational risk management system
- Prevention through design

Advanced Safety Management: Focusing on Z10 and Serious Injury Prevention,
Second Edition. Fred A. Manuele.
© 2014 John Wiley & Sons, Inc. Published 2014 by John Wiley & Sons, Inc.

- Maintenance for system integrity
- Transitions in human error prevention
- Management of change/Pre-job planning
- Incident Investigation
- The Ideal: A Systemic Socio-Technical Model for Operational Risk Management

HISTORY

Results achieved in recent years from attempts to reduce serious injuries and fatalities cannot be considered stellar. In 2007, a national forum, "Fatality Prevention in the Workplace," was sponsored by Indiana University of Pennsylvania in cooperation with the Alcoa Foundation. Attendance was great and the speakers provided good information, a large part of which suggested tweaking the elements in existing occupational risk management systems.

At about the same time, a study was in progress at ORC Worldwide to identify the characteristics of serious injuries and fatalities. ORC Worldwide is an organization whose members represent about 120 Fortune 500 companies. The intent of the study, partially achieved, was to provide member companies with information on how to improve reduction efforts.

In a 2007 press release announcing the Alcoa Foundation grant to the Foundation for Indiana University to support a national forum on fatality prevention, Lon Ferguson said:

> Reliance on traditional approaches to fatality prevention has not always proven effective. This fact has been demonstrated by many companies, including some thought of as top performers in safety and health, as they continue to experience fatalities while at the same time achieving benchmark performance in reducing less-serious injuries and illness

Ferguson's statement still applies, and let's emphasize: "Reliance on traditional approaches to fatality prevention has not always proven effective." Companies with outstanding records showing reductions on less-serious injuries may not have had similar reductions for serious injuries and fatalities.

At ORC Worldwide (now Mercer HSE Networks), about 40 companies are presently involved in an additional study to determine what can be done to reduce occupational fatalities. Such studies must be made. But it must be recognized that major innovations in safety management systems will be necessary to achieve further reductions in fatalities and serious injuries. As will be shown in this chapter, the number and rates of fatalities have plateaued.

Tweaking systems in place will not achieve the substantial improvements desired.

SCOPE OF THIS CHAPTER

For several reasons serious injuries and fatalities are treated in this chapter as a single subject. Many serious injuries could have been fatalities under slightly different circumstances. Thus, data on serious injuries (and near misses) should be analyzed because the results may provide valuable information on actions to be taken to prevent fatalities and other serious injuries.

Causal factors for serious injuries and the actions necessary to prevent them are identical to those for fatalities. A large majority of organizations do not have fatalities, but they may have serious injuries. As shown later in the chapter, data derived from analyses of serious injuries may be influential in achieving attention to incidents that have fatality potential.

SERIOUS INJURY TRENDING

A 2005 National Council on Compensation Insurance (NCCI) research brief entitled "Workers' Compensation Claim Frequency Down Again" states that "there has been a larger decline in the frequency of smaller lost-time claims than in the frequency of larger lost-time claims." The data in Table 3.1 appear in NCCI's 2005 State of the Line Report. These data pertain to the years 1999 and 2003, expressed in 2003 hard dollars.

TABLE 3.1

Value of Claim	Percent Reduction
Less than $2000	34%
>$2000 to $10,000	21%
>$10,000 to $50,000	11%
More than $50,000	7%

The reduction in cases valued from $10,000 to $50,000 is about one-third of that for cases valued at less than $2000. For cases valued over $50,000, the reduction is about one-fifth of that for less costly injuries. Thus, costly claims—those for serious injuries and fatalities—loom larger within the spectrum of all claims reported.

In the August 2011 NCCI Research Brief entitled "Workers Compensation Claims Frequency," NCCI states that from 2005 through 2009, "After accounting for wage and medical cost inflation, claims below $50,000 exhibited a greater rate of decline than those above $50,000," as shown in Table 3.2. These data are in concert,

TABLE 3.2

Value of Claim	Percent Reduction
Less than $2000	−25%
$2000 to $10,000	−22%
$10,000 to $50,000	−20%
$50,000 to $250,000	−14%
More than $250,000	−9%

generally, with the trends shown for the years 1999 through 2003. Reductions in less serious injuries are substantially more than those for more serious injuries.

In the August 2011 NCCI Research Brief, NCCI also states: "Workers compensation claim frequency for lost-time claims has increased 3% in 2010. This represents the first increase since 1997 and only the third time that frequency has increased in the last 20 years." This upward claim frequency trend is relative to the increase in fatalities for 2010 shown in Table 3.3. It is also of interest that the 2009 State of the Line Report issued by the NCCI shows that injury frequency had declined consistently for all injury types except for permanent total disabilities.

FATALITY TRENDING

For the following exhibits, the data are based on excerpts from "Accident Facts," National Safety Council (NSC), 1995, Itasca, IL and the "Census of Fatal Occupational Injuries, 1996–2011," issued by Bureau of Labor Statistics (BLS), U.S. Department of Labor, Washington, DC. In the NSC data and the BLS reports, the fatality rate is the number of fatalities per 100,000 workers.

TABLE 3.3   All Fatalities – All Occupations, 1941–2011

Year	No. of Fatalities	Fatality Rate
1971	13,700	17
1981	12,500	13
1991	9,800	8
2001	5,900	4.3
2002	5,524	4.0
2003	5,559	4.0
2004	5,703	4.1
2005	5,702	4.0
2006	5,703	3.9
2007	5,488	3.7
2008	5,214	3.7
2009	4,551	3.5
2010	4,690	3.6
2011	4,693	3.5

Reductions in the number of fatalities and fatality rates have been huge and commendable. But it must also be noted that the fatality record has plateaued in recent years. Fatality rates over the past six years are in the range 3.9 to 3.5, with an average of 3.7. Some might say that the low-hanging fruit has been picked and that achieving further substantial reductions will require exceptional effort.

For Table 3.4, years ending in 1 were chosen as a focal point beginning with 1971, so that an observation could be made of results that managements have achieved since OSHA became law. Indirectly, OSHA became a stimulus and the pace of reductions in fatalities, and fatality rates increased.

TABLE 3.4

	1941	1971	Increase	Decrease
Number of employees	48,500,000	78,500,000	63%	
Number of fatalities	18,000	13,700		24%
Fatality rate	37	17		54%

	1971	2010		
Number of employees	78,500,000	130,000,000	66%	
Number of fatalities	13,700	4,547		67%
Fatality rate	17.0	3.5		79%

DEVELOPING FATALITY DATA FOR INDIVIDUAL INDUSTRY CATEGORIES

All of the preceding data are macro; they pertain to all occupations. It is suggested that further studies be made of industry categories with respect to their trends for fatalities and fatality rates and the types of activities in which fatalities occur. Two industry categories were selected for this discussion: the North American Industry Classification System Codes for Manufacturing and Wholesale Trade, as in the BLS listing (Table 3.5).

Types of operations in these two industries and the inherent risks differ greatly. Fatality rates for the wholesale trade are over twice those for manufacturing. For examples of the studies that need to be made so that specific remedial actions can be proposed, consider the data in Table 3.6, taken from the 2010 BLS report on occupational fatalities.

TABLE 3.5

	Year	Number of Fatalities	Fatality Rate
Manufacturing	2009	304	2.2
	2010	320	2.2
	2011	322	2.2
Wholesale	2009	186	4.9
trade	2010	185	4.8
	2011	189	4.0

TABLE 3.6

	"Highway" Situations		Falls	
	Number of Fatalities	Percent of Total	Number of Fatalities	Percent of Total
BLS–totals	1044	22%	637	14%
Manufacturing	51	16%	42	13%
Wholesale trade	59	32%	15	8%

It is not surprising that 32% of the total number of fatalities in the wholesale trade group occur in highway situations, whereas only 16% of such fatalities occur in manufacturing operations. The use of motor vehicles for deliveries is greater in the wholesale trade category. And considering the nature of operations, it is logical that the percent of fatalities resulting from falls is much greater in manufacturing than in the wholesale trades.

The point is that the innovations chosen to reduce serious injuries and fatalities should pertain specifically to operational needs. While generalities can be suggested with respect to the content and order of elements in an occupational risk management system, emphasis in the application of those elements should result from studies that determine where the largest opportunities lie and where the emphasis should be placed.

ACTIVITIES IN WHICH SERIOUS INJURIES AND FATALITIES OCCUR

A statistical history supports proposing that safety professionals pay particular attention to the characteristics of incidents resulting in serious injuries, particularly with respect to the nature of the work being done and the job titles of injured personnel.

Reviews that I have made of over 1800 incident investigation reports, mostly for serious injuries and fatalities, showed that a significantly large share of incidents resulting in serious injury and fatalities occurs:

- When unusual and nonroutine work is being performed
- In nonproduction activities
- In at-plant modification or construction operations (replacing a motor weighing 800 pounds to be installed on a platform 15 feet above the floor)
- During shutdowns for repair and maintenance, and startups
- Where sources of high energy are present (electrical, steam, pneumatic, chemical)
- Where upsets occur: situations going from normal to abnormal

I made the following observations as the research progressed:

- Causal factors for low-probability/serious-consequence events are seldom represented in the analytical data on accidents that occur frequently. (Some ergonomics-related incidents are the exception.)
- Many incidents resulting in serious injury are unique and singular events, having multiple and complex causal factors that may have organizational, technical, or operational systems, or cultural origins.

Dan Petersen was an early promoter of giving particular attention to serious injury prevention. In the second edition of *Safety Management,* Petersen wrote:

If we study any mass data, we can readily see that the types of accidents that result in temporary total disabilities are different from the types of accidents resulting in permanent partial disabilities or in permanent total disabilities or fatalities. The causes are different. There are different sets of circumstances surrounding severity. Thus, if we want to control serious injuries, we should try to predict where they will happen. Today, we can often do just that. (p. 12)

The key to Petersen's message is prediction. He implies that studies should be made of serious injuries within an operation to anticipate and "predict where they will happen."

While this author listed the types of activities which research found to be prominent in serious injury and fatality reports, each entity should develop its own list, relating to its inherent risks and history.

OBTAINING INTEREST IN FATALITY PREVENTION

For this discussion, data for the manufacturing industry were selected to illustrate that annual fatality rates may be within a narrow statistical range and that the statistical probability of an organization having a fatality is low. Methods to achieve an interest in the subject are also discussed. Consider the data in Table 3.7, whose sources are reports issued by the Bureau of Labor Statistics. (Select any industry and comparable data can be produced.)

TABLE 3.7   **Manufacturing**

Year	Number of Fatalities	Fatality Rate per 100,000 Employees
2006	447	2.1
2007	392	2.4
2008	389	2.5
2009	304	2.2
2010	320	2.2
2011	322	2.2

Fatality rates fall within a very narrow range. For the six years shown in Table 3.7, the average fatality rate is 2.27%. I projected data contained in a once-in-five-years report issued by the U.S. Census Bureau for 2007 and estimated that there were about 300,000 manufacturing locations in the United States in 2010.

A Bureau of Labor Statistics report says that manufacturing had about 11,575,000 employees that year. Obviously, a very small percentage of manufacturers reported the 320 fatalities that occurred in 2010. Many have never had a fatality.

If interest is to be obtained in fatality prevention in organizations that have not have had a fatality or have had very few fatalities over a period of many years, an approach that may be successful would be to focus on the fatality potential as evidenced in the data on:

• Serious injuries that occurred

• Less-than-serious injuries that could have been serious in other circumstances

• Selected near misses that had serious injury potential

Achieving substantial reductions in the number of fatalities and the fatality rate will require developing sharp change agent skills and major innovations in operational risk management systems.

INNOVATIONS TO BE CONSIDERED

To achieve the necessary focus on serious injury and fatality prevention, the enormity of the culture changes needed must be recognized as well as how deeply some deterring premises are embedded in many companies. A list follows of innovations to be considered. Other safety professionals may want to add to the list.

- The premise that OSHA-related incidence rates are accurate measures of serious injury and fatality potential must be dislodged.
- The belief that unsafe acts of workers are the principal causes of occupational accidents must be uprooted and dislodged.
- The broadly held assumption that reducing the frequency of less-than-serious injuries will result in an equivalent reduction in serious injuries must be dislodged.
- Embedded cultural barriers that promote avoiding causal factor determination must be altered.
- Risk assessments must be recognized as the core of an operational risk management system.
- Concepts regarding prevention through design must be instituted as an element within an operational risk management system.
- An understanding must be achieved of the transition in progress concerning the prevention of human errors, which directs prevention efforts to the design of the work system and the work methods.
- Management of change/pre-job planning must be a separate and emphasized element within an operational risk management system.
- Incident investigations must be improved so that shortcomings in management systems related to serious injury and fatality potential can be identified and acted upon.
- Internally published operational risk management systems must be revised in relation to the foregoing.

Although in this chapter we focus on serious injury and fatality prevention, improving on or instituting these innovations will help reduce injuries at all severity levels.

ACHIEVING A CULTURE CHANGE WITH RESPECT TO OSHA STATISTICS

A major educational undertaking will be necessary to convince management, and subsequently all personnel, that achieving low OSHA incident rates does not indicate that controls are adequate with respect to serious injury and fatality potentials. For over 40 years, an over emphasis has been placed on achieving low OSHA incident rates, resulting in competition within companies and among companies within an industry group.

When focusing on achieving low OSHA incidence rates is deeply embedded within an entity's culture, uprooting and dislodging that concept will be difficult and will require a long-term effort. A culture change is not a one-time effort: It's a long-term journey that has to engage all members of an organization.

In the culture change process, the case should be made that recognizing and avoiding hazardous situations that have serious injury potential must be given priority attention. How to get that done must be tailored to the needs and opportunities of individual entities. Sound statistical data will help. Three courses of action are presented here for consideration.

1. All incident investigation reports for a three-year period should be collected for a sorting operation. The purpose is to select those reports describing situations for which, under slightly different circumstances, the results could have been a more serious injury or fatality. That could lead to analyses of operations in which the incidents occurred and furthering the idea that serious potential needs special consideration.

2. Have a computer run made of all workers compensation claims valued at $25,000 or more for three years. Why select this level? My experience has been that using a $25,000 cutoff level results in printouts of 6 to 8% of the total number of claims and 60 to 80% of total claims value. The number of cases in the computer outputs, even for very large companies, has been manageable and has provided good bases for analysis.

   There have been outliers. For a manufacturing company that also had a mining operation, 25% of cases valued at $25,000 or more represented 75% of total claims value. In another organization, 90% of total claims value came from 5% of the claims valued at $25,000 or more. In each case, valuable data were produced.

3. Engage employees in an information-gathering system that reports continuously on hazardous situations that have serious injury potential. The system should include near misses that could have resulted in serious consequences under slightly different circumstances. For such a process to succeed, it must be understood that employees who are encouraged to provide input are recognized as valuable resources because of their extensive knowledge regarding how the work gets done. Also, they must be respected for the knowledge and skills that they have. Feedback on employee input is a requirement.

The data collected should be mined for information that will support a proposal that additional attention be given to serious injury and fatality potential. Specifically, reviews should be made of job titles, units, or departments that are prominent in the data and the types of operations in which the injuries occurred. For example, it was found in several companies that 60 to 80% of injuries valued at $25,000 or more occurred to employees that were not making product.

Other methods must be identified to produce meaningful and convincing data relating to the inherent risks in an organization and its culture. This subject presents opportunities for safety professionals to be creative.

COUNTERING THE PREMISE: UNSAFE ACTS ARE
THE PRINCIPAL CAUSES OF OCCUPATIONAL ACCIDENTS

An article entitled "Reviewing Heinrich: Dislodging Two Myths from the Practice of Safety" was published in the October 2011 issue of *Professional Safety*. One of the myths was that unsafe acts of workers are the principal cause of occupational accidents. The paper was written based on encouragement from colleagues who encountered situations in which a speaker would quote the source of the myth—H.W. Heinrich's *Industrial Accident Prevention*—and proceed to lay out a course of preventive action focused entirely on changing worker behavior.

The article addressed moving preventive efforts from a focus on employees to: improving the work systems; the design of the work systems and work methods; the complexity of causation; and human errors occurring at organizational levels above the worker.

Response supporting the article was good, but the information developed through e-mails and phone calls indicated how deeply the Heinrichean premise—that 88% of accidents are caused by worker unsafe acts—is embedded in organizations to which safety practitioners give advice. To achieve more effective results in preventing serious injuries and fatalities, the myth that unsafe acts of workers are foremost in causal factor determination has to be dislodged from the practice of safety.

REDUCING INJURY FREQUENCY WILL NOT REDUCE
SEVERITY EQUIVALENTLY

Heinrich's premises with respect to what had become broadly known as his 300-29-1 ratios varied from the first, to the second, to the third edition of his book *Industrial Accident Prevention*. This is what appeared in the fourth edition:

> Analysis proves that for every mishap resulting in an injury there are many other similar accidents that cause no injuries whatever. From data now available concerning the frequency of potential-injury accidents, it is estimated that in a unit group of 330 accidents of the same kind and *involving the same person* [italics added], 300 result in no injuries, 29 in minor injuries and 1 in a major or lost-time injury. (p. 26)

On its face, this statement cannot be substantiated. Heinrich also wrote "that in the largest injury group—the minor injuries—lies the most valuable clues to accident causes." That became the premise from which educators taught, and many safety practitioners came to believe, that reducing accident frequency will achieve an equivalent reduction in injury severity. That is the second myth addressed in the *Professional Safety* article.

That premise is also deeply embedded in the minds of some safety practitioners and the management personnel whom they have influenced. It must be dislodged. Statistical data presented in the professional safety paper refute the premise. Long hours of research through all four editions of Heinrich's book resulted in my conclusion that the 300-29-1 ratios are not supportable.

SIGNIFICANCE OF ORGANIZATIONAL CULTURE

We refer several times in this book to an organization's culture and how it affects, favorably or unfavorably, the injury experience attained. Since causal factors for incidents resulting in serious injury are largely systemic and their presence is a reflection of an organization's culture, that subject must be explored. Comments made on organizational culture in the "August 2003 Report of the *Columbia* Accident Investigation Board" on the *Columbia* spaceship disaster are pertinent here.

In our view, the NASA organizational culture had as much to do with this accident as the foam. Organizational culture refers to the basic values, norms, beliefs, and practices that characterize the functioning of an institution. At the most basic level, organizational culture defines the assumptions that employees make as they carry out their work. It is a powerful force that can persist through reorganizations and the change of key personnel. It can be a positive or a negative force. (CAIB, Chap. 5, p. 97)

In every organization, what management does is translated directly into a system of expected performance—the reality of the culture—and that system affects, positively or negatively, decisions taken with respect to management systems, design and engineering, operating methods, and arrangement of work methods.

Consider a real-world indication of how an organization's culture—the expected and condoned system—was translated negatively into a system of expected performance. Previous to making a presentation on avoiding serious injuries and fatalities, I asked the client to provide copies of serious injury reports, from which six were selected for discussion. The causal factors in those reports focused primarily on unsafe acts of employees; and training, further education, and reinforcing employee safe practice rules was the corrective action.

During a half-hour of lecturing, the inadequacies in the investigation reports were cited and remarks were made on how an organization's culture could be a source of causal factors and could foster or impede causal factor determination. Then the investigation reports selected were distributed to discussion groups with the provision that each would choose a leader who would give a report on how the investigations could have been conducted more thoroughly. Each group was provided with a causal factor guide.

After the session, a woman said: "Mr. Manuele, I think at my location, I have the kind of a culture problem you discussed because I believe that our risks are overlooked and a lot of risk taking is accepted. I say that because all of the incident investigation reports that hit my desk put the responsibility for what happened on the worker. The reports always say things like they reinstructed the worker or discussions about safety were held with the workers or the safe practice rules are being reinforced. They don't ever really analyze the situation." Obviously, the performance expected—an aspect of the culture—accepted and condoned making superficial incident investigations. That became established as the norm.

She was asked what her job was. She said she was the plant manager. Cautiously, I was said to her: "You are the problem because you accept those shabby reports. They describe a pattern of expected performance, the culture, which has evolved in your shop over time. It has become accepted that determining and eliminating or controlling systemic causal factors is not necessary. And you can be the solution." She was sharp and quickly said: "You mean I have to convince my staff that I'm not going to accept their B.S. anymore." ..."Yes."... And her response was a pleasure to hear: "Mr. Manuele, I know how to do that."

Consider the following two examples that demonstrate how the culture accommodated a system of expected performance which condoned hazardous situations and supported excessive risk taking.

- Two workers refuse to do a job, saying that the work is too hazardous. Another worker is assigned by a supervisor to do the work, and he becomes a fatality. Speculate on the possible cultural and operational causal factors for that situation.
- There is deterioration in a tray of electrical cables, and occasionally the workers experience a minor jolt. In the maintenance department, work orders are reshuffled every Monday to establish priorities for the overly stressed maintenance personnel. Each week, the work order to make the needed repairs to the insulation in the cable tray is given a low priority. Over time, little notice is given to the jolts, and the hazard's potential is played down. An occasional jolt becomes an accepted norm. Deterioration continues. A worker makes contact that results in his electrocution. What does this sort of incident say about cultural and operational problems? Does it not establish that within the organization, excessive risk taking had become acceptable—that the system of expected performance permitted giving safety orders a lower priority?

For many incidents resulting in serious consequences, there had been, over time, a continuum of less than adequate safety decision making that resulted in a system of expected performance that condoned considerable risk taking. James Reason describes how systemic causal factors accumulate in *Managing the Risks of Organizational Accidents*. This book is recommended highly. Reason's principal research area has been human error and the way in which organizational processes and people contribute to system breakdown. He writes:

Latent conditions, such as poor design, gaps in supervision, undetected manufacturing defects or maintenance failures, unworkable procedures, clumsy automation, shortfalls in training, less than adequate tools and equipment, may be present for many years before they combine with local circumstances and active failures to penetrate the system's layers of defenses. They arise from strategic and other top-level decisions made by governments, regulators, manufacturers, designers and organizational managers. The impact of these decisions spreads throughout the organization, shaping a distinctive corporate culture and creating error-producing factors within the individual workplaces. (p. 10)

Donald A. Norman's *The Psychology of Everyday Things* is suggested as an additional resource. Norman has a background in both engineering and the social sciences. He writes:

Explaining away errors is a common problem in commercial accidents. Most major accidents follow a series of breakdowns and errors, problem after problem, each making the next more likely. Seldom does a major accident occur without numerous failures: equipment malfunctions, unusual events, a series of apparently unrelated breakdowns and errors that culminate in major disaster; yet no single step has appeared to be serious. In many cases, the people noted the problem but explained it away, finding a logical explanation for the otherwise deviant observation. (p. 128)

What Norman says about "numerous failures" being typical when major accidents occur is identical to my experience. For emphasis: As attempts are made to reduce serious injury potential, safety professionals are urged to consider seriously the following comments made by Reason and Norman:

Reason: The impact of [top-level] decisions spreads throughout the organization, shaping a distinctive corporate culture and creating error-producing factors within individual workplaces.

Norman: In many cases, the people noted the problem but explained it away, finding a logical explanation for the otherwise deviant observation.

NEEDS ASSESSMENT: AN EVALUATION OF THE CULTURE IN PLACE

Safety professionals should consider making a needs assessment, from the top down, to determine how much creative destruction and reconstruction through reeducation are needed to achieve a mindset that gives a proper place to reducing the potential for serious injury. Safety management systems that concentrate largely on the personal aspects of safety do not include activities to anticipate and identify the causal factors for low-probability/severe-consequence accidents. Nor do they include specially crafted efforts for their prevention.

James Reason has made similar comments in *Managing the Risks of Organizational Accidents*. He observes that occupational safety approaches directed largely on the unsafe acts of persons have limited value with respect to the prevention of accidents having serious consequences. (p. 239) Reason also says that:

Workplaces and organizations are easier to manage than the minds of individual workers. You cannot change the human condition, but you can change the conditions under which people work. (p. 223)

In the news release mentioned previously concerning a grant made by Alcoa to the Foundation for Indiana University of Pennsylvania (IUP) to "support a national forum on fatality prevention in the workplace," Lon Ferguson, chair of the IUP Safety Sciences Department, is quoted as saying: "The reliance on traditional approaches to fatality prevention has not always proven effective." I extend Ferguson's statement to include serious injury prevention.

ALTERING EMBEDDED CULTURE BARRIERS

All who undertake an inquiry to determine what additional actions may be taken to reduce serious injury potential will learn from R. B. Whittingham's *The Blame Machine: Why Human Error Causes Accidents*. Whittingham describes how disasters and serious accidents result from recurring but potentially avoidable human errors. He shows how such errors are preventable because they result from defective systems within a company.

From his analyses of several events, he identifies the common causes of human error and the typical system deficiencies that led to those errors. Those deficiencies were principally organizational, cultural, technical, and management systems failures. (I draw similar conclusions from my studies.)

Whittingham asserts that in some organizations, a "blame culture" exists whereby the focus in the investigation of incidents resulting in serious consequences is on individual human error, and the corrective action taken occurs at that level rather than within the system that may have enabled the human error. He stresses that placing responsibility for the incident on what a person did or did not do results in overly simplistic causal factor determination.

In his introduction, Whittingham makes a disturbing statement that derives from his investigative experience, one that safety professionals should think about as they attempt to give counsel to clients on improving serious injury prevention. He states that in many organizations, and sometimes entire industries, there is an unwillingness to look closely into error-provocative system faults. (p. xii)

In organizations where there is a reluctance to explore systemic causal factors, the incident investigation stops after addressing the individual human error—the so-called *unsafe act*. Thus, a more thorough investigation that looks into the reality of systemic causal factors is avoided. Whittingham's observation poses a serious question: In some organizations, are technical, organizational, and management systems, and cultural causal factors for incidents that result in serious injuries, glossed over when incident investigations are made?

I have not had an experience with an organization in which avoiding the reality of causal factors was an "active process" in which instructions were given to avoid the identification of systems causal factors. However, it is a certainty that avoiding the identification of system causal factors is a "passive process" in many places. In some of my studies, safety directors in several large companies provided completed incident investigation reports so that reviews could be made of their quality and to determine how deeply the investigations delved into causal factors.

On a scale of 1 to 10, with 10 being the highest possible score, some of these companies scored as low as 2. Most scored in the 4 to 5.5 range. Causal factor determination was abysmal, corrective actions were superficial, and opportunities to identify how safety management systems could be improved were overlooked.

In organizations where avoiding the identification of causal factors for incidents that result in serious injuries and fatalities is embedded within their cultures, safety professionals would have to develop superior skills as change agents to influence managements to do otherwise. A culture change is not made easily.

PROVIDING ADEQUATE RESOURCES

The following is an absolute: If adequate resources are not provided, the potential for serious injuries and fatalities increases over time. I now give more emphasis to the impact of cost reductions and organizational changes on the effectiveness of operational risk management systems. That additional emphasis is a reflection of the content of incident reports on some serious injuries and fatalities completed in companies that have experienced significant cost reductions. They indicate that:

- Equipment was operated long beyond its life expectancy.
- A significant contributing factor was a reduction in staffing.
- Unacceptable risk situations developed for such items as inadequate maintenance, inadequate employee competency, workers being overstressed beyond their mental and physical capabilities (two persons doing work for which three had previously been assigned), or a person working alone in a high-hazard situation for which the standard operating procedure calls for a work buddy.

In Z10, this subject is addressed in Section 3.1.3, "Responsibility and Authority". Top management is expected to:

A. Provide appropriate financial, human and organizational resources to plan, implement, operate, check, and review the Occupational Health and Safety Management System.

This subject is more critical when an economic downturn has been experienced that requires priority consideration. Headquarters of some companies—some very large—have required severe cost reductions, the result of which have been notable increases in situations in which risks are unacceptable. That presents difficulties for safety professionals who recognize the deterioration. As we say elsewhere, they will need to develop sharp change agent skills to have management alleviate as much of the outcome of the deterioration as practicable.

ON RISK ASSESSMENTS

I propose that risk assessments, as the core of an operational risk management system, be established as a separately identified element following soon after the first element: management leadership, commitment, demonstrated involvement, and accountability.

Europeans are in the lead in promoting risk assessments as a core value in the prevention of injuries and illnesses, but there has also been a great deal of activity throughout the world promoting risk assessments. A few examples of that activity follow.

Guidance On The Principles Of Safe Design For Work issued in 2006 by the Australian Safety and Compensation Council, an entity of the Australian government, includes a "Risk Management Process" and promotes "Integrating Risk Management into the Design Process."

Requirements for risk assessments are more explicit in the 2007 revision of BS OHSAS 18001:2007, *Occupational health and safety management systems — requirements*. Commonly spoken of as 18001, this British Standards Institution publication now says, "The organization shall establish, implement and maintain a procedure(s) for the ongoing hazard identification, risk assessment, and determination of necessary controls."

In 2008, the Health and Safety Executive in the UK issued "Five steps to risk Assessment." All employers in the UK must, by law, conduct risk assessments.

ANSI B11.0, *Safety of Machinery—General Safety Requirements and Risk Assessments*, applies to a broad range of machinery. It is an American National Standard. Note that "risk assessments' is included in the title. It describes procedures for identifying hazards, assessing risks, and reducing risks to an acceptable level over the life cycle of machinery.

In March 2011, the U.S. Pipeline and Hazardous Materials Safety Administration proposed modification of Hazardous Materials Regulations to require that risk assessments be made of loading and unloading operations.

In August 2008, the European Union launched a two-year health and safety campaign focusing on risk assessment. They said in an undated bulletin that is available on the Internet:

> Risk assessment is the cornerstone of the European approach to prevent occupational accidents and ill health. If the risk assessment process—the start of the health and safety management approach—is not done well or not at all, the appropriate preventive measures are unlikely to be identified or put in place.

It is highly significant that the European Union declared that "risk assessment is the cornerstone of the European approach to prevent occupational accidents and ill health." That statement is foundational and should be supported by safety professionals. William Johnson expressed a companion view in *MORT Safety Assurance Systems*, published in 1980:

> Hazard identification is the most important safety process in that, if it fails, all other processes are likely to be ineffective. (p. 245)

Two components must be addressed in developing a risk assessment: probability of occurrence and severity of outcome. Hazard identification and analysis establishes severity—the probable harm or damage that could result if an incident occurs. To convert a hazard analysis into a risk assessment, a probability of occurrence factor must be added. Then risk levels can be established (e.g., Low, Moderate, Serious, High) and priorities can be set.

A hazard is defined as the potential for harm. Hazards include all aspects of technology and activity that produce risk. Hazards are the generic base of, as well as the justification for the existence of, the practice of safety. If there were no hazards—no potential for harm—safety professionals need not exist. It should be understood that: *The entirety of purpose of those responsible for safety, regardless of their titles, is to manage with respect to hazards so that the risks deriving from the hazards are acceptable.*

Thus, the case can be soundly made that risk assessment should be the core of an operational risk management system. My experience has been that if employees at all levels had more knowledge and awareness of hazards and risks, there would be fewer serious injuries and fatalities. Getting the required knowledge embedded in the minds of all employees requires a major, long-term endeavor. Crafting specifically directed communication and training programs will be necessary to achieve the awareness and knowledge required—the culture change.

Literature on the many techniques available for making risk assessments is abundant. For example, in ANSI/ASSE Z690.3, *Risk Assessment Techniques*, reviews are included of 31 techniques, including primary hazard analysis, fault tree analysis, hazard and operability, studies, bowtie analysis, Markov analysis, and Bayesian statistics. Uncomplicated systems that could be introduced to supervisors and front-line employees are not as prevalent. Such a system is contained in an extension of the European Union bulletin cited previously (see Table 3.8).

TABLE 3.8   A Stepwise approach to risk assessment

1. Identify hazards and those at risk.
2. Evaluate risks.
3. Prioritize risks.
4. Decide on preventive action.
5. Take action.
6. Monitor and review results.

I emphasize strongly that having employees become capable of making risk assessments and encouraging them to adopt a mindset whereby identifying and analyzing hazards and the risks that derive from them become integral in how they approach and think about the work they do would be a major step forward in reducing serious injuries and fatalities.

Having knowledge of hazard identification and analysis and risk assessments become rooted within an organization's culture is the type of innovative action needed to further reduce serious injury and fatality potential.

PREVENTION THROUGH DESIGN

In *Guidance On The Principles Of Safe Design For Work*, comments are made about the results of a study on the "contribution that the design of machinery and equipment has on the incidence of fatalities and injuries in Australia":

Of the 210 identified workplace fatalities, 77 (37%) definitely or probably had design-related issues involved. Design contributes to at least 30% of work-related serious non-fatal injuries. (p. 6)

In review that I made of a variety of accident investigation reports (not limited to machinery), I concluded that there were implications of workplace and work methods design inadequacies in over 35%. To reduce serious injury and fatality potential, I propose that prevention through design be established as a separately identified element within an operational risk management system. To provide the necessary education for designers, safety professionals are encouraged to develop supportive data on incidents in which design shortcomings were identified and undertake a major effort to have the American National Standard ANSI/ASSE Z590.3-2011 be accepted as a design guide. The title of Z590.3-2011 is *Prevention through Design: Guidelines for Addressing Occupational Hazards and Risks in Design and Redesign Processes.*

This is from the scope section of Z590.3: "This standard provides guidance on including prevention through design concepts within an occupational safety and health management system. Through the application of these concepts, decisions pertaining to occupational hazards and risks can be incorporated into the process of design and redesign of work premises, tools, equipment, machinery, substances, and work processes, including their construction, manufacture, use, maintenance, and ultimate disposal or reuse."

This standard provides guidance for a life-cycle assessment and design model that balances environmental and occupational safety and health goals over the life span of a facility, process, or product. Z590.3 says that, in so far as is practicable, the goal shall be to assure that the design selected attains the following:

- An acceptable risk level is achieved, as defined in this standard.
- The probability of personnel making human errors because of design inadequacies is as low as reasonably practicable.
- The ability of personnel to defeat the work system and the work methods prescribed is as low as reasonably practicable.
- The work processes prescribed take into consideration human factors (ergonomics)—the capabilities and limitations of the work population.
- Hazards and risks with respect to access and the means for maintenance are at as low as reasonably practicable.
- The need for personal protective equipment is as low as reasonably practicable, and aid is provided for its use where necessary (e.g., anchor points for fall protection).
- Applicable laws, codes, regulations, and standards have been met.
- Any recognized code of practice, internal or external, has been considered.

Proposing that prevention through design be a specifically defined element in an operational risk management system is also influenced by ongoing transitions observed in the methods utilized to eliminate or reduce the occurrence of human error. As will be shown, surprisingly, some specialists in human error prevention now say that the first action to be taken when human errors are noted is to consider the design of the system in which the errors occurred.

TRANSITIONS IN PROGRESS ON HUMAN ERROR PREVENTION

On November 4 and 5, 2010, in San Antonio, ASSE sponsored a symposium entitled "Rethink Safety: A New View of Human Error and Workplace Safety." It was not a surprise at a symposium on human error that speakers commented on such as cognitive theory, the properties of human cognition, variable errors and constant errors, imperfect rationality, and mental behavioral aspects of error.

But the suggestions made by several speakers on the sources of human error and the corrective actions to be taken when human errors occur were surprising. In summary, several speakers said that:

- The first step to be taken when human errors occur is to examine the design of the workplace and the work methods.
- Managers may wish to address human error by "getting into the heads" of their employees, with training being the default corrective action—which will not be effective if error potential is designed into the work.
- It is management's responsibility to anticipate errors and to have work systems and work methods designed to reduce the potential for error.

A colleague said that to understand what is happening in the human error field, one must read Sidney Dekker's writings, which, he said, have had considerable influence. Dekker wrote *The Field Guide to Understanding Human Error*, a 2006 publication. Dekker's doctorate from Ohio State University is in cognitive system engineering. A few excerpts from Dekker's book follow.

> Human error is not a cause of failure. Human error is the effect, or symptom, of deeper trouble. Human error is … systematically connected to features of people's tools, tasks, and operating systems. (p. 15)

> Sources of error are structural, not personal. If you want to understand human error, you have to dig into the system in which people work. You have to stop looking for people's shortcomings. (p. 17)

> [Dekker quotes James Reason]: "Rather than being the main instigator of an accident, operators tend to be the inheritors of system defects created by poor design, incorrect installation, faulty maintenance and bad management decisions. Their part is usually that of adding the final garnish to a lethal brew whose ingredients have already been long in the cooking." (p. 88 in Dekker: p. 173 in Reason's *Human Error*)

> The Systemic Accident Model … focuses on the whole [system], not [just] the parts. It does not help you much to just focus on human errors, for example, or

an equipment failure, without taking into account the socio-technical system that helped shape the conditions for people's performance and the design, testing and fielding of that equipment. (p. 90)

System accidents result not from component failures, but from inadequate control or enforcement of safety-related constraints on the development, design and operation of the system. (p. 91)

This transition in the human error field, which moves from a focus on attempting to change worker behavior to an emphasis on improving the design of the system in which people work, also supports the premise that prevention through design should be a specifically defined element in an operational risk management system.

MAINTENANCE FOR SYSTEM INTEGRITY

There is a relationship between the provision to provide the necessary resources and achieving and sustaining the integrity of facilities and operating equipment. It is a given that failure to maintain systems to a high level of integrity is an invitation for serious injuries to occur.

MANAGEMENT OF CHANGE/PRE-JOB PLANNING

Management of change/pre-job planning is a process to be employed before modifications are made, and continuously throughout the modification activity, to assure that:

- Hazards are identified and analyzed and risks are assessed.
- Appropriate avoidance, elimination, or control decisions are made so that acceptable risk levels are achieved and maintained during a change process.
- New hazards are not knowingly created by the change.
- The change does not affect negatively hazards resolved previously.
- The change does not enhance the potential for harm from an existing hazard.

In the management of change (MOC) process, consideration would to be given as applicable to:

- The safety of employees making the changes
- Employees in adjacent areas
- Employees who will be engaged in operations after changes are made
- Environmental, public, product safety, and quality concerns
- Avoiding property damage and business interruption

Having an effective management of change/pre-job planning system in place would have served well to reduce the probability of serious injuries and fatalities occurring in the operational categories listed in the earlier section "Activities in Which Serious Injuries and Fatalities Occur."

Additional and substantial support for having a management of change/pre-job planning system in place comes from a study led by Thomas Krause, chairman of the board of BST. (These data were provided in a personal communication with Krause. BST is to publish a paper including this material.)

Seven companies participated in a study made in 2011 during which incidents that had serious injury or fatality potential were separated from the remainder of the reports collected. Pre-job planning shortcomings were noted in 29% of the incidents that had serious injury or fatality potential; pre-job planning inadequacies were identified in 17% of the non-serious injury potential group. Pre-job planning is another name for management of change.

Experience in the auto industry is convincing as to the value of having management of change/pre-job planning systems in place. A United Auto Workers bulletin for the period 1973 through 2007 (no longer available) indicated that 42% of fatalities occurred to skilled-trades workers, about 20% of the membership. UAW personnel provided data in 2012 through e-mail and phone correspondence indicating that from 2008 through 2011, 47% of fatalities occurred to skilled-trades workers.

The point is that skilled-trades workers in every industry are often involved in unusual and nonroutine work, at-plant modification or construction operations, shutdowns for repair and maintenance, startups, and where sources of high energy are present. The data clearly establishe that the potential for serious injuries and fatalities occurring can be diminished by having a MOC/pre-job planning system in place as a separately identified element within an operational risk management system.

INCIDENT INVESTIGATION

Incident investigation reports, done well, may include valuable data on predictive indicators pertaining to serious injury and fatality potential. But I have found that the gap between procedures established for incident investigation and what actually takes place can be huge. Even in the best safety management systems, the quality of incident investigation can be less than adequate.

In one very large organization, it was agreed that if the safety staff promoted adoption of a system as uncomplicated as the five-why technique to improve incident investigation, and achieved a B + grade in two years, a miracle would have been produced.

I suggest that internal evaluations be made of the quality of incident investigation to determine whether improvement is necessary. Such evaluations often indicate that culture problems exist and that it had become accepted practice for supervisors, management personnel above the supervisor level, and safety professionals to sign-off on shallow investigation reports.

Because of the significance of the information that can be produced, it is vital to achieve a high performance level for incident investigation, to lessen serious injury and fatality potential. Successful investigation systems for near misses that could have resulted in serious results in slightly differing circumstances may also produce critical predictive indicators.

SUMMARY: THE IDEAL—A MODEL

As a reflection of all that is discussed in this chapter, I present A Socio-Technical Model for an Operational Risk Management System. Effectively adopting the elements in this model will reduce serious injury and fatality potential—and it will have the same effect on the occurrence of all other injuries. Safety professionals are encouraged to

---

A Socio-Technical Model for an Operational Risk Management System

The Board of Directors and Senior Management Establish a Culture
for Continuous Improvement
That Requires Defining, Achieving and Maintaining Acceptable Risk Levels
in All Operations

Management Leadership, Commitment, Involvement, and the Accountability system,
Establish That the Performance
Level to be Achieved Is in Accord with the Culture Established by the Board

To Achieve Acceptable Risk Levels,
Management Establishes Policies, Standards, Procedures and Processes With Respect To:

- Providing adequate resources
- Risk assessment, prioritization and management
  - Applying a hierarchy of controls
- Prevention through design
  - Inherently safer design
  - Resiliency, reliability and maintainability
- Competency and adequacy of staff
  - Capability – skill levels
  - Sufficiency in numbers
- Maintenance for system integrity
- Management of change/pre-job planning
- Procurement – safety specifications
- Risk-related systems
  - Organization of work
  - Training – motivation
  - Employee participation
  - Information – communication
  - Permits
  - Inspections
  - Incident investigation and analysis
  - Providing personal protective equipment
- Third party services
  - Relationships with suppliers
  - Safety of contractors – on premises
- Emergency planning and management
- Conformance/Compliance assurance reviews

Performance measurement: Evaluations are made and reports are prepared for management
review to support continuous improvement and to assure that acceptable risk levels
are maintained.

---

take subjects from this model that they believe can be fit into management systems in place which are compatible with an organization's culture. Each element in the model must also be included in the ideal for an operational risk management system.

It is recognized that the model presented is a major departure from other outlines for a safety management system. But it supports the position taken in this chapter that "major and somewhat shocking innovations in the content and focus of occupational risk management systems will be necessary to achieve additional progress in serious injury and fatality prevention."

REFERENCES

Accident Facts, 1995 edition. Itasca, IL: National Safety Council.

ANSI/AIHA Z10-2012. *Occupational Health and Safety Management Systems*. Fairfax, VA: American Industrial Hygiene Association, 2012. American Society of Safety Engineers is now the secretariat. Access is at https://www.asse.org/cartpage.php?link=z10_2005.

ANSI/ASSE Z590.3-2011. *Prevention Through Design: Guidelines for Addressing Occupational Hazards and Risks in Design and Redesign Processes*. Des Plaines, IL: American Society of Safety Engineers, 2011.

ANSI/ASSE Z690.3. *Risk Assessment Techniques*. Des Plaines, IL: American Society of Safety Engineers, 2011.

ANSI B11.0. *Safety of Machinery—General Safety Requirements and Risk Assessments*. Leesburg, VA: B11 Standards, Inc., 2010.

BS OHSAS 18001:2007. *Occupational health and safety management systems—Requirements*. London: BSI Group, 2007.

Columbia Accident Investigation Report, Vol. 1. NASA, Washington, DC, 2003. At www.nasa. gov/columbia/home/CAIB_Vol1.html (Chap. 5, p. 97).

Dekker, Sidney. *The Field Guide to Understanding Human Error*. Burlington, VT: Ashgate Publishing Company, 2006.

Ferguson, Lon, Indiana, PA: Indiana University of Pennsylvania, News Release on Forum on Fatality Prevention in the Workplace, 2007.

"Five steps to risk assessment." London: The Health and Safety Executive, 2008.

Five-Why System: http://www.mapwright.com.au/newsletter/fivewhys.pdf and http://www. moresteam.com/toolbox/5-why-analysis.cfm.

Guidance On The Principles Of Safe Design For Work. Canberra, Australia: Australian Safety and Compensation Council, an entity of the Australian government, 2006.

Heinrich, H. W. *Industrial Accident Prevention*, 4th ed. New York: McGraw-Hill, 1959.

Johnson, William. *MORT Safety Assurance Systems*. Itasca, IL: National Safety Council, 1980. (Also published by Marcel Dekker, New York.)

Manuele, Fred A. "Reviewing Heinrich: Dislodging Two Myths from the Practice of Safety." *Professional Safety*, Oct. 2011.

Manufacturing employment: http://data.bls.gov/timeseries/CES3000000001?data_tool=XGtable.

"National Census of Fatal Occupational Injuries." Washington, DC: Bureau of Labor Statistics, 1997–2010. www.bls.gov/iif.

Norman, Donald A. *The Psychology of Everyday Things.* New York: Basic Books, 1998.

Petersen, Dan. *Safety Management,* 2nd ed. Des Plaines, IL: American Society of Safety Engineers, 1998.

Reason, James. *Human Error.* New York: Cambridge University Press, 1990.

Reason, James. *Managing the Risks of Organizational Accidents.* Burlington, VT: Ashgate Publishing Company, 1997.

"Risk Assessment." The European Union, 2008. http://osha.europa.eu/en/topics/riskassessment.

"State of the Line" by Dennis Mealy. National Council on Compensation Insurance News Bulletin, 2005, 2006, 2009. www.ncci.com.

"The Employment Situation—January 2011." Bureau of Labor Statistics (enter the title into a search engine).

U.S. Census Bureau for 2007. Enter "U.S. Census Bureau EC0700A1" into a search engine.

U.S. Pipeline and Hazardous Materials Safety Administration, Washington, DC: Department of Transportation, March 11, 2011.

Whittingham, R. B. *The Blame Machine: Why Human Error Causes Accidents.* Burlington, MA: Elsevier–Butterworth–Heinemann, 2004.

"Workers Compensation Claim Frequency Down Again." National Council on Compensation Insurance News Bulletin, June 2005.

"Workers' Compensation Claims Frequency" NCCI Research Brief, Aug. 2011. https://www.ncci.com/documents/2011_Claim_Freq_Research.pdf.

CHAPTER 4

HUMAN ERROR AVOIDANCE
AND REDUCTION

Several references were made in Chapter 3, "Innovations in Serious Injury and Fatality Prevention," to human errors as causal factors for accidents. It was said that many serious injuries and fatalities result from recurring but potentially avoidable human errors and that organizational, cultural, technical, and management systems deficiencies often lead to those errors.

Although emphasizing the reduction in human errors that occur above the worker level was proposed many years ago, doing so is infrequently prominent in the work of safety professionals. Fortunately, renewed interest has emerged in being able to explain why human errors occur in the occupational setting. On November 4 and 5, 2010, in San Antonio, the American Society of Safety Engineers sponsored a symposium titled "Rethink Safety: A New View of Human Error and Workplace Safety."

It was not a surprise at a symposium on human error that speakers commented on cognitive theory, the properties of human cognition, variable and constant errors, imperfect rationality, and mental behavioral aspects of error. But the suggestions made by several speakers on the sources of human error and the corrective actions to be taken when human errors occur were surprising. In summary, several speakers said that:

- The first step to be taken when human errors occur is to examine the design of the workplace and the work methods.
- Managers may wish to address human error by "getting into the heads" of their employees, with training being the default corrective action—which will not be effective if error potential is designed into the work.

Advanced Safety Management: Focusing on Z10 and Serious Injury Prevention,
Second Edition. Fred A. Manuele.
© 2014 John Wiley & Sons, Inc. Published 2014 by John Wiley & Sons, Inc.

- It is management's responsibility to anticipate errors and to have work systems and work methods designed to reduce error potential.

Several speakers referred to Sidney Dekker's work. One of his books is titled *The Field Guide to Understanding Human Error*. Dekker's Ohio State University doctorate is in cognitive system engineering. I recommend Dekker's *Field Guide* highly to all safety professionals.

So also is James Reason's book *Human Error* recommended highly. Both books support the premise that decisions made at a senior management level are primary in determining the physical aspects of a workplace and the climate in which work is done. If human errors occur at a senior management level, by commission or omission, it is likely that in many situations the result will be unacceptable risk levels.

A few excerpts from Dekker's book follow. They direct a person who gives counsel on error avoidance and reduction to try to effect the decisions made above the worker level as a primary initiative. Achieving success in such an endeavor results in a culture change.

Human error is not a cause of failure. Human error is the effect, or symptom, of deeper trouble. Human error is ... systematically connected to features of people's tools, tasks, and operating systems. (p. 15)

Sources of error are structural, not personal. If you want to understand human error, you have to dig into the system in which people work. You have to stop looking for people's shortcomings. (p. 17)

Dekker quoted from James Reason's book:

Rather than being the main instigator of an accident, operators tend to be the inheritors of system defects created by poor design, incorrect installation, faulty maintenance and bad management decisions. Their part is usually that of adding the final garnish to a lethal brew whose ingredients have already been long in the cooking. (p. 88 in Dekker; p. 173 in Reason's *Human Error*)

Dekker also wrote:

The Systemic Accident Model ... focuses on the whole [system], not [just] the parts. It does not help you much to just focus on human errors, for example, or an equipment failure, without taking into account the socio-technical system that helped shape the conditions for people's performance and the design, testing and fielding of that equipment. (p. 90)

This transition in the human error field moves the focus from attempting to change worker behavior to an emphasis on improving the design and operation of the system in which people work. Dan Petersen, whose work was influential, said as far back as March 2003 that "we need to be able to explain why human error happens and what it is." Excerpts follow from a speech given by Petersen at the "Human Error in Occupational Safety Symposium" held by the American Society of Safety Engineers.

For the last 90 years, safety has gone through many frontiers, many fads, and occasionally a true paradigm shift. Interest in Behavioral Safety has paled and some are discovering the importance of management and culture. But even in an environment with good management and a good culture, people are still being injured due to human error, their own or someone else's.

We need to be able to explain why human error happens and what it is. That knowledge will open up the next frontier in safety management.

Petersen presents an interesting proposal—that acquiring knowledge of how and where human errors occur and offering advice on human error reduction will open up the next frontier in safety management.

That ties in well with the theme of this chapter. It also relates closely to my research, which shows that *human errors at some level* are contributing factors for many incidents that result in serious injuries, particularly low-probability/serious-consequence events.

Safety professionals will do a better job in giving counsel on serious injury and fatality prevention if they are aware of human error causal factors. With a focus on improving management systems to meet Z10 provisions and reducing the potential for serious injuries to occur, in this chapter we:

- Encourage safety professionals to become more involved in human error avoidance and reduction, particularly above the worker level.
- Explore human errors as causal factors for low-probability/serious-consequence incidents.
- Direct attention to human errors that derive from:
  - Management decisions that result in unacceptable risk levels
  - Extremes of cost reduction
  - Safety management system deficiencies
  - Design and engineering decisions
  - Overly stressing work methods
  - Error-provocative operations
- Provide a selected literature and resource review.
- Comment on the relationship between behavioral safety, human error reduction, and serious injury prevention.

This chapter is not a text on human error avoidance and reduction. Selected publications are noted that provide basic knowledge on the subject that safety professionals should have. They need not be concerned over a lack of resources. Enter "Human Error Reduction" into a search engine and the screen will show that there are over 5,000,000 resources.

DEFINING HUMAN ERROR AND HUMAN ERROR REDUCTION

It seems that every author on human error has his or her own definition, and they vary somewhat. Although some are even obscure or esoteric the many definitions do have similarities. Two selected definitions of human error follow. James Reason's principal research area has been in human error and the way in which people and organizational processes contribute to the breakdown of complex, well-defended technologies. In *Human Error*, Reason offers this definition:

Error will be taken as a generic term to encompass all those occasions in which a planned sequence of mental or physical activities fails to achieve its intended outcome, and when these failures cannot be attributed to the intervention of some chance agency. (p. 9)

Reason's book was written for cognitive psychologists, human factors professionals, safety managers, and reliability engineers. His definition covers all the bases but is not quite as specific as is needed in the occupational setting.

Trevor Kletz, in *An Engineer's View Of Human Error*, gives a definition that relates more precisely to the places in which people work:

I have tried to show that so-called human errors are events of different types (slips and lapses, mistakes, violations, errors of judgment, mismatches and ignorance of responsibilities), made by different people (managers, designers, operators, construction workers, maintenance workers and so on) and that different actions are required to prevent them happening again: in some cases better training or instructions, in other cases better enforcement of the rules, in most cases a change in the work situation. (p. 181)

Kletz's definition of human error fits well with my studies of accident reports. For simplicity and to have a terse definition of human error that relates directly to the occupational setting in which the exposures to injuries and illnesses occur, I present this definition:

Human error: a decision, an oversight, or a personnel action or inaction out of which the potential arises for the occurrence of a harmful incident or exposure.

Then the goal of human error avoidance and reduction is to have the probability be as low as reasonably practicable that decisions or oversights, made individually or accumulatively, and personnel actions or inactions, will bring about the occurrence of harmful incidents and exposures.

A BIT OF HISTORY

Dan Petersen's paper "Human Error" appeared in the December 2003 issue of *Professional Safety*. Petersen offers an interesting observation on how long ago knowledge of human factors as incident causal factors has been available. He also suggests that safety professionals have been delinquent in not absorbing and utilizing that knowledge to their professional advantage:

As an industrial engineering graduate, the author studied work simplification, plant layout and motion study, not for the purpose of reducing error, but rather to increase productivity. Years later, I became acquainted with human factors concepts in graduate work in psychology. It seemed that this was a natural for the safety profession. That was in 1971, and for some reason, the profession found OSHA and its standards to be considerably more interesting. From a human factors standpoint, it seems that safety has lost 30 years of possible progress in reducing human error.

What Petersen says is true. Safety-related literature on human errors occurring at several organizational levels that become the source of causal factors for injuries dates back at least to the 1970s. Nevertheless, a large share of safety professionals ignored it. For quite some time, many safety professionals were consumed by OSHA, which took effect in 1971. In later years a form of behavioral safety that focused on improving worker behavior attracted a great deal of their attention and time.

REVIEW OF SELECTED LITERATURE

Willie Hammer's *Handbook of System and Product Safety*, published in 1971, con-tributed significantly to my developing an interest in the levels at which human errors occur. Hammer wrote this:

> Almost every mishap can be traced ultimately to personnel error, although it may not have been error on the part of the person immediately involved in the mishap. It may have been committed by a designer, a worker manufacturing the equipment, a maintenance worker, or almost anyone other than the person present when the accident occurred.
>
> It often happens that the person involved is overwhelmed by failures due to causes beyond his or her control, failures that could have been forestalled by incorporation of suitable measures in the design stage. In many instances, due consideration was not given to human capabilities and limitations and to the factors that can and may affect a human being. (p. 68)

Hammer's message was important for me. He made plain that it would be advan-tageous in the practice of safety to look for causal factors that occur above the level of a worker who may have been involved in a mishap.

Another researcher and widely published author whose work has influenced my view of incident causation is Alphonse Chapanis. He was exceptionally well known in ergonomics and human factors engineering circles. His work is often quoted, particularly as to the benefits of considering the capabilities and limitations of workers as systems are designed. Chapanis has been strong on designing to avoid error-provocative work methods.

Early in my career it became apparent that musculoskeletal injuries, particularly back injuries, were prominent in the incident experience for every client I was advising. At that time, the principal method to reduce back injuries was to conduct training programs for workers, teaching them proper lifting techniques. Very soon, it also became apparent that those methods did not achieve the results expected, for a very good reason.

Research into incident causation showed that the problem was not the worker. Study after study revealed that the problem was the design of the work methods. They were overly stressful and error-provocative for a very large share of the working population. It had to be concluded that focusing on proper lifting techniques was of little consequence if the real problem was the design of the work methods. The solution was to convince managements that, to reduce musculoskeletal injuries, workplaces and work methods had to be redesigned to reduce their overly stressful and error-provocative characteristics.

That incident causation research led into what was then mostly called *human factors engineering* and now is more often referred to as *ergonomics*. From Chapanis's writings and additional research, this observation was made with respect to management decision making for every type of occupational injury: If the design of the workplace or the work methods is overly stressful or error-provocative, you can be sure that human errors will occur.

Chapanis was the author of a chapter entitled "The Error-Provocative Situation" in *The Measurement of Safety Performance*, a 1980 publication. The following are very brief excerpts from that chapter. Note that they relate to decision-making possibilities above the worker level:

- The improvement in system performance that can be realized from the redesign of equipment is usually greater than the gains that can be realized from the selection and training of personnel.
- Design characteristics that increase the probability of error include a job, situation, or system which:
  a. Violates operator expectations
  b. Requires performance beyond what an operator can deliver
  c. Induces fatigue
  d. Provides inadequate facilities or information for the operator
  e. Is unnecessarily difficult or unpleasant
  f. Is unnecessarily dangerous

Improvement in system performance and the design of the work or operating system is principally a management responsibility, although it is wise to seek worker input in the improvement process. An appropriate goal for safety professionals is to educate decision makers so that avoiding the creation of work situations that are error-provocative or overly stressful is ingrained into their thinking. As stated several times in this book, the most economical and effective way to avoid, eliminate, reduce or control hazards (in this instance, human errors) is to address those hazards in the design and redesign processes.

James Reason's *Human Error*, first published in 1990, is also a highly recommended resource. Reason discusses the nature of error, studies of human error, performance levels and error types, cognitive underspecification and error forms, a design of a fallible machine, the detection of errors, latent errors and system disasters, and assessing and reducing the human error risk.

In a chapter on assessing and reducing human error risk, Reason acknowledges that the bulk of his book favors theory rather than practice and that this final chapter seeks to "redress the balance by focusing on remedial possibilities." He also says that this chapter was written with safety professionals and psychologists in mind.

As the title of the chapter suggests, Reason writes about several human error risk assessment techniques. One of them is the technique for human error rate prediction (THERP). Although this technique was developed in 1963 by Alan Swain, it has had staying power. Reason writes that "THERP is a technique that has attracted superlatives of all kinds" and is judged by some as "probably the best of the techniques currently available." That is a good recommendation.

Reason recognizes the major accomplishments in recent years in risk assessment and in the development of the knowledge necessary to design systems for error avoidance and resiliency. He also says that the knowledge developed is applied for most complex, high-risk systems. He does not say that the same occurs for systems having less complexity and lower risk levels.

Particular attention is given here to the *Guidelines for Preventing Human Error in Process Safety*. Although "process safety" appears in the book's title, the first two chapters provide an easily read primer on human error reduction. The content of those chapters was influenced largely by personnel with safety management experience at a plant or corporate level. Safety professionals should view them as generic and broadly applicable. This is a gutsy book in that the authors laid out plainly in the Preface and Chapter 1 where human errors occur, who commits them and at what level, and where attention is needed to reduce the potential for their occurrence. The authors state boldly that human errors arise frequently from failures at the management, design, maintenance, or technical expert levels of an organization, that the factors that directly influence error are ultimately controllable by management, that plants are particularly vulnerable to human error during shutdowns for repair and maintenance, and that there is a need to consider the organizational factors that create preconditions for errors. (To emphasize: If the design of the workplace or the work methods creates preconditions for errors, it is a near certainty that errors will occur.)

The book is an easy and informative read. A question asked in the *Guidelines* is: Why is human error neglected in the chemical process industry? The answer given is applicable in all but a few organizations.

A major reason for the neglect of human error in the chemical process industry is simply lack of knowledge of its significance for safety, reliability, and quality. It is also not generally appreciated that methodologies are available for addressing error in a systematic, scientific manner. This book is aimed at rectifying this lack of awareness. (p. 10)

Although it has been known for quite some time that the foundations for human errors may be in "failures at the management, design, or technical expert levels of an organization," offering counsel to reduce human errors at those levels is not usually a significant element within safety management systems. As the interest in serious injury prevention becomes more prominent, safety professionals will additionally be challenged to become knowledgeable about reducing human errors above the worker level.

Another of James Reason's books, *Managing the Risks of Organizational Accidents*, published in 1997, is required reading for safety professionals who want an education in human error reduction. Reason writes about how the effects of decisions *accumulate over time* and become contributing factors for incidents resulting in serious injuries or damage when all the circumstances necessary for the occurrence of a major event come together.

This book was cited in Chapter 3, "Innovations in Serious Injury and Fatality Prevention", because it stresses the need to focus on decision making above the worker level to prevent major accidents. Reason writes this:

> Latent conditions, such as poor design, gaps in supervision, undetected manufacturing defects or maintenance failures, unworkable procedures, clumsy automation, shortfalls in training, less than adequate tools and equipment, may be present for many years before they combine with local circumstances and active failures to penetrate the system's layers of defenses.
>
> They arise from strategic and other top-level decisions made by governments, regulators, manufacturers, designers and organizational managers. The impact of these decisions spreads throughout the organization, shaping a distinctive corporate culture and creating error-producing factors within the individual workplaces. (p. 10)

In addition, Reason states that the traditional occupational safety approach alone, directed largely at the unsafe acts of persons, has limited value with respect to the "insidious accumulation of latent conditions" that he notes are typically present when organizational accidents occur. (pp. 224–239)

Over and over, writers and researchers have reiterated that errors are made at the organizational, managerial, and design levels; that they form a distinctive corporate culture; and that they create overly stressing and error-producing situations within the occupational setting. Reducing the probability of such human errors occurring is the new frontier for safety professionals.

Donald A. Norman's *The Psychology of Everyday Things* is recommended as an additional and important resource. Norman's background is both in engineering and the social sciences. This book was also cited in Chapter 3 because it concentrates on "breakdowns and errors" that are causal factors for major accidents:

> Explaining away errors is a common problem in commercial accidents. Most major accidents follow a series of breakdowns and errors, problem after problem, each making the next more likely. Seldom does a major accident occur without numerous failures: equipment malfunctions, unusual events, a series of apparently unrelated breakdowns and errors that culminate in major disaster; yet no single step has appeared to be serious. In many cases, the people noted the problem but explained it away, finding a logical explanation for the otherwise deviant observation. (p. 128)

What Norman says about "numerous failures" being typical when major accidents occur is identical with my experience. It is highly recommended that the following comments by Reason and Norman be kept in mind as attempts are made to reduce serious injury potential:

> *Reason*: The impact of [top-level] decisions spreads throughout the organization, shaping a distinctive corporate culture and creating error-producing factors within individual workplaces.

> *Norman*: In many cases, the people noted the problem but explained it away, finding a logical explanation for the otherwise deviant observation.

Because of this my seagoing background, it was a pleasure to find that the U.S. Coast Guard is up to par in recognizing the sources of human error. "Human Error and Marine Safety" was written by Anita M. Robinson at the U.S. Coast Guard Research and Development Center. The following excerpts were taken from her paper, which can be found at http://www.geovs.com/_UPLOADED/Human%20Error%20and%20 Marine%20Safety.pdf.

> While human errors are all too often blamed on "inattention" or "mistakes" on the part of the operator, more often than not they are symptomatic of deeper and more complicated problems in the total maritime system. Human errors are generally caused by technologies, environments, and organizations which are incompatible in some way with optimal human performance.
>
> These incompatible factors "set up" the human operator to make mistakes. So what is to be done to solve this problem? Traditionally, management has tried either to cajole or threaten its personnel into not making errors, as though proper motivation could somehow overcome inborn human limitations. In other words, the human has been expected to adapt to the system. *This does not work.* Instead, what needs to be done is to *adapt the system to the human.*

Among many others, Rothblum also writes that the operator is not the problem. Often, it is overly stressing and error-provocative work systems that set up operators to make errors. Others have also said that expecting humans to adapt to a system does not work. The proper approach is to adapt the system to the human being.

R. B. Whittingham's book *The Blame Machine: Why Human Error Causes Accidents* was also cited and recommended in Chapter 3. Its emphasis is on human errors and defective management systems as causal factors for major accidents. From the Preface:

> *The Blame Machine* describes how disasters and serious accidents result from recurring, but potentially avoidable, human errors. It shows how such errors are preventable because they result from defective systems within a company.

W. Johnson is the author of *Human Error, Safety and Systems Development*, a 2004 publication. It is another text which indicates that a transition is taking place whereby additional emphasis is being given to human error reduction.

Recent developments in a range of industries have increased concern over the design, development, management and control of safety-critical systems. Attention has now been focused upon the role of human error both in the development and in the operation of complex systems. (p. 6)

Oliver Strater's *Cognition and Safety: An Integrated Approach to Systems Design and Assessment* was published in 2005. Strater's purpose is to promote having risk assessments made in the design process. This is a worthy goal. It fits well with the design review provisions in Z10.

Studies have shown that it is the exception when engineering students acquire knowledge about hazards, risks, and risk assessments. That sometimes results, Strater says, in designers creating constraints at the *sharp end*, which eventually lead to human errors. "Sharp end" is a British term meaning the point where the work or task is done. Strater says this in his Preface:

Safety suffers from the variety of methods and models used to assess human performance. For example, operation is interested about human error while design is aligning the system to workload or situational awareness. This gap decouples safety assessment from design. As a result, design creates constraints at the sharp-end, which eventually leads to human errors.

THE CONCEPT OF DRIFT

Jens Rasmussen was one of the first authors to comment on drift in decision making within an organization that results in the creation of an accumulation of adverse conditions and practices that could lead eventually to incidents resulting in serious injuries and fatalities. In *Risk Management in a Dynamic Society: A Modelling Problem*, he wrote:

Companies today live in a very aggressive and competitive environment which will focus the incentives of decision makers on short term financial and survival criteria rather than long term criteria concerning welfare, safety, and environmental impact. (p. 186)

James Reason expands on the concept of drift. As mentioned earlier, Reason wrote in *Managing the Risks of Organizational Accidents* about how the effects of decisions *accumulate over time* and become the contributing factors for incidents resulting in serious injuries or damage when all the circumstances necessary for the occurrence of a major event come together.

In *The Blame Machine: Why Human Error Causes Accidents*, Whittingham also writes about errors made in the decision-making process.

Failures in decision making have been identified as a principal cause of a number of major accidents.

Decision making errors are always latent and their effects can be delayed for extremely long periods of time before the errors are detected. Another problem is the difficulty in predicting and understanding how decision making errors are propagated down through an organization to, eventually, lead to an accident. (p. 31)

Particularly in difficult economic times, management decision making may result in drift and deterioration in safety-related systems. Although it may be difficult to identify the results of drift, signs of its effects become apparent when nonconformity from what had been considered appropriate practice occurs. Sidney Dekker is the author of *Drift into Failure*. A section of his book is devoted to "Normalizing Deviance, Structural Secrecy and Practical Drift". He comments on decisions that made sense and were considered favorable at the time. He then observes:

Each little decision they made, each little step, was only a small step away from what had previously been seen as acceptable risk. (p. 103)

It was easy to reach agreement on the notion that production pressure and economic constraints are a powerful combination that could lead to adverse events. (p. 105)

Dekker uses the term *normalization of deviance* to refer to situations in which there was a departure from what was considered good risk management practice and acceptance of deviations that resulted in an increase in the risk level. Each such situation is an alert for observant safety professionals and requires a good analytic and risk management approach to determine the actions to be taken—I do not say that reversing a trend toward drift will be easy.

BEHAVIORAL SAFETY, HUMAN ERROR REDUCTION, AND SERIOUS INJURY PREVENTION

The following excerpts from Reason's *Managing the Risks of Organizational Accidents* bear directly on the history of behavioral safety, human error reduction, and the prevention of serious injuries.

[A] problem that needs to be confronted is the belief held by many technical managers that the main threat to the integrity of their assets is posed by the behavioural and motivational shortcomings of those at the "sharp end." For them, the oft-repeated statistic that human errors are implicated in some 80–95 per cent of all events generally means that individual human inadequacies and errant actions are the principal causes of all accidents. What they hope for in seeking the help of a human factors specialist is someone or something to "fix" the psychological origins of these deviant and unwanted behaviours. But this—as I hope is now clear—runs counter to the main message of this book. Workplaces and organizations are easier to manage than the minds of individual workers. (p. 233)

However, a good many behavioral safety consultants built their businesses on the premise that 80% or more of occupational accidents are caused principally by the unsafe acts of workers. In 1998, the American Society of Safety Engineers held a symposium that was considered by many to be the high-water mark for behavior-based safety. Most of the big players in behavioral safety made presentations.

Some (not all) of the speakers at that symposium led their audiences to believe that worker-focused behavior-based safety was the greatest elixir ever created and that you need only apply their behavioral approaches to workers and all your problems would be solved. In their presentations, little or nothing was said of the cultural, organizational, design, engineering, and operational sources out of which many error-provocative situations arise.

In recent years, speakers and authors promoting worker-focused behavior-based safety as a cure-all, by itself, have largely been stilled. Several prominent leaders in the behavior-based safety field now speak of safety systems, workplace design, qualities for effective safety leadership, performance improvement, the need to achieve a culture change, and an organizational culture of citizenship.

Worker-focused behavior-based safety does not examine the sources of human error above the worker level and has limited impact on serious injury prevention.

Safety professionals who still base their practice on the premise that a very large percentage of occupational injuries and injuries result from the unsafe acts of workers and who still promote worker-focused behavior-based safety as the primary consideration in the safety efforts they sponsor may want to reexamine their premises in light of the foregoing. Additionally, I suggest that they consider:

- The focus of Z10, which is on reducing hazards, the risks that derive from hazards, and the deficiencies in safety management systems and process, and on identifying opportunities for improvement.
- The case made in Chapter 3 for giving special attention to serious injury prevention.
- The achievements that can derive from learning about and providing counsel on human error reduction, particularly above the worker level.

CONCLUSION

General observations can be drawn from the several sources cited here and from my experience:

- Human errors of commission or omission are factors in the occurrence of nearly all hazard-related incidents.
- Typical safety management systems do not address human error reduction above the worker level, particularly on an anticipatory basis.

- You cannot change the human condition, but you can change the conditions under which people work.
- The solutions to most human performance problems are technical rather than psychological.
- Potentials for human error derive largely from top-level decisions, and the impact of those decisions spreads throughout an organization, shaping a distinctive corporate culture and creating overly stressing and error-provocative situations.
- To avoid hazard-related incidents resulting in serious injuries, human error potentials must be addressed at the cultural, organizational, management system, design, and engineering levels, and with respect to the work methods prescribed.

All this spells opportunity for safety professionals to acquire knowledge with respect to human error avoidance and reduction and to enhance their professional status. Human error reduction may very well become the frontier for the practice of safety.

Sidney Dekker said in *The Field Guide to Understanding Human Error* that some readers may believe that his intent is to let line workers "off the hook" and that his writing "looks like a ploy to excuse defective operators." (p. 196) He assures readers that is not his intent—and that is not the intent here. If a problem in a given situation is identified as an individual employee's performance, the solution should focus on altering that person's performance. But if the employee was set up to err because of the design of the workplace or the work methods, trying to influence employee performance will not solve the problem.

REFERENCES

ANSI/AIHA Z10-2012. *Occupational Health and Safety Management Systems*. Fairfax, VA: American Industrial Hygiene Association. ASSE is now the secretariat. Available at https://www.asse.org/cartpage.php?link=z10_2005.

Chapanis, Alphonse. "The Error-Provocative Situation." In *The Measurement of Safety Performance*, edited by William E. Tarrants. New York: Garland Publishing, 1980.

Dekker, Sidney. *Drift into Failure*. Burlington, VT: Ashgate Publishing Company, 2011.

Dekker, Sidney. *The Field Guide to Understanding Human Error*. Burlington, VT: Ashgate Publishing Company, 2006.

Guidelines for Preventing Human Error in Process Safety. New York: Center for Chemical Process Safety of the American Institute of Chemical Engineers, 1994.

Hammer, Willie. *Handbook of System and Product Safety*. Englewood, NJ: Prentice-Hall, 1972.

Johnson, W. *Human Error, Safety and Systems Development*. Norwell, MA: Kluwer Academic Publishers, 2004.

Kletz, Trevor. *An Engineer's View of Human Error*, 2nd ed. Rugby, Warwickshire, UK: Institution of Chemical Engineers, 1991.

Manuele, Fred A. *On the Practice of Safety*, 4th ed. Hoboken, NJ: Wiley, 2013.

Norman, Donald A. *The Psychology of Everyday Things*. New York: Basic Books, 1988.

Petersen Dan. "Human Error." *Professional Safety*, Dec. 2003.

Petersen, Dan, "We need to be able to explain why human error happens and what it is." Speech given at the "Human Error in Occupational Safety Symposium." Des Plaines, IL: American Society of Safety Engineers. Proceedings for the Symposium, Mar. 2003.

Rasmussen, Jens. *Risk Management in a Dynamic Society: A Modelling Problem. Safety Science,* vol. 27, no. 2/3, pp. 183–213, 1997.

Reason, James. *Human Error*. Cambridge, UK: Cambridge University Press, 1990.

Reason, James. *Managing the Risks of Organizational Accidents*. Burlington, VT: Ashgate Publishing Company, 1997.

Rothblum, Anita A. "Human Error and Marine Safety." U.S. Coast Guard Research and Development Center. At http://www.geovs.com/_UPLOADED/Human%20Error%20and%20Marine%20Safety.pdf.

Strater, Oliver. *Cognition and Safety: An Integrated Approach to Systems Design and Assessment*. Burlington, VT: Ashgate Publishing Company, 2005.

Swain, A. D. "A Method of Performing a Human Reliability Analysis." Monograph SCR-685. Albuquerque, NM: Sandia National Laboratories, 1963.

Whittingham, R. B. *The Blame Machine: Why Human Error Causes Accidents*. Burlington, MA: Elsevier–Butterworth–Heinemann, 2004.

CHAPTER 5

MACRO THINKING:
THE SOCIO-TECHNICAL MODEL

To set the stage for this chapter, which promotes macro thinking versus micro thinking, citations pertaining to management systems are taken from the introduction in ANSI/AIHA Z10-2012, the *Occupational Health and Safety Management Systems* standard.

> This is a voluntary consensus standard on occupational health and safety management systems. It uses recognized management system principles to be compatible with quality and environmental management system standards such as the ISO 9000 and ISO 14000 series. The design of ANSI Z10 encourages integration with other management systems. The processes that drive implementation of the organization's management system also facilitate improved teamwork and operational performance. It places less reliance on single individuals and more emphasis on an organization's process and teamwork to maintain business functions.

In Z10's explanatory column opposite the Scope, Purpose, & Application, comment E.1.1 reads partially as follows: "The processes required in the standard are integrated for continual improvement. This system model goes beyond a simple sum of individual ... activities."

In pages 1 through 29 of Z10, the words *process*, *processes*, and *system* appear 120 times. That does not include use of the acronym OHSMS for occupational health and safety management systems, the pages prior to page 1, or the appendices.

Advanced Safety Management: Focusing on Z10 and Serious Injury Prevention,
Second Edition. Fred A. Manuele.
© 2014 John Wiley & Sons, Inc. Published 2014 by John Wiley & Sons, Inc.

The terms *process* and *system* are not included in the definitions in Z10, nor are the meanings of the terms discussed elsewhere. Since having effective processes and systems is vital in the application of Z10, they will be emphasized. In this chapter we:

- Define process and system
- Explore the integration of processes and systems and their interrelation and interdependence
- Emphasize macro thinking versus micro thinking
- Deemphasize focusing on the individual
- Suggest taking a holistic approach to risk avoidance, elimination, reduction, or control
- Explore application of the principles embodied in socio-technical theory
- Relate socio-technical theory to "A Socio-Technical Model for an Operational Risk Management System" in Chapter 3.

DEFINING *PROCESS* AND *SYSTEM*

From the many definitions of *process* and *system* available, those selected for this chapter are taken from "the Business Dictionary." Access is at http://www.business-dictionary.com/definition/system.htm.

Process: Sequence of interdependent and linked procedures which, at every stage, consume one or more resources (employee time, energy, machines, money) to convert inputs (data, material, parts, etc.) into outputs. These outputs then serve as inputs for the next stage until a known goal or end result is reached.

System An organized, purposeful structure that consists of interrelated and interdependent elements (components, entities, factors, members, parts, etc.). These elements continually influence one another (directly or indirectly) to maintain their activity and the existence of the system, in order to achieve the goal of the system.

A case can be made that the terms *process* and *system* are synonyms. How do these definitions relate to an operational risk management system?

In slightly different terms, Z10 proposes that organizations have systems that are "an organized, purposeful structure that consists of interrelated and interdependent elements", as in the definition of a system. The goal of these elements, taken as a whole, is to achieve continuous improvement and maintain acceptable risk levels.

Application of all of the elements—the processes and the systems outlined in Z10—is necessary to achieve superior results. Eliminate some elements and the probability of achieving stellar results is diminished.

For emphasis: The processes and systems in Z10 are integrated and influence each other continuously. Consider the complexity of adverse consequences on several of the processes required by Z10 if any of the following exist:

- Management leadership is absent.
- Employees are not allowed to participate.

- Risk assessments are not done or done are poorly.
- A management of change system has not been put in place.
- The procurement provision, which is to avoid bringing hazards and their accompanying risks into a workplace, has not been adopted.
- Training is not done or is inadequate.
- The design process results in hazards and unacceptable risks being brought into the work place.
- Engineers and others have not been educated with respect to hazards, risks, prevention through design, and achieving acceptable risk levels.

The terminology in Z10 promotes macro thinking—that is, thinking large about all systems and subsystems as interrelated parts of a whole—rather than taking a micro view and limiting consideration of hazards and risks and possible inadequacies in management systems as isolated components. For definitions of macro and micro, reference is made to "Dictionary Reference" on the Internet.

- *Macro* is defined as: very large in scale, scope, or capability. Access is at http://dictionary.reference.com/browse/macro.
- *Micro* is defined as: extremely small, minute in scope. Access is at http://dictionary.reference.com/browse/micro?s=t.

Taking a macro rather than a micro approach is illustrated in Figure 5.1.

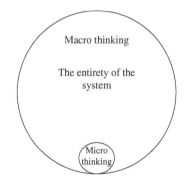

FIGURE 5.1 Taking a macro approach.

MACRO THINKING: THE SOCIO-TECHNICAL CONCEPT

Several writers on socio-technical systems say that the term was coined in the 1960s by Eric Trist and Fred Emery when working as consultants at the Tavistock Institute in London. Researchers at Tavistock said that based on their research and from their experience, what was needed for effective operations was a good fit between the technical subsystems and the social subsystems.

Trist and Emery, the writers say, postulated that the foundation of the socio-technical approach was the belief that the most effective design of a system is achieved if there is joint optimization of the subsystems, and that stellar performance can be obtained only if the interdependency of the technical and social subsystems is recognized unequivocally. Thus, in the design process, the decision makers must recognize the impact that each subsystem has on the other and design accordingly to assure that the subsystems are working in harmony.

Even though the term *socio-technical systems* is not prominent in the current literature about the organization of work, the idea it conveys is dominant in conventional thinking about the interrelationship between the technical and social aspects of operations.

Two intertwining sources promote consideration and application of a socio-technical system. A specifically stated premise in the Introduction to Z10 is: "The processes that drive implementation of the organization's management system also facilitate teamwork and operational performance." Sidney Dekker, in *The Field Guide to Understanding Human Error*, spoke of the socio-technical system as follows:

> The Systemic Accident Model ... focuses on the whole [system], not [just] the parts. It does not help you much to just focus on human errors, for example, or an equipment failure, without taking into account the socio-technical system that helped shape the conditions for people's performance and the design, testing and fielding of that equipment. (p. 90)

Dekker makes it plain that operating systems should be considered and examined as a whole to properly determine solutions to problems. Definitions of a socio-technical system vary in detail, although they maintain the substance of what the originators of the term intended. My current definition, depicted in Figure 5.2, is:

> A socio-technical system stresses the holistic, interdependent, integrated, and inseparable interrelationship between humans and machines. It fosters the shaping of both the technical and the social conditions of work in such a way that both the output goal of the system and the needs of workers are accommodated.

When offering a recommendation to improve an aspect of an operational risk management system, I propose that macro thinking—an extension of systems thinking—be employed to determine the specifics of the recommendation and how the improvement may affect other operational aspects. A composite definition of applied macro thinking follows.

> *Applied macro thinking* takes a holistic approach to analysis that focuses on the whole and its parts at the same time and the way in which a system's parts interrelate. Macro thinking contrasts with an analytical process that addresses a technical or social aspect of a system separately—micro thinking—without considering the relationship of that aspect to the system as a whole.

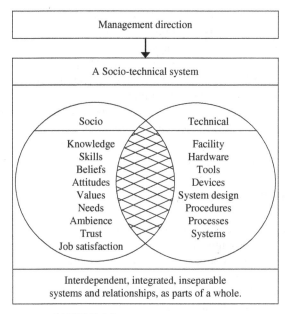

FIGURE 5.2   Socio-technical system.

In taking a socio-technical systems approach, a macro thinking approach, it would be understood that:

- The technical and social systems in an organization are inseparable parts of a whole.
- The parts are interrelated and integrated.
- The changes made in one system may have an effect on others.
- The organization and the people who work in it are not well served if when resolving a risk situation, the subject at hand is considered in isolation rather than as a part of an overall system.

Emphasis is given to the importance of the whole and the interdependence of its parts. The intent is to encourage decision makers to integrate the technical aspects with the social aspects of business practice not only for good operating results but also for the benefit of all.

The technical aspects of a system include the facility, hardware, tools, devices, system design, physical surroundings, and the procedures prescribed.

The social system consists of the knowledge and skills of employees, from the most senior management level down to the most newly hired, the attitudes that derive from their beliefs, their values, their needs, job satisfaction, respect, trust, relations with each other, the spirit and ambience of the workplace, authority structures, the reward system, and whether there is an open communications system through which the views of all can be heard.

In an effective socio-technical system, management recognizes the interdependence of the technical and social aspects of operations and integrates them, and a feedback process would be created to monitor alignment.

I believe that taken as a whole, the revision of Z10 promotes a balance within an organization of the technical and social aspects of operations and that encouragement should be given to promoting Z10 accordingly.

Among the many changes in Z10, two are prominent and deserve recognition. There is a greater emphasis on the involvement of employees within the safety management system and of an organization having an open and effective communication system. That fits well within a socio-technical system.

MACRO THINKING VS. MICRO THINKING

For the purposes of this chapter, micro thinking focuses on and considers only the direct or immediate contributing factors in a hazard/risk situation, those closely related in time and space, and does not explore decisions or shortcomings in the management systems connected to them.

The term *system deficiencies* is used frequently in Z10 to promote consideration of all of the underlying and contributing operational factors in a hazard/risk situation. In all but a few organizations, incident investigation reports are examples of a narrow and micro thought process. Typically, such reports concentrate only on the actions or inactions of the person involved and the immediate physical surroundings.

Unfortunately, too many safety practitioners are good at micro thinking and at avoiding the reality of contributing factors resulting from decisions made above the worker level. That is demonstrated when they sign shallow reports giving their approval. For a large proportion of incident investigation reports, the principal causal factor identified is the so-called "unsafe act" of the employee, and the preventive action is training or retraining, or having a group meeting to discuss and reinforce the standard operating procedure.

It is acceptable in many organizations for investigations to stop at that point. Such corrective actions are ineffective because they ignore system deficiencies that may be more important contributing factors. Examples of micro thinking follow.

- If design decisions that result in the work being overly stressful or physically demanding to an excess, or error-provocative or error-inviting or inefficient in relation to expectations, the design of the work would not be considered if micro thinking was employed—the emphasis would be on what the worker did or did not do.
- A hazardous floor condition that also had a negative effect on production existed for months—a recommendation is made to fix the floor without inquiring into the deficiencies in the interrelated management systems that allowed the situation to exist that long.

Unfortunately, micro thinking about hazards and risks is deeply embedded in the practice of safety. Mention was made in Chapter 3, "Innovations in Serious Injury and Fatality Prevention", of an article entitled "Reviewing Heinrich: Dislodging Two Myths from the Practice of Safety." One of the myths is that unsafe acts of workers are the principal causes of occupational accidents. Heinrich's approach to incident causation is a good example of micro thinking.

Discussions in the article covered such topics as moving the preventive efforts from a focus on the employee to improving the work system, the significance of the design of the work system and the work methods, the complexity of causation, and recognizing the human errors that occur at organizational levels above the worker. Although response in support of the article was good, the information developed through the communications received indicated that the Heinrichean premise—that 88% of accidents are caused by unsafe worker acts—is embedded deeply in organizations to which safety practitioners give advice.

Macro thinking is a process that considers the interrelationships and dynamics among system components and the decision making throughout the operating system. An incident investigation that applied macro thinking might include comments such as the following and promote the discussion of several of the interrelated management systems required by Z10:

- This is a design problem (the lockout/tagout station is over 400 feet away) and we were pushed hard to get the machine back into operation in a hurry.
- Pre-job planning was not done because of time pressures—we were not allowed to take the time to look into the risks.
- Our standard operating procedure was followed, but the work is risky and the risk level is high. The SOP needs to be revised.
- This incident results from our not having enough people to do the work.
- This equipment is being run beyond its expected life and needs to be replaced, as we said in our previous correspondence.
- Our work order for repair of this equipment was submitted four months ago, but the maintenance department could not get to it because they say that they are understaffed.
- Our staff was not scheduled for the training needed for this job, and we need the training.
- We used to have good personal protective equipment, but the purchasing department decided to use different specifications and we have had a frequency of hand injuries because the gloves are not substantial enough for this work.
- This is a new operation, and communication on the hazards was not a part of the indoctrination.

Comments in an investigation report of that sort—implying relationships to several interrelated contributing factors—would require inquiry into more than one facet of operations to achieve an understanding of their linkage and how they relate to and influence each other.

THE FIVE-WHY TECHNIQUE

I promote use of the five-why technique to improve the quality of incident investigation as a problem-solving method in general and to promote macro thinking. Inquiry on the Internet will reveal how broadly the technique is applied.

We discuss the technique in Chapter 22, "Incident Investigation". The five-why technique is easy to learn and easy to apply. It consists only of asking "why" five times consecutively. It is important that the first step be an identification of a "why" and not a "what." Sometimes, asking "why" only four times is sufficient. Occasionally, the inquiry needs continuation into six times.

In a particular situation, interjecting "what" or "how" may move the inquiry forward. Application of the five-why technique should identify the actions to be taken when multiple operational entities are involved. An example follows.

Operations personnel report concern over injury potential because of conditions that develop when a metal-forming machine stops when the overload trip actuates. The safety director meets with a supervisor, the person directly responsible for the work area who has had to field the complaints.

1. Why are you concerned? The electrical overload trip actuates very often when we use this new forming machine. It gets risky when it stops in midcycle and the work that has to be done to clear the partially formed metal adds risks that our employees think are more than they should have to bear. Occasionally, that's OK; often is too much.

2. Why does the overload trip actuate? This is a new problem for us. We rarely had the overload trip actuate. It started after a new order for metal was received. We are told that the purchasing department thought that it got a very good deal from a metals distributor, but it turns out that what was delivered to us did not meet our specifications for hardness. This metal is not as malleable and workable, and the metal former struggles in the forming process; so the overload trip actuates. Maintenance personnel are furious with us because we have to call on them as much as we do.

3. Why can't the amperage for the overload trip be increased for this batch of metal? Our engineers say they that don't want greater power fed into this machine.

4. Why do you have to call on maintenance so often? The rule here is that no overload trip is to be reset without a review of why it tripped and clearance from maintenance.

5. Why haven't you recommended to your operating manager that he arrange a get-together with the engineer and the maintenance manager to decide on what needs to be done to resolve the overload trip problem for this batch of metal? That's not easy for me to do at my level. But it would be good if you could find a way to get that done.

This example illustrates the interrelationships of various entities in an organization. Resolution of this problem involves future purchases, operations, engineering, and maintenance. It is often the case that risk reduction actions require participation by several interrelated and integrated functions.

Additionally, in the case cited, another matter needs attention. There is a communications problem in that a supervisor is reluctant to suggest to his or her boss that consideration be given to resolving a work situation that employees think is overly risky.

LESS RELIANCE ON SINGLE INDIVIDUALS AND MORE EMPHASIS ON PROCESSES

The phrase "less reliance on single individuals and more emphasis on processes," which appears in Z10's Introduction, has many possible implications. It suggests a movement away from a focus on what an individual employee may do or not do and a concentration on continuous improvement in the work systems—the processes—in which employees work.

One author who has, in a sense, been pleading for an understanding that an employee's actions or inactions are greatly influenced by the decisions made that determine the specifics of the work system is James Reason. In his book *Human Error*, he writes:

Rather than being the main instigator of an accident, operators tend to be the inheritors of system defects created by poor design, incorrect installation, faulty maintenance and bad management decisions. Their part is usually that of adding the final garnish to a lethal brew whose ingredients have already been long in the cooking. (p. 173)

Safety professionals should take Reason's statement seriously because it illustrates accurately the situation in which an employee is placed when the workplace is "a lethal brew whose ingredients have already been long in the cooking."

The "lethal brew" results from the decisions made at an upper management level that create the socio-technical system in which people work. To address the elements in the lethal brew properly requires applying macro thinking to many operational aspects.

Nancy Leveson is a professor of aeronautics and astronautics and engineering systems at the Massachusetts Institute of Technology. In her slide presentation entitled "Applying Systems Thinking to Risk Management in Complex, Socio-Technical Systems," she comments on safety as a system control problem rather than a component failure problem and the need for a causality model based on system theory.

On slide 8, Leveson relates to the "lethal brew … long in the cooking" when she writes:

Most major accidents arise from a slow migration of the entire system toward a state of high-risk.

She notes on the same slide that:

- Accidents arise from interactions among system components (human, physical, social) that violate the constraints on safe component behavior and interactions
- Losses are the result of complex processes, not simply chains of failure events.

A good number of safety practitioners will consider the idea of less reliance on single individuals and more emphasis on processes both disturbing and threatening. If safety practitioners have based the advice they give principally on the premise that unsafe acts of workers are the principal causes of occupational injuries, they would have to make major changes in their thinking if they tried to implement Z10.

Such safety practitioners would do themselves, the profession, and society a major good if they adopted the premises on which Z10 is based: focusing on integrated processes and systems as a whole to identify and analyze hazards so that appropriate actions can be taken to achieve acceptable risk levels.

To repeat for emphasis, the words *processes* and *systems* dominate Z10. There is a huge difference between:

- Applying macro thinking to focus on interrelated and interdependent processes and systems in a continuous improvement effort, and
- Applying micro thinking and focusing principally on the actions of individual employees and their immediate surroundings.

Another group who will probably find the thrust of Z10 threatening are the safety practitioners, whose endeavors concentrate mostly on worker-directed, behavior-based safety and who give little attention to contributing factors for accidents that result from decisions that affect the workplace and the work methods.

ON WORKER-FOCUSED BEHAVIOR-BASED SAFETY

It is typical for the various fields within the practice of safety to be in transition with respect to knowledge needs and the counsel given. Not surprisingly, that element within the practice of behavior-based safety is in the process of evolution.

In the May 27, 2002 issue of ISHN's *E-News*, the editor asks "What's next?" and says the question "came up lately because signs point to the decline of behavior-based safety, which has been hailed as the latest, greatest elixir for easing safety headaches." And behavior-based safety, which stressed giving attention principally to the behavior of the employee, was touted by some safety practitioners as the ultimate elixir that would solve all problems. But in recent years, speakers and writers engaged in behavior-based safety don't propose nearly as often that the proper course of action to reduce injuries is to focus on the at-risk behavior of employees—to the exclusion of all else.

Some of the well-known behavior-based safety consultants have shifted or expanded their message. In articles written by a few, the word *behavior* does not appear at all. Terms such as the following dominate: qualities for effective safety leadership, culture change and motivation, performance improvement, organizational citizenship culture, and transformational leadership. In the application of a behavior-based initiative, this would be the typical procedure:

- Data would be gathered on at-risk behaviors through interviews with employees, through safety surveys, and through analyses of accident reports.
- At-risk behavioral inventories would be produced.
- Worker groups would be trained to give each other positive reinforcement for safe behavior.
- The intent being that peer involvement would reduce at-risk behavior.

Behaviorists write that immediate, frequent, and positive rewarding of proper behaviors will promote repetition of those desired behaviors. A good part of the endeavor concentrates on what is called the "at-risk" behavior of individual workers as though their behavior was the principal element to be corrected when an injury occurred.

Many of the concerns expressed by highly capable safety professionals over the worker-focused, behavior-based safety approach pertain to the fact that causal factors deriving from the work environment and work practices in place are ignored. Hardly ever in the relevant literature is there recognition that the antecedents for at-risk behavior should be analyzed to determine their sources. It is not generally recognized by behaviorists that antecedents may derive from the work environment created by management.

Management is responsible for the processes that result in the design of the work-place and the design of the work systems. If the antecedents for at-risk behavior derive, for example, from material-handling ergonomics problems that overstress a large percentage of the employee population, the focus for reducing at-risk behavior logically has to be on redesigning or revising the material-handling methods.

Although interest in behavior-based safety has diminished, it is not going to go away. Thinking practitioners in behavior-based safety will recognize that the marketability of the advice they give will be improved if they broaden their concepts and include the sound processes and systems on which Z10 is based.

All safety practitioners who choose to adopt the premises in Z10 that promote consideration of the interrelationships between processes as they give advice toward reducing injuries and illnesses will benefit from an exposure to macro thinking and socio-technical concepts.

SAFETY CLIMATE – SAFETY CULTURE

A critic of a draft of this chapter suggested that comments should be included on the subject referred to in the literature as a "safety climate" because it has a relationship to the "socio" aspects of a socio-technical system and has attained a significant status within the practice of safety. That subject has been researched. Unfortunately, a literature review indicates that there is considerable confusion about what a safety climate really is, and also about the term *safety culture*. The terms may be interchanged. *Safety culture* appears often in the literature on *safety climate* and is used as a synonym, but they are not the same.

An excellent reference on these subjects is a study made by the Aviation Research Lab, Institute of Aviation, at the University of Illinois for the Federal Aviation Administration. The title is "A Synthesis of Safety Culture and Safety Climate Research" (Technical Report ARL-02-3FAA-02-2). In that paper, 12 definitions of safety climate are given. Excerpts from those definitions follow:

- Snapshot of employees' perceptions of the current environment
- Perceptions of safety systems, job factors, and individual factors
- Employee attitudes about the current state of safety initiatives at their workplace
- Safety climate: the perceived state of safety of a particular place at a particular time
- Employee perceptions of safety-related policies, procedures, and rewards

And as the proponents on safety climate say, if efforts are focused on improving the safety climate as perceived by employees, injuries will be reduced on a large scale. It seems that such efforts are to concentrate on a part of one aspect of a socio-technical system—the socio aspect. That implies micro thinking rather than promoting a holistic view of the operational system. This approach seems to have a relationship to the principles on which behavior-based safety is based.

Jim Howe, who chaired the committee that developed the 2012 version of Z10, critiqued this chapter and wrote: "The emphasis has to be on system thinking. That means improvement cannot be made on either the socio or the technical aspects of the system independently." Then he wrote of his own experience as follows (reproduced with his permission).

A department made critical parts out of specialty steel. Each day, machine operators were required to fill out production sheets listing the number of acceptable parts made and the number of defects. If they reported defects, they would be disciplined and sent home. So, they did not report defects. Yet, at the end of the day, scrap tubs were full of defective parts. That cost the company millions of dollars a year.

A Deming trained engineer was assigned to the department. When he asked machine operators why they did not report the defects, they told him that they wanted to avoid the punishment they would receive if they did. He told the machine operators that they would not be punished if they stopped the machine immediately when defects were recognized and reported the stoppage to him.

Machine operators did not believe him and they continued to not report defects. But, a few of them decided to test the new engineer. They stopped their machines when it was apparent that a defective part was being produced and reported the stoppage. They were not punished. They encouraged other machine operators to report defects. They were not punished and their fear of being punished disappeared. (One of Deming's principles—drive out fear.)

Being Deming trained and knowing that to obtain significant performance improvement the focus has to be on improving the process, the new engineer commenced a macro thinking review—a systems thinking review—that included participation and input from machine operators. They found:

- excessive material variations from suppliers
- inadequate material testing on receipt
- measurement tools so worn that extreme differences resulted when measurements were taken with respect to quality specifications
- machines that were so worn—gears, bearings—that they were not capable of producing quality parts consistently
- et cetera

If machine operators had not been convinced that they need not fear being punished if they reported defects, it's unlikely that the good outcome would have been achieved. Fear was driven out. It became apparent to them that their knowledge and experience was being respected as their input was sought in the macro thinking—the systems thinking—process.

The result was employees engaged and working collaboratively with management and engineers for continual improvement and millions of dollars of savings each year.

Several aspects of a good socio-technical system can be found in the foregoing narrative. On the technical side: Specifications for material received from a supplier were applied rigidly; a random testing system was instituted for material received; new measurement tools were purchased; and machinery was replaced. On the socio side (*socio* is used here as an acronym for *sociological*), respect was gained for the experience and knowledge of machine operators; an atmosphere of trust was created by the engineer; and fear was driven out.

Consultants and some organizations proposing an emphasis on improving a safety climate have developed tools to produce a measure of the safety climate at a given moment. I am unable to find significant differences between those tools and that age-old tool—perception surveys.

Wiegman et al., the authors of "A Synthesis of Safety Culture and Safety Climate Research," list 13 definitions of safety culture. Excerpts from them follow.

- Safety culture is defined as the shared values, beliefs, assumptions, and norms which may govern organizational decision making.
- Safety culture reflects attitudes, beliefs, perceptions, and values that employees share in relation to safety.

- Safety culture refers to entrenched attitudes and opinions which a group of people share
- Safety culture is defined as the attitudes, values, norms and beliefs which a particular group of people share with respect to risk and safety.

As in the foregoing and in many other definitions of safety culture, terms such as the following appear: shared beliefs, attitudes, values, and norms of behavior; shared assumptions; individual and group attitudes about safety; and entrenched attitudes and opinions that a group of people share. Although I have used such terms, on reflection, they do not relate to the reality of how a safety culture develops. An organization's safety culture—a subset of its overall culture—derives from the decisions made at the board of directors' level and at the senior management level, which result in acceptable or unacceptable operational risk levels.

Management establishes the culture. An organizational culture with respect to occupational safety, environmental safety, the safety of the public, and product safety is the outcome of decisions made by management. Those decisions determine what the technical aspects and the social aspects at a facility are to be. In some organizations, where employers believe that operational risks are acceptable, employees may conclude that some of the risks are unacceptable and that protection is inadequate. Thus, employees may not share the views and beliefs held by management with respect to safety and operational risk levels.

A question occasionally arises about the level of management where decisions are made with respect to acceptable or unacceptable risk levels. For a hugely huge majority of the time, that level resides in the board of directors and executive management.

Occasionally, a manager from outside the organization is put in charge at a location where accident experience has been poor and she observes that the risk levels considered acceptable, reflecting the culture in place, are contrary to what she thinks is appropriate. She finds a way to display that she has taken the leadership role to reduce injuries (such as placing the review of incident experience at the top of the agenda for her management meetings, insisting on improved incident reports, and personally following some situations to a conclusion).

After a time of doubt, the normal resistance to change, and with sufficient demonstrations by what she does that she is serious about reducing risks, the culture shows improvement, even though she has the same workforce and the same tools and machinery. How much improvement in the culture can be attained will depend largely on the constraints or lack of constraints on the budgets she is allowed.

In the paper cited previously, Wiegman et al. say:

Although the debate over the definition of safety culture has not reached unanimous agreement, a similar term "safety climate" has been used frequently in the literature and has added to the confusion. The literature has not presented a generally accepted definition of safety climate either. (p. 8)

The authors conclude, as a result of the studies made:

Safety culture, as defined in the literature, is commonly viewed as an enduring characteristic of an organization that is reflected in its consistent way of dealing with critical safety issues. On the other hand, safety climate is viewed as a temporary state of an organization that is subject to change depending on the features of the specific operational or economic circumstances. (p. 11)

Note in that observation that a safety culture is organizational and enduring. Management controls the organization, and the outcomes of its decisions may be enduring. The culture created by management is the dominant and governing factor with respect to the risk levels attained: acceptable or unacceptable.

A precaution is offered: Focusing principally and largely on improving a safety climate to the exclusion of other aspects of a socio-technical system—that is, taking a micro view—is not good operational risk management.

SOCIO-TECHNICAL CONCEPTS: A MODEL OPERATIONAL RISK MANAGEMENT SYSTEM

In Chapter 3, "Innovations in Serious Injury and Fatality Prevention", we proposed adopting "A Socio-Technical Model for an Operational Risk Management System" as a substantially different means of improving on serious injury and fatality prevention. The model is duplicated here. In proposing the model, the hope is that socio-technical concepts would be foundational:

- At a board of directors level, where establishing a culture begins
- In the decisions made at a senior management level to demonstrate its involvement to achieve the culture
- In the policies, standards, and procedures established
- Throughout the administration of all the individual aspects of the operational risk management system listed in the model

It is close to the top of the probability scale that acceptable risk levels can be achieved and maintained if such a model is the base upon which an operational risk management system is built and that superior results can be achieved as well.

CONCLUSION

This chapter is not intended to be an in-depth commentary on macro thinking or the ideas that are basic to a socio-technical system. It introduces the subjects for those who want to become more familiar with them.

A Socio-Technical Model for an Operational Risk Management System

The Board of Directors and Senior Management Establish a Culture
for Continuous Improvement
That Requires Defining, Achieving and Maintaining Acceptable Risk Levels
in All Operations

Management Leadership, Commitment, Involvement, and the Accountability system,
Establish That the Performance
Level to be Achieved Is in Accord with the Culture Established by the Board

To Achieve Acceptable Risk Levels,
Management Establishes Policies, Standards, Procedures and Processes With Respect To:

- Providing adequate resources
- Risk assessment, prioritization and management
  - Applying a hierarchy of controls
- Prevention through design
  - Inherently safer design
  - Resiliency, reliability and maintainability
- Competency and adequacy of staff
  - Capability – skill levels
  - Sufficiency in numbers
- Maintenance for system integrity
- Management of change/pre-job planning
- Procurement – safety specifications
- Risk-related systems
  - Organization of work
  - Training – motivation
  - Employee participation
  - Information – communication
  - Permits
  - Inspections
  - Incident investigation and analysis
  - Providing personal protective equipment
- Third party services
  - Relationships with suppliers
  - Safety of contractors – on premises
- Emergency planning and management
- Conformance/Compliance assurance reviews

Performance measurement: Evaluations are made and reports are prepared for management
review to support continuous improvement and to assure that acceptable risk levels
are maintained.

I believe that the state of the art in the practice of safety would be enhanced if safety professionals recognized and acted on the following premises:

- Technical and social aspects evolve from decisions made by management that determine the work environment.
- The technical and social aspects are inseparable parts of a whole.

- Their parts are interrelated.
- Making changes in one of its parts may have an effect on others.
- The organization and the people who work in it are not well served if when giving counsel to resolve a risk situation the subject at hand is addressed in isolation.
- Macro thinking—systems thinking—is required to resolve issues that effect several operational elements.

REFERENCES

ANSI/AIHA Z10-2012. American National Standard, *Occupational Health and Safety Management Systems*. Fairfax, VA: American Industrial Hygiene Association, 2012. ASSE has become the secretariat. Available at   https://www.asse.org/cartpage. php?link=z10_2005.

"Business Dictionary" on the Internet at http://www.businessdictionary.com/definition/ system.htm.

Dekker, Sidney. *The Field Guide to Understanding Human Error*. Burlington, VT: Ashgate Publishing Company, 2006.

"Dictionary Reference" at http://dictionary.reference.com/browse/macro and http://dictionary. reference.com/browse/micro?s=t.

Johnson, Dave. "Behavior-Based Safety" in the May 27, 2002 issue of ISHN's *E-News*.

Leveson, Nancy. "Applying Systems Thinking to Risk Management in Complex, Socio-Technical Systems" can be found at http://us.yhs4.search.yahoo.com/yhs/search?p=Applying+System+ Thinking+to+Risk+Management+in+Complex%2C+Socio-Technical+systems&hspart=att &hsimp=yhs-att_001&type=att_lego_portal_home.

Manuele, Fred A. "Reviewing Heinrich: Dislodging Two Myths from the Practice of Safety." *Professional Safety*, Oct. 2011.

Reason, James. *Human Error*. Cambridge, UK: Cambridge University Press, 1990.

Trist, E. L. *The evolution of socio-technical systems: A conceptual framework and an action research program*. Toronto, Canada: Ontario Quality of Working Life Center, Occasional Paper No. 2, 1981.

The Five-Why System: http://www.mapwright.com.au/newsletter/fivewhys.pdf and http:// www.moresteam.com/toolbox/5-why-analysis.cfm.

Wiegman, Douglas A., Hui Zhang, Terry von Thaden, Gunjan Sharma, and Alyssa Mitchel. "A Synthesis of Safety Culture and Safety Climate Research". Technical Report ARL-02-3/ FAA-02-2. Prepared for the Federal Aviation Administration, June 2001 (enter the title of the study into a search engine for access).

OTHER READING

"Safety Climate Measurement." Undated. At http://www.lboro.ac.uk/departments/sbe/down-loads/pmdc/safety-climate-assessment-toolkit.pdf.

"Socio-Technical Systems in Professional Decision Making", by William Frey. At http://cnx. org/content/m14025/latest/?collection=col10396/latest.

"Socio-Technical Systems Theory and Environmental Sustainability." At http://sprouts.aisnet.org/11-3/.

"Socio-Technical Systems: Team Building" by John W. Aldridge. At http://www.argospress.com/Resources/team-building/socitechnisistem.htm.

"Socio-Technical Systems: Information on Socio-Technical System" by John W. Aldridge, 2004. At http://www.argospress.com/Resources/team-building/socitechnisistem.htm.

"Socio-Technical Theory." At http://istheory.byu.edu/wiki/Socio-technical theory.

Weisbord, M. R. *Productive Workplaces: Organizing and Managing for Dignity, Meaning and Community*. San Francisco: Jossey-Bass, 1987.

Yule, Steven. *Safety Culture and Safety Climate: A Review of the Literature*, 2003. At http://www.efcog.org/wg/ism_pmi/docs/Safety_Culture/Feb08/safety_culture_and_safety_climate_a_review_of_the_literature.pdf.

CHAPTER 6

SAFETY PROFESSIONALS AS CULTURE CHANGE AGENTS

This chapter promotes the idea that the overarching role of a safety professional is that of a culture change agent. The case will be made that every proposal made by a safety professional for improvement in an occupational safety management system pertains to a deficiency in a system or process. And the deficiency can be corrected only if there is a modification in an organization's culture—a modification in the way things get done, a modification in the *system of expected performance*. Thus, the primary role for a safety professional is that of a culture change agent.

What does the term *overarching* mean? A composite definition, as found in dictionaries, is: encompassing everything; embracing all else; including or influencing every part of something. This premise—that the overarching role of a safety professional is that of a culture change agent—applies universally to all who give advice on improving safety management systems. There are no exceptions. Definitions of a change agent are numerous. This definition is a composite that fits well with the safety professional's position.

> A *change agent* is a person who serves as a catalyst to bring about organizational change. Change agents assess the present, are controllably dissatisfied with it, contemplate a future that should be, and take action to achieve the culture changes necessary to achieve the future desired.

To make the case that the overarching role of a safety professional is that of a culture change agent, in this chapter we:

Advanced Safety Management: Focusing on Z10 and Serious Injury Prevention,
Second Edition. Fred A. Manuele.
© 2014 John Wiley & Sons, Inc. Published 2014 by John Wiley & Sons, Inc.

- Comment on a safety culture within an organization's overall culture and how the advice given by safety professionals affects the culture
- Recognize the difficulties when the safety culture is negative
- Provide resources with respect to safety professionals as culture change agents
- Test the premise that safety professionals are cultural change agents against:
  - The requirements of Z10
  - The provisions in "The Socio-Technical Model for an Operational Risk Management System"

THE SIGNIFICANCE OF MANAGEMENT LEADERSHIP AND ORGANIZATIONAL CULTURE

From whence does the culture derive that safety professionals are to influence as they promote continuous improvement? As management provides the leadership (Z10 sections 3.0 and 3.1) and makes decisions and takes actions directing the organization, the outcome of those decisions establishes the reality of its safety culture as a subset of its overall culture. The outcome could be positive or negative. Safety is culture driven, and management establishes the culture.

An organization's culture with respect to occupational safety, environmental safety, the safety of the public, and product safety is determined by the outcome of decisions made by management as measured by the risk levels attained in the technical and social aspects of a facility's operations. The culture created by management is the dominant factor with respect to the risk levels attained, whether they are acceptable or unacceptable. Over the long term, the injury, illness, environmental damage, and property damage experience attained are a direct reflection of an organization's safety culture.

Management owns the culture. An organization's culture is translated into a *system of expected performance*. Strong emphasis is given to the phrase *a system of expected performance* because it defines the staff's beliefs with respect to what management wants done in reality. Although an organization may issue commendable safety policies, manuals, and operating procedures, the staff's perception of what is expected of them and the performance for which they will be measured—its *system of expected performance*—may differ from what is written.

Colleagues remind me of having written years ago that management is what management does, which may differ from what management says. What management does defines the actuality of an organization's safety culture and its commitment or noncommitment to safety. Employee perceptions of management's position on safety are, in effect, their reality. Realistic or unrealistic, employee perceptions are their truths.

To achieve superior results, only top management can provide the leadership and direction needed to "establish, implement and maintain an occupational health and safety management system" (Z10 Section 3.1.1). Major improvements in safety—being free from unacceptable risks—will be achieved only if a culture change takes place—only if major changes occur in the *system of expected performance*. That must be understood by safety professionals in their role as culture change agents.

THE ROLE OF SAFETY PROFESSIONALS WITH RESPECT TO THE SAFETY CULTURE

What is the safety professional's role with respect to an organization's safety culture? Assume that safety is a core value in an organization and that the board of directors and senior management are determined to achieve and maintain acceptable risk levels in all operations. Usually, the environmental, health, and safety professionals in those organizations are well qualified, they have stature, and the advice they give is well received and seriously considered.

But even in organizations where safety is a core value and superior safety management systems are in place, change—favorable or unfavorable—is a constant. Information will be developed through the safety management processes, indicating that improvements can be made in certain safety-related processes.

Then, acting from a sound professional base, the advice given by safety professionals in their role as culture change agents is generally welcomed. Their role requires:

- Diligent data gathering and analysis to identify the process shortcomings
- Proposing arrangements for hazard identification and analyses and risk assessments to be made
- Giving advice on prioritizing risks
- Recommending the management system actions that should be taken for improvement

But all organizations do not have superior safety management systems. In such entities, the role of the safety professional as a culture change agent can be more difficult, particularly if senior management takes the position that all is well and believes that there is little need to make changes. The operating base for a safety professional is the same as in the previously bulleted items.

The skill level required to be a successful culture change agent in such situations can be exceptionally demanding. Patience is required. Satisfaction may derive principally from small steps forward. But the goal remains the same: to attempt to influence the safety culture positively as continual improvement is proposed.

Now, assume that the organization's culture has always been negative or is drifting into a negative state. I have observed that the concept of drift needs discussion because of the economics in recent years and the significant expense reductions required by executive management in some companies.

In "Risk Management in a Dynamic Society: A Modelling Problem," Jens Rasmussen writes:

The scale of industrial installations is steadily increasing with a corresponding potential for large-scale accidents. Companies today live in a very aggressive and competitive environment which will focus the incentives of decision makers on short term financial and survival criteria rather than long term criteria concerning welfare, safety and environmental impact. (p. 186)

The word drift has been attached to Rasmussen's premise. For example, in a book by Sidney Dekker entitled *Drift into Failure*, he cites Rasmussen's work. Dekker writes:

Drift occurs in small steps. This can be seen as decrementalism, where continuous adaptation around goal conflicts and uncertainty produces small, step-wise normalizations of what was previously judged as deviant or seen as violating some safety constraint. (p. 15)

When a safety professional senses drift resulting from the decisions made that result in "violating some safety constraint" or, more likely, several safety constraints, the challenges faced as a culture change agent require an approach that attempts to deter or slow the pace of drift. With the same diligent data gathering and analysis to identify process shortcomings and risk prioritization as mentioned previously, counsel can be given to management on the facts and pace of the "drift to danger".

The intent is to have management become aware that the organization is putting in place the elements that increase the potential for a 'large-scale accident' and to encourage that the pace of deterioration in processes be slowed down or stopped. Prioritizing risks and emphasizing the growing potential for the occurrence of a low-probability/severe-consequence event acquires a high degree of importance.

Safety personnel at locations where the safety culture has drifted into a negative state or the culture has always been negative probably have to deal with situations in which resources are greatly limited. Their communication should provide management with information on which safety-related decisions can be made on a priority basis so that the limited resources available can be applied to achieving the greatest good. That, again, particularly, requires priority setting and focusing on preventing low-probability/serious-consequence events.

I do not claim that achieving success by safety professionals in such situations will be easy. But the probability of being successful as a culture change agent is enhanced if a safety professional attains the status of an integral member of the business team. That will result from giving well-supported, substantiated, and convincing technical and managerial risk management advice that is perceived as serving the business interests. Admittedly, convincing management that safety should be one of the organization's core values may be difficult to achieve. A short list of professional skills that safety professionals should strive to obtain and refine includes:

- A general understanding of business and financial terms and language
- Use of basic financial analysis tools such as benefit/cost analysis
- Oral presentation and writing skills
- An enhanced ability to influence business decision makers
- A positive attitude and problem-solving approach

Notice that some of these skills and attributes are business-related. To be successful as an agent of change, the safety professional must operate within the business framework in the organizations to which they give counsel.

SELECTED RESOURCES ON SAFETY PROFESSIONALS AS CULTURE CHANGE AGENTS

In 2008, Jim Spigener and Don Groover published a paper entitled "Staying Relevant: The Emerging Role of the Safety Professional—Becoming a Change Agent." They say:

> Staying relevant as an organization changes means learning how to leverage your knowledge, skills and experience in new ways. If you are a technical expert in environmental health, and safety, the good news is that you already have the skills and knowledge to contribute to safety strategy. The hard part will be gaining fluency in organizational change management.

Spigener and Groover write about the core competencies of a change agent, and their comments have a kinship to that of other writers. They, and others, say that change agents:

- Are forever inquisitive and never-ending learners
- Advance performance by identifying what ought to be, deciding how to get there, and influencing decision makers to adopt their ideas
- Do not leave their expertise behind
- Leverage their knowledge and experience to develop strategies to positively influence actions that result in a higher performance level
- Recognize that to be influential in achieving change, they must acquire change management skills
- Become aware of the culture in place and learn how to manage within it to effect change
- Recognize the effect of management decisions and actions on the culture
- Find ways to tactfully inform management when they believe those decisions and actions may have negative results

Chapter 4 in book *Safety Through Design* is entitled "Achieving the Necessary Culture Change." Steven I. Simon was the author. Its opening paragraph reads as follows:

> A full explanation of what culture change is and is not, who is involved, why it is necessary and can achieve world class safety through design, and how to make it happen is provided in this chapter.

Although the focus of the chapter is on safety through design, it is largely generic with respect to attaining a culture change.

In Volume 41, Issue 1, February 2003 of the magazine *Safety Science*, there is an article entitled "The safety adviser/manager as agent of organisational change: a new challenge to expert training". Paul Swuste and Frank Arnoldy are the authors. They are attached to the Safety Science Group at the Delft University of Technology. This is the Abstract for their article:

There is a great need for health and safety advisers/managers to act as agents of change, both in respect to the technology of the company and the design of its workplaces, and in the organisation of the company health and safety management system. This article reports on the development of training to meet these increasing needs. The postgraduate masters course 'Management of Safety, Health and Environment' of the Delft University of Technology has now introduced a course-module of 1 week, addressing the issue of the learning organisation and the specific role of the safety adviser/manager.

The course-module starts from the assumption that for a health and safety adviser/manager his or her personal effectiveness and ability to influence and stimulate others are qualities as important to a company as the quality of a safety and health management system. This paper will describe the development in the role of the safety adviser/manager and the mainstream thinking on change management and training. The consequence for the content and programme features of the course-module is presented as well as the results of the evaluation of its effectiveness.

For emphasis, we repeat a sentence from this abstract: "The course-module starts from the assumption that for a health and safety adviser/manager his or her personal effectiveness and ability to influence and stimulate others are qualities as important to a company as the quality of a safety and health management system."

John Kello, who is connected with Davidson College, had an article published in Volume 6, Number 4, of *The International Journal of Knowledge, Culture & Change Management* entitled "Changing the Safety Culture: Safety Professionals as Change Agents." Excerpts from that article follow.

The Safety Change-Agent

So what is the proper role of the safety professional in the Total Safety Culture, to which many organizations today aspire? It is definitely not the same old technical expert role, even with a broader bandwidth. It is fundamentally, qualitatively different in its approach.

In the field of Organization Development, OD practitioners have been referred to as "change agents" from the very beginnings of the discipline. My central thesis is that, whether they normally think of it in these terms or not, to be truly effective in the flexible, team-based High Performance Organization, safety professionals must perform as change agents too. In my view, much like organization development consultants, safety professionals encourage and help people make constructive behavior change, to do things differently, to challenge longstanding habits and to get out of their "comfort zone". Further, and also like the OD consultant, they are almost always more of an influencer than a director.

Kello goes on to say that "modern safety professionals are agents for positive change in their organizations. They are trying to build deep working relationships that allow them to effect constructive change through influence, even when the client system may not want to change."

The next reference is not an article or a book. It is a slide presentation pertaining to patient safety that was developed for the U.S. Department of Defense. The title is "Change Management: How to Achieve a Culture of Safety."

I have tried to avoid including a "how to" dissertation in this chapter. However, as the title of the slide series says, this is a how-to piece of work. Its first objective, as shown below, is to identify and discuss the eight steps of change. Recognition is given to John Kotter's book as the source of the eight steps, about which more will be said later. Objectives to be achieved are:

- To identify and discuss the Eight Steps of Change.
- To describe the actions required to set the stage for organizational change.
- To identify ways to empower team members to change.
- To discuss what is involved in creating a new culture.
- To begin planning for the change in the organization.

If safety professionals presented this slide series and attempted to influence others on how to achieve organizational change, they would be serving as culture change agents. A review of this slide series is recommended for its informative value.

John P. Kotter is the author of *Leading Change*. His book has been well received and critiqued positively. Many authors and consultants refer to Kotter's work; it should be considered foundational. For a good and thorough review of the content of *Leading Change*, safety professionals could profitably spend some time with the slide presentation cited above: "Change Management: How to Achieve a Culture of Safety." Also, as will be noted later, Kotter has a lengthy paper on the Internet that presents his views. I have been greatly impressed with the thought pattern in *Leading Change*, a good part of which pertains to what was learned by Kotter from practical applications.

WHY CULTURE CHANGE INITIATIVES FAIL

Kotter makes it clear that many change initiatives fail. Many other authors agree and have posted relative material on the Internet. As culture change agents, safety professionals should be well informed on how change initiatives succeed and fail and how success and failure are measured. Several references on initiative failures follow.

- At http://EzineArticles.com/26088 you will find five reasons why leaders fail to create successful change.

- At http://andybrazier.blogspot.com/2007/04/seven-reasons-why-organisational-change.html, an author presents seven reasons why organizational change fails.
- With the source given as *Leading at a Higher Level* by Ken Blanchard, you will find "Why Change Efforts Typically Fail: 15 Predictable Reasons & Situations to Avoid" at http://a80.cci.fsu.edu/flalib/www/conference_2010/round_table_documents/Why%20Change%20Fail%20or%20succeed.pdf.
- John P. Kotter has a paper entitled "Leading Change: Why Transformation Efforts Fail" at http://www.stratafrica.com/media/2a8f73ccf40110beffff80be7f000101.pdf. This is a multipage presentation on "The Eight Stage Process" to transform an organization. Although the wording in Kotter's paper is not absolutely identical to the content of the book *Leading Change*, the paper is only a little short of a verbatim excerpt. This is a good and valuable read. Safety professionals are encouraged to download it, as it has significant informational value.

Reflecting on my own experience, a few of the reasons why change initiatives fail are recorded here. The first is the most important.

1. The culture in place and how to work within it are largely ignored.
2. The leadership and commitment necessary at sufficiently high levels to achieve change may not exist in reality because the change agent has not invested the time necessary to achieve the commitment.
3. Decision makers are not seriously enthused about the change proposed because the supporting data are shallow and not convincing.
4. The importance of becoming aware of the power structure and determining how to work within it have not been sufficiently well recognized.
5. Team building, which is vital to success, has been inadequate.
6. Preparing for the typical resistance to change at all levels comes up short.
7. Communication to all personnel levels that would be affected by the change is not as thorough as needed.
8. Management personnel who are assigned responsibility for the change may not be held accountable for progress by the people to whom they report, and in time, the urgency and importance for the change that may have been established diminishes.
9. Assumptions are made that a change in a process or system has occurred without determining that it has. Others refer to this as declaring victory too quickly. Some authors say that a change in a system or process should not be considered a success as a culture modification until the change has been in place for at least a year. Too often, operators revert to previous methods if supervisers allow such a reversion.
10. Change agents are not sufficiently aware that achieving a culture change may take several years.

A BASIC GUIDE

The second edition of *Environmental Management Systems: An Implementation Guide for Small and Medium-Sized Organizations* was prepared by NSF International with funding through a cooperative agreement with the U.S. Environmental Protection Agency. It is on the Internet and downloadable. Although the text is devoted to environmental management, many of its parts are generic. A few excerpts follow.

I believe that the contents of the *Guide* are basic to almost all change initiatives. This publication can be a good addition to a library titled Safety Professionals as Culture Change Agents.

Objectives and Targets:

Establishing goals for environmental management ISO 14001

Objectives and targets help an organization translate purpose into action. These environmental goals should be factored into your strategic plans. This can facilitate the integration of environmental management with your organization's other management processes.

You determine what objectives and targets are appropriate for your organization. These goals can be applied organization-wide or to individual units, departments or functions – depending on where the implementing actions will be needed.

In setting objectives, keep in mind your significant environmental aspects, applicable legal and other requirements, the views of interested parties, your technological options, and financial, operational, and other organizational considerations.

There are no "standard" environmental objectives that make sense for all organizations. Your objectives and targets should reflect what your organization does, how well it is performing and what it wants to achieve.

Hints

- Setting objectives and targets should involve people in the relevant functional area(s). These people should be well positioned to establish, plan for, and achieve these goals. Involving people at all levels helps to build commitment.
- Get top management buy-in for your objectives. This should help to ensure that adequate resources are applied and that the objectives are integrated with other organizational goals.
- In communicating objectives to employees, try to link the objectives to the actual environmental improvements being sought. This should give people something tangible to work towards.
- Measureable objectives should be consistent with your overall mission and plan and the key commitments established in your policy (pollution prevention, continual improvement, and compliance). Targets should be sufficiently clear to answer the question: "Did we achieve our objectives?"

- Be flexible in your objectives. Define a desired result, then let the people responsible determine how to achieve the result.
- Objectives can be established to maintain current levels of performance as well as to improve performance. For some environmental aspects, you might have both maintenance and improvement objectives.
- Communicate your progress in achieving objectives and targets across the organization. Consider a regular report on this progress at staff meetings.
- To obtain the views of interested parties, consider holding an open house or establishing a focus group with people in the community. These activities can have other payoffs as well.
- How many objectives and targets should an organization have? Various EMS implementation projects for small and medium-sized organizations indicate that it is best to start with a limited number of objectives (say, three to five) and then expand the list over time. Keep your objectives simple initially, gain some early successes, and then build on them.
- Make sure your objectives and targets are realistic. Determine how you will measure progress towards achieving them.

A TEST OF THE PREMISE

The source for the following quote is the planning section, section 4.0 in Z10. "The planning process goal is to identify and prioritize occupational health and safety management issues (defined as hazards, risks, management system deficiencies and opportunities for improvement)."

The table of contents for Z10 is duplicated here, excluding the "Scope, Purpose, & Application," section 1.0; and "Definitions," section 2.0. Readers are asked to review every section and subsection to consider whether they can locate exceptions to the following premises.

If hazards, risks and management system deficiencies exist, the solution is to change and improve the relative management systems and, if the changes are successful, a culture change will have been achieved. Achieving a culture change is the ultimate goal.

3.0 Management Leadership and Employee Participation
    3.1 Management Leadership
        3.1.1 Occupational Health and Safety Management System
        3.1.2 Policy
        3.1.3 Responsibility and Authority
    3.2 Employee Participation
4.0 Planning
    4.1 Initial and Ongoing Reviews (Error: For 2012 version, substitute "Review Process")
    4.2 Assessment and Prioritization

I cannot conceive of a situation in which a hazard, a risk, or a management system deficiency exists for which the remedy would not require a revision in a process and the *system of expected performance*. If the process revision is successful, acceptable risk levels will be the result and a culture change will have been achieved. A precaution expressed previously needs emphasis: Revisions in a process, and thereby a culture change, should be examined over time. The purpose is to assure that:

- Personnel have not reverted to previous practices after what seems to be a success in the short term.
- The modification made is effective in delivering the outcomes expected.
- No unintended consequences are created that increase risk.

A TEST IN RELATION TO A SYSTEM TO ENHANCE SERIOUS INJURY PREVENTION

In Chapter 3, "Innovations in Serious Injury and Fatality Prevention", it was said that "major and somewhat shocking innovations in the content and focus of occupational risk management systems will be necessary to achieve additional progress in fatality and serious injury prevention." Such innovations were included in my socio-technical model for an operational risk management system. That model is presented here in sections to illustrate the various levels of approaches to be taken to achieve a culture change.

As was done for the requirements of Z10, readers are asked to review each of the elements in the model to find exceptions to the premise that if hazards, risks, or deficiencies in management systems are identified, a process change is necessary that results in improved performance with respect to the management system, thereby achieving a culture change. Thus, changing the culture is the ultimate goal.

The first two sections of the model pertain to establishing a safety culture by the board of directors and to management making decisions to assure that the culture expected by the board is established. Those sections follow.

The board of directors and senior management establish a culture for continuous improvement that requires defining, achieving, and maintaining acceptable risk levels in all operations.

Management leadership, commitment, involvement, and the accountability system establish that the performance level to be achieved is in accord with the culture established by the board.

Changes in policies established at a board of directors' level and applied by senior management can be achieved if sound, logical, convincing presentations are made up through the organizational structure. Nevertheless, that's the managerial level from which an organization's safety culture derives. Having culture changes made at the board and executive levels should be the goal of safety professionals. Achieving policy revisions at that level will require superior skills as a culture change agent.

In the model, the next step by management is to establish policies, standards, and procedures to promote adoption of the safety policies set at the board level. Several subjects are listed here as a collection because they exhibit a similarity—a large majority of organizations do not have comparable written processes in place. The next section of the model follows.

To achieve acceptable risk levels, management establishes policies, standards, procedures, and processes with respect to:

- Providing adequate resources
- Risk assessment, prioritization, and management
  - Applying a hierarchy of controls
- Prevention through design
  - Inherently safer design
  - Resiliency, reliability, and maintainability
- Competency and adequacy of staff
  - Capability—skill levels
  - Sufficiency in numbers
- Maintenance for system integrity
- Management of change/pre-job planning

For most of the foregoing, having written and established processes in place will be the exception. Having new systems established for such items as risk assessment or prevention through design would be a major undertaking and not easily achieved. Having top-level skills in management of change would be necessary.

If the elements listed above became a part of an operational risk management system, it is a near certainty that the probability of serious injuries and fatalities occurring would be reduced. That is a good reason for safety professionals trying to apply culture change methods to have them instituted.

Organizations that employ safety professionals will have procedures in place for most of the following safety-related elements. The exception is "Procurement—safety expectations." It is understood that getting safety specifications included in procurement orders or contracts is not easy to do (see Chapter 20). "The Procurement Process". Those elements follow.

- Safety-related systems
  - Organization of work
  - Training—motivation
  - Employee participation
  - Information—communication
  - Permits
  - Inspections
  - Incident investigation and analysis
  - Providing personal protective equipment
- Third-party services
  - Relationships with suppliers
  - Safety of contractors—on premises
  - Procurement—safety specifications
  - Emergency planning and management
  - Conformance/compliance assurance reviews

Since improvements suggested by safety professionals often are to refine existing systems, the culture change exercise may be easier—but not necessarily easy. In every case, the planning, communication, and consideration of people relationships that are a part of a good culture change process must be at a high level.

The last element of the model requires periodic performance measurements upon which sound decisions can be made in support of continuous improvement. That element follows.

Performance measurement: Evaluations are made and reports are prepared for management review to support continuous improvement and to assure that acceptable risk levels are maintained.

Even at this level, achieving improvement in reporting systems (such as audits) that overlook hazards, risks, and management system deficiencies requires the employment of culture change principles.

CONCLUSION

A sound case can be made in support of the idea that the overarching role of a safety professional is that of a culture change agent. In my role as the managing director for a safety and fire protection consultancy, it became evident early on that the only product that we had to sell was advice—and that our success would be determined by whether clients believed that our advice provided value. Think about it. Are most safety professionals not primarily providers of advice to achieve change?

Safety professionals are most often in staff positions and their role is to provide advice to decision makers on hazards, risks, and deficiencies in management systems so that modifying actions can be taken to achieve acceptable risk levels.

Actions taken to attain acceptable risk levels require culture changes. Although it has been apparent only to a few, safety professionals have always been culture change agents. If they recognize the reality of their role, they should become more competent in achieving changes in systems and processes.

Many big ideas on organizational change have failed because the prevailing culture was ignored. Nearly every author who writes on organizational change makes it clear that achieving permanent change is difficult. Safety professionals must become aware of the importance of giving attention to an organization's existing culture, how deeply certain practices are embedded within processes, who presumes to have ownership of them, and the need to plan and communicate to achieve change.

Safety professionals must be perceived to be members of the management team. Change methods they adopt should, if practicable, be in concert with procedures with which the management team is comfortable so that major conflict can be avoided. A good place to start in applying culture change agent skills is with "Management of Change", Chapter 18 in this book.

REFERENCES

ANSI/AIHA Z10-2012. American National Standard, *Occupational Health and Safety Management Systems*. Fairfax, VA: American Industrial Hygiene Association, 2012. ASSE is now the secretariat. Available at https://www.asse.org/cartpage.php?link=z10_2005.

"Change Management: How to Achieve a Culture of Safety." U.S. Department of Defense. Available at http://us.yhs4.search.yahoo.com/yhs/search?p=Change+Management%3A+How+to+Achieve+a+Culture+of+Safety&hspart=att&hsimp=yhs-att_001&type=att_lego_portal_home.

Christensen, Wayne and Fred Manuele, Editors. *Safety Through Design*. Itasca, IL: National Safety Council, 1999.

Dekker, Sidney. *Drift into Failure*. Burlington, VT: Ashgate Publishing Company, 2011.

Environmental Management Systems: An Implementation Guide for Small and Medium-Sized Organizations, 2nd ed. Available at http://us.yhs4.search.yahoo.com/yhs/search?p=Environmental+Management+Systems%3A+An+Implementation+Guide+for+Small+and+Medium+Sized+Organizations%2C+Second+Edition&hspart=att&hsimp=yhs-att_001&type=att_lego_portal_home. The copyright is held by NSF International Strategic Registrations, Ltd., Ann Arbor, MI.

Kello, John. "Changing the Safety Culture: Safety Professionals as Change Agents." *The International Journal of Knowledge, Culture & Change Management*, Vol. 6, No. 4. Also available for a $5.00 download charge at http://us.yhs4.search.yahoo.com/yhs/search?p=Jo hn+Kello+Changing+the+Safety+Culture%3A+Safety+Professionals+as+Change+Agent s&hspart=att&hsimp=yhs-att_001&type=att_lego_portal_home.

Kotter, John P. *Leading Change*. Boston: Harvard Business Review Press, 1996.

Rasmussen, Jens. "Risk Management in a Dynamic Society: A Modelling Problem." *Safety, Science*, Vol. 27, No. 2/3, pp. 183–213, 1997.

Spigener, Jim and Don Groover. "Staying Relevant: The Emerging Role of the Safety Professional—Becoming a Change Agent," 2008. It is available at http://www.bstsolutions. com//resources/knowledge-resource/staying-relevant-the-emerging-role-of-the-safety -professional-part-2-becoming-a-change-agent.

Swuste, Paul and Frank Arnoldy. "The Safety Adviser/Manager as Agent of Organisational Change: A New Challenge to Expert Training," *Safety Science*, Vol. 41, Issue 1, 2003, pp. 15–27. The authors are attached to the Safety Science Group, Delft University of Technology, Delft, The Netherlands. An abstract of the article is available at http://www.sciencedirect.com/ science/article/pii/S0925753501000509.

CHAPTER 7

THE PLAN-DO-CHECK-ACT CONCEPT (PDCA)

In the Introduction to ANSI/AIHA Z10-2012, the *Occupational Health and Safety Management Systems* standard, it is stated that the design of ANZI Z10 encourages integration with other management systems to facilitate organizational effectiveness using the elements of Plan-Do-Check-Act (PDCA) Model as the basis for continual improvement. Prominence is given in this chapter to the application of PDCA concepts as an asset in continuous improvement.

Vic Toy, the vice chair for the committee that wrote the Z10 standard, wrote an article entitled "Let Your OHS Management System Do the Work: How the New Z10 Adds Even Better Value." What Toy wrote also relates to continuous improvement.

> The beauty of an Occupational Health and Safety Management System (OHSMS) is that it provides health and safety management in an integrated, interconnected, organic way to maintain focus on continual improvement. The Z10 standard provides a systematic framework and the tools required for continual improvement.

In discussions with James Howe, chair of the Z10 committee, he also emphasizes that the intent of the committee was to have the application of PDCA concepts be a major influence on continuous improvement.

Within Toy's article, which is recommended reading, there was a depiction of the PDCA concept, which includes emphasis on continual improvement. With his

Advanced Safety Management: Focusing on Z10 and Serious Injury Prevention,
Second Edition. Fred A. Manuele.
© 2014 John Wiley & Sons, Inc. Published 2014 by John Wiley & Sons, Inc.

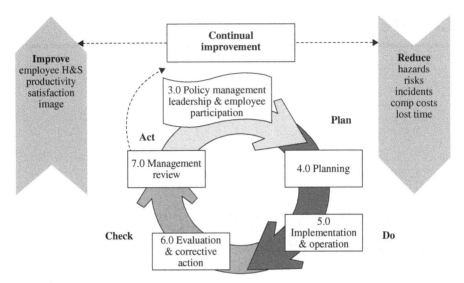

FIGURE 7.1   PDCA model. (Reprinted from *Synergist*, Feb. 2012; with permission.)

concurrence and that of Ed Rutkowski, representing the American Industrial Hygiene Association, the depiction is shown here as Figure 7.1.

To provide guidance on the PDCA concept and its application in problem-solving initiatives, in this chapter we:

- Discuss the origin and substance of the PDCA concept.
- Comment on the process, systems, and continual improvement aspects of PDCA.
- Discuss variations in PDCA applications.
- Relate the PDCA concept to basic problem-solving techniques.
- Give guidance on initiating a PDCA process.

ORIGIN AND SUBSTANCE OF THE PDCA CONCEPT

In *Out of the Crisis*, W. Edwards Deming provided a diagram designated as the Shewhart cycle. This is what Deming said about it:

> The perception of the cycle shown came from Walter A. Shewhart. I called it in Japan in 1950 and onward the Shewhart cycle. It went into immediate use in Japan under the name of the Deming cycle, so it has been called ever since. (p. 88)

Deming became world-renowned for his successful approaches to quality management. Deming's depiction of the Shewhart cycle predates all other diagrams that I have been able to locate that are comparable to what is now known as the *PDCA concept*.

Walter A. Shewhart was a Bell Laboratories scientist and friend and mentor of Deming. Shewhart is credited with having developed a statistical process control method in the late 1920s. Thus, the origin of the PDCA concept lies in statistical process control, a methodology developed to address the need for improvement in product quality.

The emphasis in the application of the PDCA concept with respect to product quality is on process control and continual improvement. That is also the case in Z10. The words *process* and *processes* appear in Z10 more than 120 times.

Deming's original depiction of the Shewhart cycle is a six-step, numerically identified process. Following are the six steps as they appear in *Out of the Crisis*. Keep in mind that this is a quality improvement process:

1. What would be the most important accomplishments of this team? What changes might be desirable? What data are available? Are new observations needed? If yes, plan a change or test. Decide how to use the observations.
2. Carry out the change or test decided upon, preferably on a small scale.
3. Observe the effects of the change or test.
4. Study the results. What did we learn? What can we predict?
5. Repeat step 1, with knowledge accumulated.
6. Repeat step 2, and onward. (p. 88)

The foregoing outline represents good thinking to achieve continual improvement. As Deming wrote, the Shewhart cycle became known as the Deming cycle, and it metamorphosed into the PDCA form.

Out of the Crisis was published in 1982. The second edition of Deming's *The New Economics for Industry, Government, Education* was published in 1994. What Deming referred to as "the Shewhart cycle for learning and improvement" was still close to the depiction of the PDCA model in Figure 7.1, but one revision was made. Deming referred to his amended model as the Plan-Do-Study-Act (PDSA) cycle. That replacement term—*study*—has not been adopted broadly. Most of the literature on the continual improvement process refers to the PDCA model.

DEFINING PDCA

A variety of descriptive terms appear in the literature on PDCA, and safety professionals should adopt the language that best suits their purposes. In the following examples, emphasis is indicated by underscores:

- In *Out of the Crisis*, Deming wrote that "the Shewhart cycle will be helpful as a procedure to follow for improvement of any change." (p. 88)
- In *The New Economics*, Deming said that "the PDSA cycle is a flow diagram for learning, and for improvement of a product or a process." (p. 132)
- On the Internet, a paper issued by the North Carolina Department of Environmental and Natural Resources is entitled "Plan–Do–Check–Act—A Problem Solving Process." (p. 1)

- The U.S. Environmental Protection Agency has issued an "Implementation Guide for the Code of Environmental Management Principles for Federal Agencies." They state: "The approach incorporates the "plan–do–check–act" cycle and the emphasis on continuous improvement." (p. 2)
- In ANSI/ISO/ASQ Q9001-2008, the American National Standard *Quality Management Systems—Requirements*, reference is made to "The model of a process-based quality management system, and the methodology known as 'Plan–Do–Check–Act. (p. ix) This standard gives a brief but adequate definition of the PDCA processes, as follows:
  - *Plan*: Establish the objectives and processes necessary to deliver results in accordance with customer requirements and the organization's policies.
  - *Do*: Implement the processes.
  - *Check*: Monitor and measure processes and product against policies, objectives, and requirements for the product and report the results.
  - *Act*: Take actions to continually improve process performance. (p. x)

So, PDCA is a concept, cycle, procedure, flow diagram, process, methodology, and model for continual improvement. To relate directly to Z10 and the work of safety professionals who give counsel on its implementation, the following definition is offered, *striving for simplicity*.

The PDCA concept presents a sound problem-solving and continual-improvement model.

ON PROCESSES, SYSTEMS, AND CONTINUOUS IMPROVEMENT

Throughout Z10 there is frequent repetition of the premise that "the organization shall establish and implement *processes* [emphasis added] to ensure" that the elements of the OHSMS are established and implemented. The word *system* appears as often as does *processes*.

What is a process? What is a system? From the many definitions of process and system available, those selected for this chapter are taken from the "Business Dictionary" on the Internet.

Process: sequence of interdependent and linked procedures which, at every stage, consume one or more resources (employee time, energy, machines, money) to convert inputs (data, material, parts, etc.) into outputs. These outputs then serve as inputs for the next stage until a known goal or end result is reached.

System: an organized, purposeful structure that consists of interrelated and interdependent elements (components, entities, factors, members, parts, etc.). These elements continually influence one another (directly or indirectly) to maintain their activity and the existence of the system, in order to achieve the goal of the system.

The emphasis given to processes is appropriate and important. Having effective processes—management systems—is necessary in fulfillment of the PDCA concept. Furthermore, stressing the need for effective processes appropriately puts the focus in determining contributing factors for injuries and illnesses on the adequacy or inadequacy of the processes, that is, the management systems. In accord with the PDCA concept, Z10 is a process standard.

Focusing on deficiencies in management systems is particularly advantageous in an attempt to identify and select the actions to be taken to reduce the probability of incidents or exposures occurring that may result in serious injuries or illnesses.

Deming emphasized making system improvements to achieve significant advances in product quality rather than directing efforts on what employees do. In her book *Deming: Management at Work*, Mary Walton recorded how Deming expressed the need to focus on improving the system:

> In the American style of management, when something goes wrong the response is to look around to blame or punish or to search for something to "fix" rather than to look at the system as a whole for improvement. The 85–15 Rule holds that 85 percent of what goes wrong is with the system, and only 15 percent with the individual person or thing.

In applying the PDCA model, the goal is to improve processes. Taken as a whole, the processes make up the management system. In applying Z10, processes are to be established and implemented to create an OHSMS. Although Z10 does state that "employees shall assume responsibility for aspects of health and safety over which they have control, including adherence to the organization's health and safety rules," the focus of the standard is on improving processes that are controlled by management.

Each process is dependent on the others for the overall management system to achieve its goals. In application of the PDCA concept, the impact that making a change in one process may have on another process must be considered. Deming addresses the interdependence of systems and processes in *The New Economics*:

> A system is a network of interdependent components that work together to accomplish the aim of the system. A system must have that aim. Without an aim, there is no system. (p. 95)

> The greater the interdependence between components, the greater will be the need for communication and cooperation between them. (p. 96)

Consider the following as examples.

1. Changes or alterations in the operating system may require:
   - New job hazard analyses
   - Revisions in the standard operating procedures
   - Adjustments in the content of training programs

- Rescheduling for maintenance
- Extensions in inspection details
2. Changes in design specifications may have an impact on the safety-related specifications to be included in purchasing documents.

The aim of the OHSMS, in accord with the PDCA concept, is clearly established: *To provide a management tool to reduce the risk of occupational injuries, illnesses, and fatalities.* Each continual improvement process is an integral part of the system to achieve the desired control of risks. To reduce the risk of injuries and illnesses, the goal in applying every element in the PDCA concept in the continual improvement process must be to achieve acceptable risk levels.

MEASUREMENT SYSTEMS

A continual improvement process requires that measurement systems be in place to observe progress toward achieving stated goals. Section 6.0 in Z10, "Evaluation and Corrective Action," requires that processes be in place to evaluate the performance of the OHSMS through monitoring, measurement, assessment, incident investigations, and audits. Their purpose is to determine whether the processes are functioning as designed. The process measurements chosen must be compatible with an organization's size, operations, and other measurement systems in use.

Valid statistical measures such as control charts are convincing and I encourage their use. In some situations, such as initiating the culture change necessary to call attention to serious injury prevention, the occurrence data on such incidents that would be placed on a control chart will not be available since low-probability/serious-consequence events do not occur frequently. Cost data for such events can be influential. For an additional reference, see "Measurement of Safety Performance" in *On the Practice of Safety, 4th edition.*

VARIATIONS ON A THEME

Entering Plan-Do-Check-Act into a search engine on the Internet will bring up thousands of variations of the concept. There are many adaptations of the concept because, through its use, it has proven to be a sound problem-solving and continual-improvement model. Comments are made here on two PDCA innovations in which many safety practitioners have an interest. They are six sigma and the U.S. EPA's environmental management system (EMS).

Six sigma is principally a management strategy used to achieve a product or service defect level of 3.4 or fewer defects per million opportunities. Writers on six sigma refer to the system as a parallel to Deming's PDCA cycle. As Deming does in the 14 Points of Management as presented in *Out of the Crisis*, six sigma emphasizes the importance of designing processes and systems to enable a staff to achieve six-sigma performance levels. Quoting from Deming in *Out of the Crisis*:

A theme that appears over and over in this book is that quality must be built in at the design stage. It may be too late once plans are on the way. (p. 49)

The same principle applies for six sigma as for OHSMS. In six-sigma programs, the adaptation of Deming's PDCA is called "Define, Measure, Analyze, Improve, and Control" (DMAIC). Achieving a six–sigma quality level is a major accomplishment. How does it relate to occupational injury and illness performance?

OSHA's recordable incident rate is based on 200,000 hours worked. Project the 200,000 base to a million by multiplying by 5. If an organization had an OSHA recordable incident rate of 0.68, it would be operating at the six-sigma level. Star performers in some industries achieve recordable incident rates lower than 0.68. But for most organizations, that performance level is beyond comprehension.

A search for easily available literature on the PDCA concept that would be helpful to safety professionals, particularly for smaller and medium-sized organizations, did not result in extensive success. However, a publication issued by the Environmental Protection Agency entitled *Environmental Management Systems: An Implementation Guide for Small and Medium-Sized Organizations, 2nd edition*, is a good reference for more than one purpose.

1. It is a valuable guide.
2. EPA recommends that the environmental management system be combined with existing management systems. In that context, reference is made in the text to a Total Quality Management System and an Occupational Safety and Health Management System.
3. The environmental management system proposed is built on the "Plan, Do, Check, Act model introduced by Shewhart and Deming and endorses the concept of continuous improvement." (p. 8)
4. A variation of the PDCA model is included in most of its pages.
5. EPA recognizes the difficulty in achieving a culture change and gives good guidance on the subject which is applicable to all hazard/risk situations. They say:

   The EMS approach and an organization's culture should be compatible. For some organizations, this involves a choice: (1) tailoring the EMS to the culture, or (2) changing the culture to be compatible with the EMS approach. Bear in mind that changing an organizations culture can be a long-term process. Keeping this compatibility issue in mind will help you ensure that the EMS meets your organization's needs. (p. 9)

6. The preceding comments, in item 5, have a direct relationship to the institution of a PDCA system.
7. Hints given, beginning on page 29, for the institution of an EMS are quite good and also apply to the initiation of a PDCA system.

This appears in the *Guide*: "This Guide has been copyrighted by NSF to preserve all rights under U.S. Copyright law and Copyright laws of Foreign Nations. It is not the intent of NSF to limit by this Copyright the fair use of these materials. Fair use shall not include the preparation of derivative works. Published by NSF International: E-mail: information@nsf-isr.org Web: www.nsf-isr.org."

RELATING PDCA CONCEPTS TO BASIC PROBLEM-SOLVING TECHNIQUES

Safety professionals who have an understanding of basic problem-solving techniques and how they are applied to hazard identification and analysis and risk assessments, and actions to avoid, eliminate, reduce, or control risks will have a relatively easy time adapting to the PDCA concept. Consider the alignments of problem-solving techniques and the Plan-Do-Check-Act idea shown in Table 7.1. The lists under the headings "Problem-Solving Methodology" and "Plan-Do-Check-Act" are composites of the methods shown in several texts.

In applying either the problem-solving methodology or the PDCA concept to prevent injuries and illnesses, the process is as follows.

1. Hazards are identified and analyzed.
2. The risks that derive from the hazards are assessed.
3. Alternative solutions for risk avoidance, elimination, reduction, or control are determined in accord with a prescribed hierarchy of controls.
4. Remedial risk management methods are selected and actions are taken to implement them.
5. Review processes are to determine whether the desired risk reduction was achieved and whether the residual risk is acceptable or unacceptable.
6. If the residual risk is unacceptable, start over.

The only important difference in the literature on the PDCA concept in relation to the writings on age-old, problem-solving techniques is that more emphasis is given in the PDCA literature to the continual improvement of processes. That Z10 is a process, and a continual improvement standard cannot be stated often enough. Improvements

TABLE 7.1

Problem-Solving Methodology	Plan–Do–Check–Act
Identify the problem.	*Plan*: Identify the problem.
Analyze the problem.	*Plan*: Analyze the problem.
Explore alternative solutions.	*Plan*: Develop solutions.
Select a plan and take action.	*Do*: Implement solutions.
Examine the effects of the actions taken.	*Check*: Evaluate the results.
If results are not acceptable, start over.	*Act*: Adopt the change, abandon it, or start over.

in the processes as the PDCA concept is applied are to address the OHSMS issues. Those issues are defined in the standard as "hazards, risks, management systems deficiencies, and opportunities for improvement."

Giving due recognition to the emphasis in the standard on PDCA processes and continual improvement, I believe nevertheless, that safety professionals who are schooled in basic problem-solving techniques can take comfort in knowing that they need make only minor adjustments in their thinking to adapt to the PDCA concept.

INTRODUCING THE PDCA CONCEPT

Assume that a safety professional wants to take the leadership to apply the PDCA concept for an organization's safety and health management system. To begin with, an assessment should be made of the opportunity for success. After reviewing the situation at hand, the safety professional would bring to bear the technical, small-group leadership, planning, and communication skills necessary to motivate management's buy-in to the PDCA concept. The goal is to create strategies for success. There cannot possibly be a one-size-fits-all approach. Consider these extremes:

- The organization has been certified with respect to the ISO 9000 (quality) and ISO 14,000 (environmental) standards. The management staff is familiar with the PDCA concept and welcomes with enthusiasm the fact that the proposals being made to achieve continual improvement in the safety management systems are in accord with the PDCA applications in other standards. Even then, a few successful hands-on demonstrations that lead to process improvements related to hazards, risks, and deficiencies in the safety management system will be beneficial.
- A safety professional is employed in a 500-employee organization in which the management knows little about the PDCA concept and is resistant to change. That does not preclude the safety professional from framing risk management proposals in a way to favorably affect the processes set forth in Z10 and to seek continual improvement in those processes. In that respect, the principles on which PDCA is based—good problem-solving techniques—would be applied. Achieving successes in that manner may be convincing and gain the respect of management for the PDCA concept.

To move forward in applying the process and continual improvement concept, change management techniques must be brought to bear. The participation of technically qualified people in all departments will be necessary, and the knowledge and capabilities of the people doing the work should be sought. Help with respect to change management can be found in Chapters 6, "Safety Professionals As Culture Change Agents" and chapter 19, "Management of Change".

The starting point in undertaking a continual improvement initiative can be as narrow as addressing a particular hazard or as broad as installing a process that does not exist (e.g., risk assessment.) Two real-world indicators of micro and macro applications of the PDCA concept follow.

In the first, workers complain that although the standard operating procedure (SOP) requires that they lockout/tagout electric power when a setup job is being done in the production line, the distance to the lockout station makes doing so a time-consuming nuisance. It is at the opposite end of the building—a walk of about 400 feet. They say that the inconvenience promotes taking excessive risks and ignoring the SOP and that they have done so under pressure to get the production line working again. A safety professional is brought into the discussion. He looks into the lockout/tagout design systems throughout the location and finds several that are comparable to the error-provocative work situation described by the workers. He promotes a meeting of the design, operating, and maintenance personnel at his location.

- Hazards are identified and the risks are judged to be excessive.
- It is agreed that when error-provocative work situations and methods exist, errors will probably occur, and that in this instance, the injury potential could be a fatality.
- A plan of action is agreed upon to study and correct all situations where a lockout/tagout station is not readily available.
- Responsibilities are assigned, with target dates for completion.
- The plan is put into effect.

It was determined that the action plan achieved its purpose. The risks were reduced to acceptable levels. More important, a recognition developed throughout the departments involved of the fact that hazards and their accompanying risks should be given greater attention in the design of work systems.

In another multilocation operation, a fatality resulting from electrocution occurred in a situation comparable to that just described. The CEO took charge and she quickly recognized that shortcomings in the design of lockout/tagout systems result in unacceptable risks and that those risks may be pervasive. The safety professional was asked to work with operating executives and put together a Plan.

It was recognized that the electrical hazards had to be identified and corrected. Under the direction of the division vice presidents, each location manager was required to Do, that is, to create study groups consisting of all levels of employment to identify potential electrocution situations, outline the actions to be taken to correct them, assign responsibilities and resources, and monitor the actions taken.

Location managers had to report that a Check had been made to assure that the hazard identification and risk assessment actions had been thorough and that the necessary risk reduction measures have been taken. Furthermore, location managers were expected to Act by providing the necessary resources and demonstrations of leadership to reinforce the culture change that had been achieved. It became understood that during safety audits, all levels of employees would be interviewed concerning the effect of the actions taken.

Throughout this activity, which took several months, the safety professional was available to personnel at the many locations for consulting. He was quite busy.

CONCLUSION

Drafters of Z10 did well in basing the *Occupational Health and Safety Management Systems* standard on the PDCA concept. Doing so makes the standard compatible with other internationally accepted standards and facilitates, melding the provisions in Z10 with other business practices.

The PDCA concept is a sound problem-solving and continual improvement model with which safety practitioners should be familiar. Fortunately, those safety professionals who have knowledge of basic problem-solving methods are well equipped to adopt the PDCA concept.

REFERENCES

ANSI/ISO/ASQ Q9001-2008. *American National Standard; Quality Management Systems—Requirements*. Milwaukee, WI: American Society for Quality, 2008.

"Business Dictionary." Access is at http://www.businessdictionary.com/definition/system.htm.

Deming, W. Edwards. *Out of the Crisis*. Cambridge, MA: Massachusetts Institute of Technology for Advanced Engineering Study, 1982.

Deming, W. Edwards. *The New Economics for Industry, Government, Education*, 2nd ed. by the W. Edwards Deming Institute. Cambridge, MA: MIT Press, 1994.

Manuele, Fred A. *On the Practice of Safety*, 4th ed. Hoboken, NJ: Wiley, 2013.

North Carolina Department of Environmental and Natural Resources, "Plan–Do–Check–Act—A Problem Solving Process." At http://quality.enr.state.nc.us/tools/pdca.htm.

Toy, Vic. "Let Your OHS Management System Do the Work: How the New Z10 Adds Even Better Value." Synergist, February 2012.

U.S. Environmental Protection Agency. "Environmental Management Systems: An Implementation Guide for Small and Medium-Sized Organizations," 2nd ed. At http://yosemite.epa.gov/water/owrccatalog.nsf/e673c95b11602f2385256ae1007279fe/7c3822f9d489db0185256d83004fd7cb!OpenDocument.

U.S. Environmental Protection Agency. "Implementation Guide for the Code of Environmental Management Principles for Federal Agencies (CEMP)." At http://www.epa.gov/compliance/resources/publications/incentives/ems/cempmaster.pdf.

Walton, Mary. *Deming Management at Work*. New York: Putnam Publishing Group, 1990.

CHAPTER 8

MANAGEMENT LEADERSHIP AND EMPLOYEE PARTICIPATION: SECTION 3.0 OF Z10

In Chapter 1, "An Overview of ANSI/AIHA Z10-2005", it was stated that Section 3.0, which addresses "Management Leadership and Employee Participation", is the most important section in the *Occupational Health and Safety Management Systems* standard. Having superior management leadership is an absolute requirement—a sine qua non—if the goal is to achieve superior results. With respect to this very important section of Z10, in this chapter we:

- Discuss the significance of management direction as it influences an organization's safety culture
- Comment on the role of safety professionals with respect to the safety culture
- Set forth the absolutes needed in management leadership to attain stellar results
- Acknowledge the impact that the current business environment may have on achieving or maintaining a superior safety culture
- Comment on specific elements in Section 3.0:
  - 3.1　Management Leadership
  - 3.1.1 Occupational Health and Safety Management System
  - 3.1.2 OHS Policy
  - 3.1.3 Responsibility and Authority
  - 3.2　Employee Participation

Advanced Safety Management: Focusing on Z10 and Serious Injury Prevention,
Second Edition. Fred A. Manuele.
© 2014 John Wiley & Sons, Inc. Published 2014 by John Wiley & Sons, Inc.

- Relate management leadership to serious injury prevention
- Describe a case of inadequate management leadership and employee participation that resulted in catastrophe
- Propose that an internal analysis of the safety culture be made, give an outline for such an analysis, and comment on a case study

THE SIGNIFICANCE OF MANAGEMENT LEADERSHIP AND ORGANIZATIONAL CULTURE

As management provides the leadership (Sections 3.0 and 3.1) and makes decisions directing the organization, the outcomes of those decisions establish its culture with respect to safety. Safety is culture driven, and management establishes the culture. *Safety* is defined as freedom from unacceptable risks.

In many definitions of safety culture, terms such as the following appear: *shared beliefs, attitudes, values, and norms of behavior*; *shared assumptions*; *individual and group attitudes about safety*; and *entrenched attitudes and opinions* which a group of people share. This author has used such terms. Note the frequent use of the term *shared* which may be inappropriate in some instances.

But such definitions need examination. An organization's safety culture, a subset of its overall culture, derives from the decisions made at the board of directors' level, at the senior management level, and occasionally at a location management level. An organization's culture with respect to occupational safety, environmental safety, the safety of the public, and product safety is determined by the outcome of decisions made by management as measured by the risk levels in the technical and social aspects at a facility. The culture created by management is the dominant factor with respect to the risk levels attained—acceptable or unacceptable.

Management owns the culture. An organization's culture is represented by the reality of application of its goals, performance measures, and sense of responsibility to its employees, to its customers, and to its community—all of which are translated into a *system of expected performance*. Over the long term, the injury and illness experience attained are a direct reflection of an organization's safety culture.

Strong emphasis is given to the phrase *system of expected performance* because it defines what the staff believes: in reality, what management wants done. Although organizations may issue safety policies, manuals, and standard operating procedures, the staff's perception of what is expected of them and the performance for which they will be measured—its *system of expected performance*—may differ from what is written.

Colleagues remind me of having written years ago that what management does rather than what management says defines the actuality of an organization's safety culture and its commitment or noncommitment to safety. Occasionally, there is a difference between what management says and what management does.

To achieve superior results, only top management can provide the leadership and direction needed to "establish, implement and maintain an occupational health and safety management system" (3.1.1). Major improvements in safety—toward being free from unacceptable risks—will be achieved only if a culture change takes place—only if major changes occur in the *system of expected performance.*

THE ROLE OF SAFETY AND HEALTH PROFESSIONALS WITH RESPECT TO THE SAFETY CULTURE

What is the safety and health professional's role with respect to the safety culture? In an organization where safety is a core value and management at all levels "walks the talk" and demonstrates by what it does that it expects the safety culture to be superior, the role of the safety and health professional is easier in the role of a culture change agent as he or she gives advice that supports and maintains the culture.

In a large majority of organizations, an advanced safety culture does not exist. Then the role of the safety and health professional as a culture change agent has greater significance and requires more diligence as attempts are made to influence management to move toward achieving a superior culture.

The possibility of being successful in that endeavor is enhanced if the safety professional attains the status of an integral member of the business team. That will result from giving well-supported, substantial, and convincing risk reduction advice that serves the business interests.

Admittedly, convincing management that safety should be one of an organization's core values may not be easily achieved. Safety and health professionals should understand that as steps forward are taken by management to improve on management system deficiencies, the result in each instance is a culture change. And the requirements to achieve a permanent culture change should be intertwined into each proposal made to improve on a management deficiency.

ABSOLUTES FOR MANAGEMENT TO ATTAIN SUPERIOR RESULTS

During a review of statements made in annual reports on safety, health, and environmental controls issued by five companies that consistently achieve outstanding results, a pattern became evident that defines the absolutes necessary to attain such results. Elements in that pattern follow.

- Safety considerations are incorporated within the company's culture, within its expressed vision and core values and its *system of expected performance.*
- The board of directors and senior management lead the safety initiative and make clear by what they do that safety is a fundamental within the organization's culture.
- There is a passion for and a sense of urgency for superior safety results.
- Safety considerations permeate all business decision making, from the concept stage for the design of facilities and equipment through to their disposal.
- An effective performance measurement system is in place.
- All levels of personnel are held accountable for results.

Whatever the size of an organization—10 employees or 100,000—the foregoing principles apply to achieve superior results. Safety is culture driven, and the board of directors and senior management define the culture and the *system of expected performance*. Where there is a passion for superior results, management insists that its hazard and risk problems be identified and resolved. This insistence by management at all levels to be informed of the reality of management system deficiencies is vital to achieving stellar results.

Companies with superior results demonstrate their insistence on having evidence-based management for all purposes, through their commitment to finding, facing, and acting on the facts—no matter how unpleasant those facts might be. That commitment applies to "hazards, risks and management system deficiencies," referred to as "OSHMS issues" in Section, 4.0.

To achieve superior results, management must establish open communications so that knowledge of hazards and risks flows upward to decision makers. Proof of management's wanting to know about hazards and risks is demonstrated by the actions taken to eliminate or control them.

Safety professionals should be working toward convincing management that it is in their best interest to have processes in place to find, face, and act on hazards and the risks that derive from system shortcomings. Yet, realism with respect to management practices in some companies must be acknowledged. Unfortunately, in actuality, what Whittingham wrote in *The Blame Machine: Why Human Error Causes Accidents* is sometimes true.

> Organizations, and sometimes whole industries, become unwilling to look closely at the system faults which caused the error. Instead the attention is focused on the individual who made the error and blame is brought into the equation. (Preface)

Readers will not find a statement in this book indicating that the role of the safety professional in favorably influencing an organization's culture is easily fulfilled. Yet the endeavor is worth undertaking, and attaining positive results, perhaps in small steps, can be rewarding.

THE BUSINESS ENVIRONMENT

It is possible that the prevailing business environment makes it more difficult for safety professionals in some organizations to influence its safety culture favorably. Consider these excerpts from the "Report of the OECD Workshop on Lessons Learned from Chemical Accidents and Incidents, 21–23 September 2004, Karlskoga, Sweden." OECD is the international Organization for Economic Cooperation and Development, the directorate of which is in Paris. Although this report was published more than eight years ago, the premise cited is still a matter of concern.

> The concept of "drift" as defined by Rasmussen as "the systematic organizational performance deteriorating under competitive pressure, resulting in operation

outside the design envelop where preconditions for safe operation are being systematically violated" was generally agreed as being a far too common occurrence in the current business environment. (p. 8)

There are other references in the OECD report indicating that the effects of pressures to maintain high profit levels and reduce costs may be among the contributing factors for incidents that have low probability but result in serious consequences. In such cases, safety requirements may be compromised and the safety culture deteriorates. Although the OECD report pertains to the chemical process industries, similar observations are made with respect to the negative impact of bottom-line pressures in other industries.

Later in this chapter we comment on a catastrophe in which management acknowledged in its own internally prepared report that its safety culture, over time, was allowed to deteriorate. In Chapter 18, "Lean Concepts: Emphasizing the Design Process", reference is made to safety concerns being over ridden as lean concepts are applied. In discussions with several safety directors, it is readily agreed that everyone is expected to do more with less and that bottom-line pressures are weighty.

It is appropriate, then, to acknowledge that where the business environment results in management decision making that affects the safety culture negatively, convincing management that safety should be one of the organization's core values will not be easy to achieve. Yet the safety professional has an obligation to be professional, factual, and complete in the recommendations made to keep risks at an acceptable level.

OHS POLICY: SECTION 3.1.2

Z10 says that "the organization's top management shall establish a documented occupational health and safety policy." Appendix A provides guidance on what a policy statement should contain and provides two samples. An organization's policy statement should be tailored particularly to its needs and written in the language that the issuer would normally use. The policy statement has to be believable. Consideration of the following may be helpful when drafting a policy statement. The policy statement should:

1. State clearly management's position on safety, health, and the environment, and indicate that avoiding injury and illness to employees and to the public from operations or from products sold, as well as damage to the environment are organizational values.
2. Bear the signature of the senior executive or manager.
3. Be appropriate to the nature of the organization's operations and their scope.
4. Be current, reviewed at least annually, and prominently displayed.
5. State a commitment to comply with all applicable legislation and standards.
6. Affirm that all safety, health, and environmental policies that are in place are to be followed.

7. Make clear that employees are to participate actively in all elements of the safety and health management system.

8. Pledge to a continual improvement process to reduce risks further.

If additional examples of policy statements are desired, they can be found in the safety, health, and environmental reports issued by companies such as:

- Bayer at http://www.annualreport2011.bayer.com/en/sustainability-strategy.aspx
- DuPont at http://www2.dupont.com/personal-protection/en-us/dpt/safety-protection. html
- Intel at http://www.intel.com/content/www/us/en/corporate-responsibility/environ-mental-health-safety-policy-otellini.html

RESPONSIBILITY AND AUTHORITY: SECTION 3.1.3

Section 3.1.3 of Z10 requires that management define roles, assign responsibilities and authority, provide the necessary resources (financial and human), and establish accountability. If a management accountability system for safety, health, and environmental results is not in place, management commitment to attaining superior results is questionable. Accountability without consequences is not accountability.

In the Introduction to the standard, it is made clear that it was drafted to be compatible with other business processes. That thought is reinforced in this section. Management is to provide the leadership and assume responsibility for "integrating the occupational health and safety management system into the organization's other business systems and processes and assuring the organization's performance review, compensation, reward and recognition systems are aligned with the OHS policy and the OHSMS performance objectives." Doing so is a goal worthy of achievement. Getting that done will interweave safety and health processes into and be supportive of the organization's endeavors.

While management has responsibilities for safety, so do employees. As the standard indicates: "Employees shall assume responsibility for aspects of health and safety over which they have control, including adherence to the organizations health and safety rules and requirements."

Appendix B, "Roles and Responsibilities," provides an excellent reference from which excerpts can be taken to "define roles, assign responsibilities, establish account-ability, and delegate authority" as suitable to an entity's needs. The data cover these employment categories:

- President, Chief Executive Officer, Owner
- Executive Officers, Vice Presidents, and other Senior Leadership
- Directors, Managers, and Department Heads
- Supervisors
- Employees
- Health and Safety Department.

Defining responsibilities and establishing accountabilities is an important step. It must be done for safety and health management systems to be effective and to provide a basis for performance and accountability reviews.

EMPLOYEE PARTICIPATION: SECTION 3.2

Not only are employees to assume responsibility for aspects of health and safety over which they have control, but they are also to have an opportunity to participate in every aspect of the occupational health and safety management system. They are to have the mechanisms, time, and resources necessary to participate.

A statement made in the standard's advisory column next to employee participation is close to one that I have made and believe to be fundamentally true. If an employer does not take advantage of the knowledge, skill, and experience of workers close to the hazards and risks, opportunities are missed to improve safety management systems and reduce injury and illness potential.

Employers improve their prevention efforts if they recognize the creativity of the workers doing the jobs. The purpose of reducing risk is well served if the culture makes it clear that the knowledge of workers is valued and respected and that they are to participate in the ownership of the safety management system.

Although the provisions in the standard for employee participation are not changed substantially from the earlier issue:

- Comments in the advisory column are more extensive, including reference to W. Edwards Deming's principles : drive our fear; break down barriers; and eliminate exhortations.
- Contents of Appendix C, "Encouraging Employee Participation," fills about twice as much space as was given to this subject in the earlier version of Z10. These topics are covered in Appendix C:
  - Introduction
  - Organizational Readiness and Effective Leadership
  - Employee Participation
  - Other Issues to Consider for Effective Employee Participation
  - Barriers to Participation.

Extensions made in the advisory material for employee participation are a major change in the standard. The material provided is a good reference source. Two examples of outstanding contributions to risk reduction made by hourly workers come to mind.

In a maker of heavy machinery, the innovations of tool and die makers in redesigning work situations to reduce ergonomics risks were so creative that visitors to the plant were often shown their inventions as a matter of pride.

In a space industry location, it became the standard practice for design engineers to seek the opinions of hourly workers before proceeding to manufacture what was designed. They learned through experience that the suggestions made by hourly workers resulted in risk avoidance, particularly human factors design errors, and resulted in improved efficiency in the production process.

Comments made in the advisory column (E3.2C) on obstacles or barriers to meaningful employee participation include lack of response to employee input or suggestions, and reprisals that penalize or discourage participation. Both of these examples define a negative safety culture.

RELATING MANAGEMENT LEADERSHIP
TO SERIOUS INJURY PREVENTION

My analysis of over 1,800 incident investigation reports indicates that a significantly large share of incidents resulting in serious injuries and fatalities occurs:

- When unusual and nonroutine work is being performed
- In nonproduction activities
- In at-plant modification or construction operations (replacing a motor weighing 800 pounds to be installed on a platform 15 feet above the floor)
- During shutdowns for repair and maintenance and during startups
- Where sources of high energy (electrical, steam, pneumatic, chemical) are present
- Where upsets occur: situations going from normal to abnormal

It was also determined that:

- Causal factors for low-probability/high-consequence events are seldom represented in the analytical data on accidents that occur frequently. (Some ergonomics-related incidents are the exception.)
- Many incidents resulting in serious injuries are unique and singular events, having multiple and complex contributing factors that may have technical, operational systems, or cultural origins.

These studies reveal that often, over time, there had been an accumulation of shortcomings in safety and health management resulting from decisions made that reflected adversely on the safety culture. Other writers have written similarly.

Incidents that result in serious injuries are often low-probability events that result from what James Reason refers to as an accumulation of latent technical conditions and operating practices that are built into a system and shape an organization's culture. He discusses the long term impact of a continuum of less-than-adequate management leadership and decision making in *Managing the Risks of Organizational Accidents*, from which the following excerpt is taken.

Latent conditions, such as poor design, gaps in supervision, undetected manufacturing defects or maintenance failures, unworkable procedures, clumsy automation, shortfalls in training, less than adequate tools and equipment, may be present for many years before they combine with local circumstances and active failures to penetrate the system's layers of defenses. They arise from strategic and other top-level decisions made by governments, regulators, manufacturers, designers and organizational managers. The impact of these decisions spreads throughout the organization, shaping a distinctive corporate culture and creating error-producing factors within the individual workplaces. (p. 10)

As the impact of less-than-adequate decision making by management spreads throughout an organization, employees at all levels respond to the negative safety culture that develops, and risky work practices become common. Such a situation, once recognized, presents a challenge to safety professionals in that giving advice to reduce the probability of incidents occurring that result in serious injuries must become a principal goal for them.

Although not easy to do, they must prepare data to try to convince management to recognize the possible systemic causal factors that have accumulated and take action to reduce them. Thus, to achieve a significant reduction in the potential for low-probability/serious-consequence incidents occurring, a different mindset and a culture change have to be achieved. In that respect, safety professionals are culture change agents.

A CASE OF INADEQUATE MANAGEMENT LEADERSHIP AND EMPLOYEE PARTICIPATION

Data follow to illustrate a situation in which deterioration in safety management leadership and employee participation resulted in a catastrophic incident. Negative safety decision making resulted in a deteriorating safety culture, failure to adequately involve employees in the safety process, and poor communication.

A positive safety culture results from management leadership and direction that produces the opposite of what is described in this case. It is suggested that readers ask whether similar situations occur in the operations to which they give counsel.

A Catastrophe in Texas City, 2005

Excerpts are taken from an internally produced report on an incident that occurred on March 23, 2005 at a BP Products North America owned and operated refinery. A fire and explosion resulted in 15 deaths, 170 injuries, and extensive property damage. The report, *Fatal Accident Investigation Report, Isomerization Unit Explosion Final Report, Texas City, Texas, USA*, was approved for release by J. Mogford, the investigation team leader and an employee of BP. The website through which this report can be accessed is listed in the reference section of this chapter.

The Executive Summary highlights the content of the report. As the following excerpts from it are read, keep the safety culture, management leadership, accountability, and employee participation implications in mind.

[The] underlying causes are identified as follows:

- Over the years, the working environment had eroded to one characterized by resistance to change, and lacking of trust, motivation, and a sense of purpose. Coupled with unclear expectations around supervisory and management behaviors this meant that rules were not consistently followed, rigor was lacking and individuals felt disempowered from suggesting or initiating improvements. Process safety, operations performance and systematic risk reduction priorities had not been set and consistently reinforced by management.
- Many changes in a complex organization had led to the lack of clear accountabilities and poor communication, which together resulted in confusion in the workforce over roles and responsibilities.
- A poor level of hazard awareness and understanding of process safety on the site resulted in people accepting levels of risk that are considerably higher than comparable installations. One consequence was that temporary office trailers were placed within 150 feet of a blowdown stack which vented heavier than air hydrocarbons to the atmosphere without questioning the established industry practice.
- Given the poor vertical communication and performance management process, there was neither adequate early warning system of problems, nor any independent means of understanding the deteriorating standards in the plant. (p. iii)

A statement in the first bulleted item in the preceding excerpts is significant in understanding the positive development of, or deterioration in, a safety culture. Changes in a safety culture, for better or worse, don't occur quickly. Note that in this situation: "Over the years, the working environment had eroded to one characterized by resistance to change, and lacking of trust, motivation, and a sense of purpose." The time element is further recognized as follows in the Executive Summary.

It is evident that [the causal factors] had been many years in the making and will require concerted and committed actions to address. (p. iii)

The authors of the report recognized that reversing the cultural trend will take a long time, considering the nature of the recommendations made. They pertain to:

- People and procedures
- Control of work and trailer citing
- Design and engineering (p. iii)

The following excerpts were taken from the body of the report. They relate specifically to inadequate participation by the hourly workforce, poor motivation, a safety culture that accepted high risk taking, and the failure of senior management to hold people accountable for following the "defined processes/procedures." For emphasis, we repeat: This is an internally produced report by BP personnel.

- There was a failure by leadership to hold employees at all levels accountable for executing defined processes/procedures. A workplace environment characterized by poor motivation, unclear expectations around supervisory/management behaviors, no clear system of reward and consequences, and high distrust between leadership and the workforce, had developed over a number years within the site. The working relationships between leadership and workers, and employees and contractors were poor. (p. 153)
- Specifically, the team observed that the relationship between operations and engineering was inconsistent and fractured; evidence of this was the fact that engineering was not called during the startup problems; nor did the team find evidence that engineering rounds were being conducted. In addition, the "independent" functions, such as training and Process Safety Management, had become under-resourced and lacked the influence to ensure that standards were met. (p. 166)
- The fourth cultural issue is the inability to see risks and, hence, toleration of a high level of risk. This is largely due to poor hazard/risk identification skills throughout management and the workforce, exacerbated by a poor understanding of process safety. (p. 167)

To describe a positive safety culture that results from good management leadership and employee participation, start by turning the negatives in the foregoing into affirmatives.

PROPOSING AN INTERNAL ANALYSIS OF THE SAFETY CULTURE

Assume that management responded favorably to a suggestion made by a safety professional that an internally conducted survey of the organization's safety culture would be beneficial. The purpose would be to gather the perceptions of all levels of employment on the quality of the safety management system in place. It should be understood that for those who participate in the exercise, their perceptions are their reality. The result of such an exercise will be a culture survey.

The self-analysis provides data on the positive and negative effects of management leadership, the extent of employee participation, and whether an accumulation has developed of latent technical conditions and operating practices that could be the causal factors for low-probability incidents having severe consequences.

For such a self-analysis, a survey mechanism is necessary. An outline of a basic survey guide follows. For the survey mechanism to relate to the hazards and risks in a particular operation, it is necessary that management, assisted by a safety professional, add or delete items. Also, a scoring system for each item, compatible with practices applicable in the organization, should be included in the revision of the guide so that a compilation of results can be made. In many situations, a simple "yes," "no," or "not applicable" scoring system will suffice. It must be emphasized—this guide is not offered as a one-size-fits-all mechanism.

Safety Management System Survey Guide

1. Is the safety management system in place in our organization effective?
2. Does management demonstrate by what it does that safety is a core value in our organization?
3. Is there a significant gap between what management says and what management does?
4. Has the staff reporting directly to the senior manager been held accountable, in reality, for a high level of safety decision making?
5. Is this a safe place to work?
6. Are you asked to participate effectively in safety discussions and meetings?
7. Are you asked to provide input on safety matters that affect you directly?
8. Is your input on safety matters respected and considered valuable?
9. Do you believe that some of the equipment you operate or the work methods you are required to follow are hazardous or overly risky?
10. Do you believe that you are free to report hazardous conditions and practices without reprimand?
11. Are you encouraged to report hazardous conditions and practices?
12. Does your supervisor effectively give safety a high priority?
13. Is accident investigation in sufficient depth to identify the reality of causal factors (organizational, cultural, design and engineering, technical, procedural)?
14. Is there a broadly held belief that unsafe acts of workers are the principal causes of accidents?
15. Is safety often relegated to a lower status and overlooked when there are production pressures?
16. Have you been given adequate training on hazards, risks, and safe operating procedures?
17. Does the organization's culture accept gradually escalating risk?
18. Does the organizational structure enhance or dissuade adequate safety decision making?
19. Are there organizational barriers that prevent effective communication on safety, up and down?
20. Have streamlining and downsizing conveyed a message that efficiency and being on schedule are paramount, and that safety considerations can be overlooked?
21. Is staffing adequate in your group adequate so that work can be done safely?
22. Are you discouraged from reporting injuries?
23. Have technical and operational safety standards been at a sufficiently high level?
24. Has it been the practice to accept safety performance at a lesser level than the Standard Operation Procedures prescribed?
25. Have known safety problems, over time, been relegated to a "not of concern" status and, thereby become "acceptable risks"?
26. Has safety-related hardware or software become obsolete?

27. Are certain operations continued with the knowledge they are unduly hazardous?
28. Have budget constraints had a negative effect on safety decision making?
29. Has inadequate maintenance resulted in an accumulation of hazardous situations that have gone unattended (e.g., is the detection equipment adequate, maintained, and operable; are basic safety-related repairs postponed unduly)?
30. Has adequate attention been paid to "near miss" incidents that could, under other circumstances, have resulted in a major accident?
31. Are safety personnel encouraged to be aggressive when expressing their views on hazards and risks, even though their views may differ from those held by others?
32. Has there been an overreliance on outside contractors (outsourcing) to do what they cannot do effectively with respect to safety?
33. Are purchasing and contracting procedures in place to limit bringing hazards into the workplace?

CASE STUDY

A safety director in a very large organization with about 13,000 employees read an article of mine in which emphasis was given to the necessity of having a positive safety culture to achieve superior performance levels. That organization's work is considered "high hazard", and fatalities and serious injuries occur.

The safety director had concluded that the senior executive in his organization, to whom he reported, was somewhat removed from the leadership necessary to reduce fatalities and serious injuries and that he did not hold the staff reporting to him accountable for their incident experience.

The safety director sought help. During our discussions it was agreed that he would approach his boss to convince him that the organization would be well served if the opinions of the staff were solicited on the quality of the safety management system in place. He did so, and it worked.

A Safety Management System Survey Guide comparable to that shown previously was sent to the safety director. He worked up a version of it that fit the high-hazard operation with which he is involved. The survey guide was sent to a statistically adequate sampling of the staff at all employment levels. Over 70% of those who received of the survey guide responded. They took the survey seriously.

When the safety director analyzed the results, he found that the same shortcomings in the safety management system were recorded, largely, by all levels of employment; and there were many shortcomings. But most important, some of the staff members reporting directly to the senior executive who authorized the survey said that for the department as a whole, safety was not a high-level value.

During a meeting that I attended with the safety director, his boss, and other interested persons, the senior executive was well prepared with questions about how superior safety results were achieved elsewhere. He was surprised by the results of the culture survey. He learned that every level of the organization said that they wanted safety to be given a higher status.

As the discussions proceeded, I asked the senior executive to draw an organizational chart starting with his position and showing the positions of all who reported to him. He was quick to acknowledge that if he wanted risks of injury and fatalities to be reduced, he would have to provide the leadership and hold the staff reporting to him accountable for results.

Subsequently, the senior executive convened his staff and laid out what he expected of them. A bulletin was issued throughout the organization setting forth the safety policy and the procedures to implement it. Safety is now an agenda item for several levels of management meetings, division heads are holding the staffs reporting to them accountable for results, and management is encouraging input and involvement from all levels of employees.

Communication has improved downward and upward. Safety-related suggestions get quicker attention than formerly.

CONCLUSION

Management Leadership and Employee Participation is the most important section in the Z10 standard. Safety is culture driven, and leaders create the culture. It is the responsibility of leadership to change the safety culture when it is deficient.

Top administrators must take responsibility for risk management by remaining alert to the effects their decisions have on the work system. Leaders are responsible for establishing the conditions and the atmosphere that lead to their subordinates' successes or failures.

As top management makes decisions directing the organization, its safety culture is established and that culture is translated into *a system of expected performance.* Where safety management systems are most effective, there is a commitment to:

- Find the facts about hazards, risk, and deficiencies in management systems, regardless of any unpleasantness that may arise in the discovery process
- Take actions to achieve acceptable risk levels

A safety culture is unsound if employees do not have opportunities to participate in every aspect of the occupational health and safety management system and if they do not have the mechanisms, time, and resources necessary to participate. Furthermore, if an employer does not take advantage of the knowledge, skill, and experience of those workers close to the hazards and risks, opportunities are missed to reduce injury and illness potential and for improvement in safety and health management systems.

Too much cannot be made of the importance of concentrating on having risks be at acceptable levels, particularly at what James Reason calls the "sharp end," where the work gets done.

In Section 4.0, the planning section, Z10 provides a focus for all that is expected of management and employees. The planning process goal is to identify occupational health and safety management issues, which are defined as "hazards, risks, management system deficiencies, and opportunities for improvement."

Consider this premise: The entirety of purpose of those responsible for safety, regardless of their titles, is to manage their endeavors with respect to hazards so that the risks deriving from those hazards are at an acceptable level.

REFERENCES

ANSI/AIHA Z10-2012. *Occupational Health and Safety Management Systems.* Des Plaines, IL: American Society of Safety Engineers, which is the secretariat. Available at https://www.asse.org/shoponline/products/Z10_2005.php.

Bayer, for a safety, health, and environmental report. Available at http://www.annualreport2011.bayer.com/en/sustainability-strategy.aspx.

DuPont, for a safety, health, and environmental report Available at http://www2.dupont.com/personal-protection/en-us/dpt/safety-protection.html.

Fatal Accident Investigation Report, Isomerization Unit Explosion Final Report, Texas City, Texas, USA. Date of accident: March 23, 2005. Date of report: December 9, 2005. Approved for release by J. Mogford, investigation team leader. http://www.rootcauselive.com/Files/Past%20Investigations/BP%20Explosion/texas_city_investigation_report.pdf.

Intel, for a safety, health, and environmental report. Available at http://www.intel.com/content/www/us/en/corporate-responsibility/environmental-health-safety-policy-otellini.html.

Manuele, Fred A. *On the Practice of Safety*, 4th ed. Hoboken, NJ: Wiley, 2013.

Reason, James. *Managing the Risks of Organizational Accidents.* Burlington, VT: Ashgate Publishing Company, 2001.

"Report of the OECD Workshop on Lessons Learned from Chemical Accidents and Incident, 21–23 September 2004, Karlskoga, Sweden." Paris, France: Organization for Economic Cooperation and Development, 2005. Available at http://www.oecd-ilibrary.org/economics/oecd-series-on-chemical-accidents-no-14_oecd_papers-v5-art15-en.

Whittingham, R. B. *The Blame Machine: Why Human Error Causes Accidents.* Burlington, MA: Elsevier–Butterworth–Heinemann, 2004.

CHAPTER 9

PLANNING: SECTION 4.0 OF Z10

In the Plan-Do-Check-Act model, the purpose of the plan step is to identify and analyze problems. Establishing and implementing the processes set forth in the planning Section (4.0) of Z10 will serve problem identification and analysis purposes as well as to develop solutions and implementation plans with respect to the problems identified.

As is said in the standard, the "planning process goal is to identify and prioritize occupational health and safety management system issues." Those issues are "defined as hazards, risks, management system deficiencies, and opportunities for improvement." Also: objectives are to be established that offer the greatest opportunities for improvement and risk reduction; occupational health and safety issues are to be prioritized; and plans are to be formulated to accomplish the prioritized objectives.

REVIEW PROCESS: SECTION 4.1

This section states that the organization shall establish continuing review processes to identify the differences between the safety and health management systems in place and the requirements of Z10. This statement, in E4.1A, provides significant guidance on what is expected. *System reviews are focused on management systems and are not specific to operations.* Note the emphasis on management systems. While the reviews are "not specific to operations," hazards and risks in operations may indicate that there are shortcomings in management systems that need review.

These system reviews will fulfill the problem identification and analysis steps and the solution consideration steps of the PDCA concept. Information gathered on hazards,

Advanced Safety Management: Focusing on Z10 and Serious Injury Prevention, Second Edition. Fred A. Manuele.
© 2014 John Wiley & Sons, Inc. Published 2014 by John Wiley & Sons, Inc.

risks, and deficiencies in safety management systems is to be analyzed to provide a decision and priority base to be considered by an organization as it creates "the processes necessary to establish or improve its management systems."

The ongoing review process is to be supplemented by information arising from application of other sections in the standard, such as "Management Leadership and Employee Participation," "Implementation and Operation," "Evaluation and Corrective Action", and "Management Review."

To move forward on continual improvement, information is to flow to the decision makers on how the elements in the safety and health management systems can be improved. That knowledge flow is necessary to accomplish the needs of the management review process outlined in Section 7.

For the planning section to be accomplished successfully, an organization must demonstrate by its culture—its *system of expected performance*—that management is committed to being informed on the facts about hazards, risks, and deficiencies in its safety management systems, regardless of any unpleasantness that may arise in the discovery process, and to taking actions to achieve acceptable risk levels.

Since very few organizations meet all of the requirements of Z10—a state-of-the-art standard—it is a near certainty that shortcomings in existing systems will be identified. The result should be prioritizing the shortcomings and developing action plans to improve the existing safety and health management system.

For the review process, the standard suggests taking into consideration: the relevant business management systems; hazards, risks, and controls; allocation of resources (funding, personnel, equipment); regulations and standards; assessments; and other relevant activities.

SHORTCOMINGS EXPECTED IN THE REVIEW PROCESSES

When making the reviews required by this standard, it is suggested that particular emphasis be given to those processes in Section 5.0, "Implementation and Operation," for which most organizations will be found wanting. Those processes include:

- Prevention through design, which includes safety design reviews
- Risk assessments and prioritization
- Management of change
- A prescribed hierarchy of controls
- Safety specifications being included in purchasing documents

ASSESSMENT AND PRIORITIZATION: SECTION 4.2

This section requires that processes be in place to assess and prioritize the occupational health and safety management issues identified in Section 4.1, the ongoing review processes. The intent is to "assess the impact on health and safety of OHSMS issues and assess the level of risk for identified hazards" and to "establish priorities based on factors such as the level of risk, etc."

In addition, the processes shall "identify the underlying causes and other contributing factors related to system deficiencies that lead to hazards and risks." As the "system deficiencies that lead to hazards and risks" are identified and planning is begun to eliminate them, consideration must be given to the probability that the system deficiencies define safety culture inadequacies. If so, culture change mechanisms would be utilized.

Section 4.3, "Objectives," and Section 4.4, "Implementation Plans and Allocation of Resources," logically follow the assessment and prioritization requirements. They require, principally, that processes be established and implemented to:

- Set documented objectives based on the priorities developed with respect to the hazards and risks identified and the shortcomings found in the safety and health management system (Section 4.3).
- Establish a documented implementation plan to achieve those objectives. That plan is to define resources, responsibilities, time frames, and appropriate measures of progress (Section 4.4).
- Allocate resources to achieve the objectives established (Section 4.4).

The goal of the risk assessment and prioritization process is to provide input for decision makers as they attempt to attain an occupational environment in which the risks are judged to be acceptable. Safety professionals must recognize the real world of economics with respect to resource allocation and setting priorities so as to achieve the best probable good from the expenditure of those resources. In that context, prioritization needs special comment.

- Some risks are more significant than others.
- Usually, significance is determined judgmentally through a review of estimates made of the probability of event or exposure occurrence and severity of outcome potential.
- Safety professionals must be capable of distinguishing the more important from the less important.
- Resources will always be limited, and staffing and money are never adequate to attend to all risks.
- The greatest good to employees, to employers, and to society is attained if resources available for risk avoidance, elimination, reduction, or control are applied effectively and economically.
- On a priority basis, consideration is given, first, to those risks that have the greatest potential for serious harm or damage.

Although Appendix D, "Planning—Identification, Assessment and Prioritization" (Section 4), is of general interest to all organizations, it begins with guidance to entities that are implementing a formal occupational safety and health management system for the first time. It is suggested that such entities perform a baseline or gap analysis "that identifies differences between the requirements of this standard and any pre-existing management systems."

Help is given on the procedures to be followed and the information to be gathered and reviewed to identify occupational safety and health management system needs. Guidance comments are provided on assessment and prioritization factors and system and operational issue examples.

Appendix E, "Objectives/Implementation Plans" (Sections 4.3 and 4.4), provides reference material on managing safety and health and planning and setting objectives. This appendix includes a helpful outline for "Managing Safety and Health: Planning and Setting Objectives."

CONCLUSION

Success of an occupational safety and health management system is largely contingent on how thorough the provisions in the planning process are applied. They are to identify and prioritize the issues, which are "defined as hazards, risks, management system deficiencies, and opportunities for improvement." As that process moves forward, it should be understood that the entirety of purpose of those responsible for safety, regardless of their titles, is to manage their endeavors with respect to hazards so that the risks deriving from those hazards are acceptable.

This planning section requires that shortcomings in existing safety management systems be identified in relation to the requirements of Z10. Existence of hazards and risks is evidence of system shortcomings. Once those inadequacies are known, priorities are to be set, objectives are to be established for improved risk control, and actions are to be outlined in a documented plan for continual improvement. Those processes represent good management as respects applying the PDCA concept.

CHAPTER 10

IMPLEMENTATION AND OPERATION: SECTION 5.0 OF Z10

INTRODUCTION

Provisions in the planning section of Z10 (Section 4.0) pertain to the first three steps in the Plan-Do-Check-Act (PDCA) *process*:

Plan: Identify the problem.
Plan: Analyze the problem.
Plan: Develop solutions.
Do: Implement solutions.
Check: Evaluate the results.
Act: Adopt the change, abandon it, or start over.

In the planning processes, occupational health and safety management systems issues are identified, analyzed, and prioritized. Those issues are defined in the standard as *hazards, risks, management system deficiencies and opportunities for improvement* (italic type is used for material taken directly from the standard). After the identification and analysis process, objectives are then to be established that offer the greatest opportunities for improvement. The following steps are to draft a documented implementation plan to allocate the necessary resources and to achieve the objectives.

As the Implementation and Operating section of Z10 is applied, the activity moves into the "Do" element of the PDCA process. The standard says that *this section* [5.0]

Advanced Safety Management: Focusing on Z10 and Serious Injury Prevention,
Second Edition. Fred A. Manuele.

defines the operational elements that are required for implementation of an OHSMS and that these elements *provide the backbone of an OHSMS and the means to pursue the objectives from the planning process.*

Particular emphasis is given to this "Do" element. Although all the sections in Z10 have significance with respect to achieving an effective safety and health management system, several implementation and operation requirements in Section 5.0 have particular significance. They are, as the standard says, *the backbone of an occupational health and safety management system.*

Section 5.1 provides general guidance with respect to establishing an implementing the operational elements of a safety and health management system, and says that the organization shall integrate the operational elements into the organization's management system.

Section 5.1.1, "Risk Assessment," is a vital and highly significant addition to the 2012 version of Z10. It requires that:

The organization shall establish and implement a risk assessment process(es) appropriate to the nature of hazards and risks.

This provision recognizes the worldwide trend toward giving risk assessment a prominent place within safety and health management systems. Addendum A in Chapter 11, "A Primer on Hazard Analysis and Risk Assessment", provides a list of actions that have been taken in many countries to promote the inclusion of risk assessments as an integral part of safety endeavors.

Brief comments, only, are made in this chapter on sections of 5.0 that are abundantly addressed in occupational safety literature. They are: "Contractors," 5.1.5; "Emergency Preparedness," 5.1.6; "Education, Training and Awareness," 5.2; "Communication," 5.3; and "Document and Record Control Process," 5.4. Comments are also made regarding 5.1.3.1, "Applicable Life Cycle Phases," and 5.1.3.2, "Process Verification."

Separate chapters follow on the requirements in Z10 for which the literature is not quite as prevalent and which are vital in achieving superior results. Those separate chapters are on:

- Risk assessment, Section 5.1.1: Chapter 11 – "A Primer on Hazard Analysis and Risk Assessment"
- Risk assessment – 5.1.1 – Chapter 12 – "Provisions for Risk Assessments in Standards and Guidelines"
- Risk assessment – 5.1.1 – Chapter 13 – "Three and Four Dimensional Risk Scoring Systems"
- The hierarchy of controls, Section 5.1.2: Chapter 14 – "Hierarchy of Controls"
- Safety design reviews, 5.1.3: Chapter 15 – "Safety Design reviews"
- Management of change, 5.1.3: Chapter 19 – "Management of Change"
- Procurement, 5.1.4: Chapter 20 – "The Procurement Process"

Applied lean concepts as discussed in this book relate to the safety design review provisions (Section 5.1.3) in Z10. Chapter 18, Lean Concepts: Emphasizing the

Design Process, contains a description of a system in which lean safety and environmental concerns were addressed in the design process.

Contractors: Section 5.1.5

Relations with contractors who do work at an organization's site are addressed in this section. An organization is to have processes in place to protect the organization's employees from the risks that may be presented by the contractor's work or the activity of the contractor's employees, and to protect the contractor's employees from the organization's activities and operations. These requirements are stated briefly.

A good reference on contractor selection procedures and the key safety, health, and environmental protection provisions to which contractors should adhere are discussed in the first chapter in *Construction Safety Management and Safety Engineering*. Also, entering "contractor safety requirements" into an Internet search engine will bring up a large number of downloadable examples of procedures established by a variety of entities with respect to contractor selection and the safety performance expected of contractors while on an organization's premises.

Appendix J pertains to "Contractor Safety and Health (Section 5.1.5)." It states: *The purpose of this appendix is to provide organizations guidance with respect to exposures and control measures required to maintain the business by contracting for expertise and/or services.* Examples are given of factors to be considered when contractors or service providers are engaged. Separate lists are included in the appendix of measures to be taken into consideration to control losses associated with on-site activities related to low-risk, medium-risk, and high-risk operations.

Six publications are listed as resources on contractor safety in Appendix O, "Bibliography and Reference". Two are standards approved by the American National Standards Institute (ANSI). One pertains to multiemployer projects; the other covers basic safety management elements in construction activities.

Two American Petroleum Institute (API) references give guidance on implementing a contractor safety and health program. The International Association of Oil and Gas Producers (OGP) publication provides guidelines for working together in a contract environment. An American Industrial Hygiene Association publication pertains to the health and safety requirements in contract documents.

Emergency Preparedness: Section 5.1.6

With respect to emergency preparedness, an organization is to have processes in place to *identify, prevent, prepare for, and/or respond to emergencies.* This is a subject that has been much written about since the 9/11 catastrophe, the 2005 Katrina hurricane, and the 2012 Sandy hurricane. A good basic reference on the subject can be found in Chapter 18 "Emergency Preparedness" in the National Safety Council's *Accident Prevention Manual: Administration & Programs*, 13th edition.

Enter "emergency preparedness" into an Internet search engine and the number of references available for review is well up into the millions. In Appendix O, five references for emergency preparedness are given: Federal Emergency Management

Agency (FEMA), National Fire Protection Association (NFPA), Department of Transportation (DOT), and two issued by the Occupational Safety and Health Administration (OSHA). They serve well as resources.

EDUCATION, TRAINING, AWARENESS, AND COMPETENCE: SECTION 5.2

For these four subjects, an organization is to have processes in place to establish competency needs for employees and contractors, see that they are educated and trained in a language that they understand, ensure that all involved are aware of applicable occupational health and safety management systems and their importance, ensure that the trainers are competent, and provide training on an ongoing and timely basis. In the advisory column, it is said that training should be given, for example, to:

a. *Engineers in safety design (e.g., hazard recognition, risk assessment, mitigation, etc.).*
b. *Those conducting incident investigations.*
c. *Those conducting audits.*
d. *Procurement personnel on the impact of purchasing decisions.*
e. *Others involved with the identification of issues, methods of prioritization, and controls.*

An Internet search will reveal millions of resources on training and awareness, but not as many on establishing competence levels. These chapters in the *Accident Prevention Manual* are helpful:

• Motivation, Chapter 28
• Safety and Health Training, Chapter 29
• Safety Awareness Programs, Chapter 31

Appendix O lists four publications pertaining to training, awareness, and competence. Of particular interest is a publication issued by the International Association of Oil and Gas Producers (OGP), Reference 6.78/292. entitled "Competence Assessment Training Guidelines for the Geophysical Industry." Despite the title, the document is largely generic and the guidelines apply broadly to competence assessment training. Enter the title into a search engine and you will find that the document can be downloaded without charge.

COMMUNICATION: SECTION 5.3

The standard's communication provisions require that all levels of the organization are informed about the OHSMS and the implementation plan, injuries and illnesses are reported promptly, employees be encouraged to recommend improvements on safety and health matters, that there be consultation with contractors and other interested parties when changes are made that affect the occupational health and safety management system, and any barriers to communication on hazards and risks and safety management deficiencies be eliminated.

DOCUMENT AND RECORD CONTROL PROCESS: SECTION 5.4

The organization is allowed some discretion with respect to its document and record control process. This is stated in the advisory column: *The type and amount of formal documentation that is necessary to effectively manage an OHSMS should be commensurate with the size, complexity and risks of an organization.*

Nevertheless, records shall be kept to *demonstrate or assess performance with the requirements of this standard.* In several provisions in Z10, statements are made indicating that documentation is necessary (e.g., health and safety policy, objectives, implementation plan, audits, and management reviews). For many of Z10's other provisions, activity cannot be managed properly without adequate documents.

REFERENCES

Accident Prevention Manual: Administration and Programs, 13th ed. Itasca, IL: National Safety Council, 2009.

Competence Assessment Training Guidelines for the Geophysical Industry. International Association of Oil and Gas Producers (OGP) Reference 6.78/292, 1999. Available at http://info.ogp.org.uk/Geophysical/reports/M3.pdf.

Hill, Darryl C, Editor. *Construction Safety Management and Engineering*. Des Plaines, IL, American Society of Safety Engineers, 2004.

CHAPTER 11

A PRIMER ON HAZARD ANALYSIS AND RISK ASSESSMENT: SECTIONS 4.2 AND 5.1.1 OF Z10

Requirements for risk assessment and prioritization within an occupational safety and health management system are introduced in Section 4.2 of Z10: *Processes are to be in place to: assess management system issues and assess the level of risk for identified hazards; establish priorities based on factors such as the level of risk; and identify underlying causes and other contributing factors related to system deficiencies that lead to hazards and risks.*

In the advisory column at E4.2, it is said that the method of assessment should be selected *based on the type of issue, nature of risk, or operations and that various methods may be used in determining the level of risk imposed.* Organizations are advised that priorities are to be set based on several factors, such as *issues requiring immediate attention; opportunities having the greatest potential for improvement or risk reduction; organizational, resource, participation, or accountability issues; and issues with the highest impact or severity.*

E4.2 encourages employers to give consideration not only to hazards that relate to the high probability of incident occurrence but also to hazards that relate to low-probability/serious-consequence events.

It is stated clearly that assessments conducted for Section 4.2 are for the prioritizing of occupational health and safety issues and that they may not be sufficient for the requirements of Section 5.1.1, "Risk Assessments," or to determine the appropriate hazard controls as outlined in Section 5.1.2, "Hierarchy of Controls." Appendix D provides guidance on assessment and prioritization.

Advanced Safety Management: Focusing on Z10 and Serious Injury Prevention,
Second Edition. Fred A. Manuele.
© 2014 John Wiley & Sons, Inc. Published 2014 by John Wiley & Sons, Inc.

An important step forward was taken by adding a provision in the 2012 version of Z10 requiring that risk assessments be made. The requirement is simply stated and the advisory comments are brief. Section 5.1.1 states in its entirety that:

> *The organization shall establish and implement a risk assessment process(es) appropriate to the nature of hazards and level of risk.*

Little is said about risk assessments in the advisory column at E5.1.1. Organizations are informed that:

> *Assessing risks can be done using quantitative (numeric) or qualitative (descriptive) methods. There are many methods of risk assessment. The organization should select methods appropriate to the hazards and type of process.*

Appendix F, "Risk Assessment," has been lengthened considerably and now includes a broader overview of risk assessment processes and data on several risk assessment methods. The bibliography, Appendix O, also lists several risk assessment resources.

Safety professionals can expect that being able to make documented risk assessments will be necessary for their job retention and career enhancement. That premise has acquired weight because of the more frequent inclusion in safety standards and guidelines of provisions requiring that hazards be identified and analyzed and that risk assessments be made. Examples of recently approved examples follow.

- ANSI/ASSE Z590.3-2011: *Prevention Through Design: Guidelines for Addressing Occupational Hazards and Risks in Design and Redesign Processes*
- ANSI/AIHA Z10-2012: *Occupational Health and Safety Management Systems*
- CSA Z1002-12: *Occupational Health and Safety—Hazard Identification and Elimination and Risk Assessment and Control,* issued by the Canadian Standards Association

Other standards and guidelines having such provisions are discussed in Chapter 11, "Provisions for Risk Assessments in Standards and Guidelines".

Since safety professionals are to analyze hazards and assess the risks that derive from them, the question logically follows: What do they need to know? The intent of this chapter is to provide a primer that will serve many of the hazard analysis and risk assessment needs that safety professionals will encounter. In this chapter we:

- Define the terms that must be understood in the hazard analysis and risk assessment process
- Establish the parameters for a hazard analysis
- Indicate how a hazard analysis is extended into a risk assessment
- Establish that the goal is to achieve acceptable risk levels
- Include a hazard analysis and risk assessment guide

- Give examples of the terms used in risk assessment matrices, and their variations
- Present examples of basic two-dimensional risk assessment matrices
- Describe several of the most commonly used hazard analysis and risk assessment techniques

DEFINING HAZARD, HAZARD ANALYSIS, RISK, AND RISK ASSESSMENT

When a safety professional identifies a hazard and its potential for harm or damage and decides on the probability that an injurious or damaging incident can occur, a subjective risk assessment has been made. In doing so, for the simpler and less complex hazards and risks, the assessment may be based entirely on a priori knowledge and experience, without documentation. Making informal risk assessments has been an integral part of the practice of safety and health professionals from time immemorial.

Recent developments take the risk assessment subject to a higher level within the practice of safety. By formalizing the hazard analysis and risk assessment process, a better appreciation of the significance of individual risks is achieved. As risk levels are categorized and prioritized, more intelligent decisions can be made with respect to their elimination or reduction. For the hazard analysis and risk assessment process, it is necessary that agreement be reached on the definitions of hazards, hazards analyses, risks, and risk assessments.

ANSI/AIHA Z10 is an occupational health and safety management systems standard, and its definitions of hazard and risk are, understandably, worker injury and illness related. They do not include considerations for injury to the public, possible damage to the environment or damage to property, or business downtime. This chapter has a broader purpose than Z10, as will be seen.

- A *hazard* is defined as the potential for harm to people, property, or the environment. If there is no potential for harm, injury or damage cannot occur. (In Z10, a hazard is defined as *a condition, set of circumstances, or inherent property that can cause injury, illness, or death.*) The dual nature of hazards must be understood. Hazards encompass all aspects of technology or activity that produce risk. Hazards include the characteristics of things and the actions or inactions of people.
- A hazard analysis is made to estimate the severity of harm or damage that could result from a hazard-related incident or exposure. The hazard analysis process need not include an estimate of incident or exposure probability. Examples of hazards analyses that do not include probability indicators are the estimates made by fire protection engineers of maximum foreseeable loss or maximum probable loss for insurance purposes, and the hazard analysis requirements of *OSHA's Rule For Process Safety Management Of Highly Hazardous Chemicals,* 29 CFR 1910.119.

- Whether hazardous situations are simple or complex, the process for making a hazard analysis will address the following questions:
  1. Is there potential for harm deriving from aspects of the technology or activity, the characteristics of things, or the actions or inactions of people?
  2. Can the potential be realized?
  3. Who and what are exposed to harm or damage?
  4. What is the frequency of endangerment?
  5. What will the consequences be (i.e., the severity of harm or damage) if the potential of a hazard is realized?
- Making a hazard analysis is necessary to and precedes making a risk assessment. William Johnson said this about hazard analysis in *MORT Safety Assurance Systems*: "Hazard analysis is the most important safety process in that, if that fails, all other processes are likely to be ineffective." Johnson's premise is stated soundly.
- *Risk* is defined as an estimate of the probability of a hazards-related incident or exposure occurring and the severity of harm or damage that could result. (Z10 defines risk as *an estimate of the combination of the likelihood of an occurrence of a hazardous event or exposure(s), and the severity of injury or illness that may be caused by the event or exposures.*)
- *Probability* is defined as the likelihood of a hazard's potential being realized and initiating an incident or exposure that could result in harm or damage for a selected unit of time, events, population, items, or activity being considered.
- *Severity* is defined as the extent of harm or damage that could result from a hazard-related incident or exposures.
- *Risk assessment* is a process that begins with hazard identification and analysis, followed by an estimate of the probable extent of severity of harm or damage if an incident or exposure occurs, concluding with an estimate of the probability of the incident or exposure occurring.

In a statement indicating risk level, both probability of occurrence and severity of outcome must be included. After determining the severity of expected damage or harm through a hazard analysis, estimating the probability of an incident or exposure occurring is the additional and necessary step in concluding a risk assessment.

These excerpts from the *Framework for Environmental Health Risk Management* issued by The Presidential/Congressional Commission on Risk Assessment and Risk Management are an indication of the widespread adoption of the foregoing definitions.

What Is "Risk"

Risk is defined as the probability that a substance or situation will produce harm under specified conditions. Risk is a combination of two factors:

- The *probability* that an adverse event will occur;
- The *consequences* of the adverse event.

Risk encompasses impacts on public health and on the environment, and arises from *exposure* and *hazard*. Risk does not exist if exposure to a harmful substance or situation does not or will not occur. Hazard is determined by whether a particular substance or situation has the potential to cause harmful effects.

MAKING A HAZARD ANALYSIS/RISK ASSESSMENT

For many hazards and the risks that derive from them, knowledge gained by safety professionals through education and experience will lead to proper conclusions on how to attain an acceptable risk level without bringing teams of people together for discussion. For the more complex situations, it is vital to seek the counsel of experienced personnel who are familiar with the work or process. Reaching group consensus is a highly desirable goal. Sometimes, for what a safety professional considers obvious, achieving consensus is still desirable so that buy-in is obtained for the actions to be taken. A general guide follows on how to make a hazard analysis and how to extend the process into a risk assessment.

This guide is applicable to every phase of the life cycle of facilities, equipment, processes, and materials (e.g., design, construction, operation, maintenance, and disposal).

Whatever the simplicity or complexity of the hazard/risk situation, and whatever analysis method is used, the following thought and action process is applicable.

A Hazard Analysis and Risk Assessment Guide.

1. Select a risk assessment matrix. A risk assessment matrix provides a method to categorize combinations of probability and severity, thus establishing risk levels. A matrix helps in communicating with decision makers and influencing their decisions on risks and the actions to be taken to ameliorate them. Also, risk assessment matrices can be used to compare and prioritize risks and to effectively allocate mitigation resources.

Definitions of the levels of probability and severity used in risk assessment matrices vary greatly. That reflects the differences in the perceptions people have of risk. Since a risk assessment matrix is a management decision tool, management personnel at appropriate levels must agree on the definitions of the terms to be used. In so doing, management establishes the levels of risk that require reduction and those that are acceptable.

Repeating for emphasis: Safety professionals must understand that definitions of terms for incident probability and severity and for risk levels vary greatly. Thus, they should tailor a risk assessment matrix to suit the hazards and risks and the management tolerance for risk with which they deal. Examples of definitions used for incident probability and severity are presented here as well as for risk categories and risk assessment matrices.

They are to provide safety professionals with a broad base of information from which choices can be made in developing the matrix considered appropriate for their clients' needs.

The breadth of possibilities in drafting a risk assessment matrix is extensive. Matrices have been developed that display only one or a combination of several of the following injury or damage classes: employees, members of the public, facilities, equipment, product, operation downtime, and the environment.

For this primer, two-dimensional risk assessment matrices are discussed. They are displays of variations for two categories of terms: the *severity* of harm or damage that could result from a hazards-related incident or exposure, and the *probability* that the incident or exposure could occur. They also show the *risk levels* that derive from the various combinations of severity and probability.

A review of three-and four-dimensional risk assessment systems is given in Chapter 13, "Three and Four Dimensional Numerical Risk Scoring Systems".

2. **Establish the analysis parameters.** Select a manageable task, system, process, or product to be analyzed, establish its boundaries and operating phase (standard operation, maintenance, startup), and define its interface with other tasks or systems, if appropriate. Determine the scope of the analysis in terms of what can be harmed or damaged: people (the public, employees), property, equipment, productivity, the environment.

3. **Identify the hazards.** A frame of thinking should be adopted that gets to the bases of causal factors, which are hazards. These questions would be asked: What are the aspects of technology or activity that produces risk? What are the characteristics of things or the actions or inactions of people that present a potential for harm? Depending on the complexity of the hazardous situation, some or all of the following may apply.

- Use intuitive engineering and operational sense: This is paramount throughout.
- Examine system specifications and expectations.
- Review codes, regulations, and consensus standards.
- Interview current or intended system users or operators.
- Consult checklists.
- Review studies from similar systems.
- Consider the potential for unwanted energy releases.
- Take into account possible exposures to hazardous environments.
- Review historical data: industry experience, incident investigation reports, OSHA and National Safety Council data, manufacturers' literature.
- Brainstorm.

4. **Consider the failure modes.** Define the possible failure modes that would result in realization of the potentials of hazards. Ask: What circumstances can arise that would result in the occurrence of an undesirable event? What controls are in place that mitigate against the occurrence of such an event or exposure? How effective are the controls?

5. **Determine the frequency and duration of exposure.** For each harm or damage category selected for the analysis (people, property, business interruption, etc.), estimate the frequency and duration of exposure to the hazard. This is a very important part of this exercise. For example, in a workplace situation, ask: How often is a task performed, how long is the exposure period, and how many people are exposed? More judgments than one might realize will be made in this process.

6. **Assess the severity of consequences.** The purpose is to determine the magnitude of harm or damage that could result. Informed speculations are made to establish the consequences of an incident or exposure: the number injuries or illnesses and their severity, and fatalities; the value of property or equipment damaged; the time for which productivity will be lost; and the extent of environmental damage. Historical data can be of value as a baseline but should not be treated as the primary or sole determinant. Judgment of knowledgeable and competent personnel should prevail.

   On a subjective basis, the goal is to decide on the worst credible consequences should an incident or exposure occur, not the worst conceivable consequence.

   When the severity of the outcome of a hazard-related incident or exposure is determined, a hazard analysis has been completed.

7. **Determine occurrence probability.** Extending the hazard analysis into a risk assessment requires the one additional step of estimating the likelihood, the probability, of a hazardous event or exposure occurring. Unless empirical data are available, and that would be a rarity, the process of selecting incident or exposure probability is subjective. As is the case when determining severity of consequences, historical data can be of value as a baseline but should not be the sole determinant. Judgment of knowledgeable and competent personnel, controls in place and their capability, and lessons learned with respect to similar systems should prevail.

   For the more complex hazardous situations, it should be considered a necessity that brainstorming sessions be held with knowledgeable people and that consensus be reached in determining occurrence probability.

   To be meaningful, probability has to be related to an interval base of some sort, such as a unit of time or activity, events, units produced, or the life cycle of a facility, equipment, process, or a product.

8. **Define the initial risk.** Conclude with a statement that addresses the probability of a hazard-related incident or exposure occurring, the expected severity of

adverse results, and a risk category (e.g., – high, serious, moderate, or low). Using a risk assessment matrix for that purpose assists in communicating on the risk level.

9. **Risk prioritization.** A risk ranking system should be adopted so that priorities can be established. Since the risk assessment exercise is subjective, the risk ranking system would also be subjective. Prioritizing risks gives management the knowledge needed for intelligent allocation of resources with respect to risk avoidance, elimination, or control.

10. **Select and implement risk reduction and control methods.** When the initial risk assessment indicates that elimination or reduction measures are to be taken to achieve acceptable risk levels:
    - Alternative proposals for the design and operational changes necessary would be recommended.
    - In their order of effectiveness, the actions outlined in Chapter 14, "Hierarchy of Controls", would be the basis upon which remedial proposals are made.
    - If informal judgments on costs are considered inadequate for a risk situation, formal action would be taken to establish remediation cost for each proposal and an estimate would be given of its effectiveness in achieving risk reduction to an acceptable risk level.
    - Risk avoidance, elimination, or reduction methods would be selected and implemented.
    - A tracking system should be put in place to assure completion of the actions undertaken.

11. **Assess the residual risk.** *Residual risk* is defined as the risk remaining after preventive measures have been taken. No matter how effective the preventive actions, there will always be residual risk if a facility exists or an activity continues. Attaining zero risk is not possible.
    - If the residual risk is not acceptable, the action outline set forth in this hazard analysis and risk assessment process would be applied again.
    - The risk assessment process is to continue until an acceptable risk level is attained.
    - If an acceptable risk level cannot be achieved, operations should not be continued, except in unusual and emergency circumstances or as a closely monitored and limited exception and with the approval of the person who has the authority to accept the risk.
    - Information on residual risks is to be communicated to relevant downstream stakeholders so that they can use this information in their own risk assessments.

    Even though the residual risk may be acceptable, management should consider taking additional risk reduction measures if the cost is reasonable, particularly if so doing resolves concerns of employees.

12. **Risk acceptance decision making.** Risk acceptance decisions shall be made at the appropriate management levels.

   - Temporary acceptance of high and serious risks shall be made at the top management level.
   - Responsibility for moderate or low risks may be assigned to lower management levels.

13. **Documentation.** To the extent appropriate for a given risk situation, documentation should include the:

   - The names and job titles/qualifications of the persons who did the risk assessment
   - The risk assessment method(s) used
   - Hazards identified
   - Risks deriving from the hazards
   - Avoidance, elimination, reduction, and control measures taken to attain acceptable risk levels
     In certain cases, consideration would be given to additional documentation that includes relevant information about the use, effectiveness of design, problems during startup and continued use, and changes to designs over the life cycle of the design.

     The foregoing applies whether the documentation is done directly under the direction of management or by a supplier of services.

14. **Reassess the risk:** follow up on actions taken. This step is a part of all effective problem-solving exercises. Follow-up activity would determine that the:

   - Problem was resolved, only partially resolved, or not resolved, and that acceptable risk levels were or were not achieved.
   - Actions taken did or did not create new hazards.

If acceptable risk levels were not achieved or new hazards were introduced, the risk is to be reevaluated and other countermeasures are to be proposed and taken.

DESCRIPTIONS: PROBABILITY AND SEVERITY

Examples follow in Tables 11.1 to 11.5 to show variations in the terms and their descriptions as used in a variety of applied risk assessment processes for the probability of occurrence and severity of consequence. There is no single correct method of selecting probability and severity categories and their descriptions. Tables 11.4 to 11.6 show how severity of harm or damage categories can be related to several types of adverse consequences and levels of harm or damage.

TABLE 11.1   Example A: Probability Descriptions

Descriptive Word	Probability Descriptions
Frequent	Likely to occur repeatedly
Probable	Likely to occur several times
Occasional	Likely to occur sometime
Remote	Not likely to occur
Improbable	So unlikely that one can assume occurrence will not be experienced

TABLE 11.2   Example B: Probability Descriptions

Descriptive Word	Probability Descriptions
Frequent	Occurs often, experienced continuously
Probable	Occurs several times
Occasional	Occurs sporadically, occurs sometimes
Seldom	Remote chance of occurrence; unlikely but could occur sometime
Unlikely	Can assume that incident will not occur

TABLE 11.3   Example C: Probability Descriptions

Descriptive Word	Probability Descriptions
Frequent	Could occur annually
Likely	Could occur once in two years
Possible	Would occur no more than once in five years
Rare	Would occur no more than once in 10 years
Unlikely	Would occur no more than once in 20 years

TABLE 11.4   Example A: Severity Descriptions for Multiple Harm and Damage Categories

Catastrophic	Death or permanent total disability, system loss, major property damage, and business downtime
Critical	Permanent, partial, or temporary disability in excess of three months, major system damage, significant property damage, and downtime
Marginal	Minor injury, lost-workday accident, minor system damage, minor property damage, and little downtime
Negligible	First aid or minor medical treatment, minor system impairment

TABLE 11.5   Example B: Severity Descriptions for Multiple Harm and Damage Categories

Catastrophic	One or more fatalities, total system loss, chemical release with lasting environmental or public health impact
Critical	Disabling injury or illness, major property damage and business downtime, chemical release with temporary environmental or public health impact
Marginal	Medical treatment or restricted work, minor subsystem loss or damage, chemical release triggering external reporting requirements
Negligible	First aid only, nonserious equipment or facility damage, chemical release requiring only routine cleanup without reporting

TABLE 11.6   Example C: Severity Descriptions for Multiple Harm and Damage Categories

Category: Descriptive Word	Facilities; People: Employees, Public	Product or Equipment Loss	Operations Downtime	Environmental Damage
Catastrophic	Fatality	Exceeds $3 million	Exceeds 6 Months	Major event, requires more than 2 years for full recovery
Critical	Disabling injury or illness	500,000 to $3million	4 weeks to 6 months	Significant event, injury or requires 1 to 2 years for full recovery
Marginal	Minor injury or illness	50,000 to 500,000	2 days to 4 weeks	Recovery time is less than 1 year
Negligible	Injury requires only first aid	Less than 50,000	Less than 2 days	Minor damage, easily repaired, little time for recovery

EXAMPLES OF RISK ASSESSMENT MATRICES

Five examples of risk assessment matrices follow. Shown here in Table 11.7 is an adaptation of the "Risk Assessment Matrix" in MIL-STD-882E, the Department of Defense, *Standard Practice for System Safety*. (p. 11) MIL-STD-882, first issued in 1969, is the grandfather of risk assessment matrices. All of the over 30 variations of matrices that I have collected include the basics found in the 882 standard. They include event probability categories, severity of harm or damage ranges, and risk gradings.

TABLE 11.7 Risk Assessment Matrix

Occurrence Probability	Severity of Consequence			
	Catastrophic	Critical	Marginal	Negligible
Frequent	High	High	Serious	Medium
Probable	High	High	Serious	Medium
Occasional	High	Serious	Medium	Low
Remote	Serious	Medium	Medium	Low
Improbable	Medium	Medium	Medium	Low
Eliminated	This category is used only for identified hazards that are totally removed			

TABLE 11.8 Risk Scoring System: B155.1-2011[a]

Probability Level	Severity Category			
	Catastrophic	Critical	Marginal	Negligible
Frequent	High	High	Serious	Medium
Probable	High	High	Serious	Medium
Occasional	High	Serious	Medium	Low
Remote	Serious	Medium	Medium	Low
Improbable	Medium	Medium	Low	Low

[a]Identified in B155.1-2011 as MIL STD 882: Two-Factor Risk Model [4×5].

The risk scoring system shown in Table 11.8 appears in the ANSI/PMMI B155.1-2011 standard titled *Safety Requirements for Packaging Machinery and Packaging-Related Converting Machinery* (p. 53). It is shown here for two reasons. It is an indication of the validity of the concepts on which the risk assessment matrices in MIL-STD-882 are based and why so many developers of matrices use 882 as a reference.

Table 11.8 is almost identical with the 882 version shown in Table 11.7 except for the addition of an "eliminated" probability category in the 882 model. But there is one other slight difference: People who drafted B155.1 made a change in one risk severity category.

As noted previously, people who develop risk assessment matrices work their own risk perceptions into them. And that's great. Safety professionals should feel free to do so too.

Table 11.9 shows a risk assessment matrix that combines types of severity categories and uses alphabetical risk gradings.

Table 11.10 is close to a risk assessment matrix shown in Annex E of the original version of Z10. Its uniqueness is that it was a broadly published matrix that includes not only probability and severity categories and descriptive data for each category but also advice on decision making and remedial action for each risk category.

As the matrix in Table 11.10 was developed, one of the participants thought that the risk levels presented were one step too high in two places. In the discussion that

TABLE 11.9   Risk Assessment Matrix: Alphabetical Risk-Level Indicators: Probability That Something Will Go Wrong[a]

Severity Categories	Frequent: Likely to occur immediately or soon	Likely: Quite likely to occur often	Occasional: May occur in time	Seldom: Not likely to occur but possible	Unlikely: Unlikely to occur
Catastrophic: death, multiple injuries, severe property or environmental damage	E	E	H	H	M
Critical: serious injuries, significant property or environmental damage	E	H	H	M	L
Marginal: may cause minor injuries, financial loss, negative publicity	H	M	M	L	L
Negligible: minimum threat to persons or damage to property	M	L	L	L	L

[a]E, extremely high risk; H, high risk; M, moderate risk; L, low risk.

followed, it was agreed that changes could be made. What is the significance of this? Risk assessment is more art than science. Since establishing risk levels is largely a matter of judgment, people will come to different conclusions in a given situation. Nevertheless, the ultimate goal needs to be kept in mind, which is to be satisfied that the residual risk that exists after risk reduction measures are implemented is acceptable.

There are no restrictions or rules with respect to the terms used to establish qualitative risk levels. But at a minimum, a matrix should show probability and severity categories and risk gradings. Tables 11.7 through 11.10 and Figure 11.1 show a general acceptance of a group of terms for incident probability and severity and for risk categories. However, safety professionals should draft matrices with which they are comfortable. Since risk assessment matrices are valuable communication tools, the terms used in them must be agreed upon and the education time necessary to achieve an understanding of them must be allocated.

The risk assessment matrix shown in Figure 11.1 is a composite of matrices that include numerical values for probability and severity levels that are transposed into qualitative risk gradings. It was originally developed for people who prefer to deal with numbers rather than qualitative indicators. To provide a one-page tool that could be used by safety professionals who wanted to get shop floor personnel involved in

TABLE 11.10   The Risk Assessment Matrix Example in Z10

Example of a Risk Assessment Matrix				
Likelihood of	**Severity of Injury or Illness Consequence and Remedial Action**			
OCCURRENCE or EXPOSURE For selected Unit of Time or Activity	**CATASTROPHIC** Death or permanent total disability	**CRITICAL** Disability in excess of 3 months	**MARGINAL** Minor injury, lost workday accident	**NEGLIGIBLE** First Aid or Minor Medical Treatment
Frequent Likely to occur repeatedly	**HIGH** Operation not permissible	**HIGH** Operation not permissible	**SERIOUS** High priority remedial action	**MEDIUM** Take Remedial action at appropriate time
Probable Likely to occur several times	**HIGH** Operation not permissible	**HIGH** Operation not permissible	**SERIOUS** High Priority remedial action	**MEDIUM** Take Remedial action at appropriate time
Occasional Likely to occur sometime	**HIGH** Operation not permissible	**SERIOUS** High Priority Remedial action	**MEDIUM** Take Remedial Action at appropriate time	**LOW** Risk Acceptable: Remedial Action discretionary
Remote Not likely to occur	**SERIOUS** High Priority Remedial action	**MEDIUM** Take Remedial action at appropriate time	**MEDIUM** Take Remedial action at appropriate time	**LOW** Risk Acceptable: Remedial Action Discretionary
Improbable Very unlikely – may assume exposure will not happen	**MEDIUM** Take Remedial action at appropriate time	**LOW** Risk Acceptable: Remedial Action Discretionary	**LOW** Risk Acceptable: Remedial Action Discretionary	**LOW** Risk Acceptable: Remedial Action Discretionary

risk assessments, extensions were made to include severity and probability descriptions and action levels.

In two instances, shop floor personnel said to safety professionals that relating numbers to each other first—such as 6 to 12—was a big help in understanding whether the risk category was moderate or serious. If using a risk assessment matrix in which numbers are used to begin with to establish risk levels makes the process more understandable and acceptable for operating personnel, that should be encouraged.

As noted in Figure 11.1, numbers in the matrix were derived intuitively: They are qualitative, not quantitative. Thus, the numbers have value only in relation to each other. And that is the case for all risk scoring systems that are not based on hard probability and severity numbers, which are rarely available.

On Acceptable Risk

The theoretical goal should always be to eliminate all hazards, but that may not be feasible. When a hazard cannot be eliminated, the associated risk should be reduced to a level as low as reasonably practicable and acceptable, considering cost constraints

Severity levels and values	Occurrence probabilities and values				
	Unlikely (1)	Seldom (2)	Occasional (3)	Likely (4)	Frequent (5)
Catastrophic (5)	5	10	15	20	25
Critical (4)	4	8	12	16	20
Marginal (3)	3	6	9	12	15
Negligible (2)	2	4	6	8	10
Insignificant (1)	1	2	3	4	5

Numbers were intuitively derived. They are qualitative, not quantitative. They have meaning only in relation to each other.

Incident or Exposure Severity Descriptions

Catastrophic: One or more fatalities, total system loss and major business down time, environmental release with lasting impact on others with respect to health, property damage or business interruption.

Critical: Disabling injury or illness, major property damage and business down time, environmental release with temporary impact on others with respect to health, property damage or business interruption.

Marginal: Medical treatment or restricted work, minor subsystem loss or property damage, environmental release triggering external reporting requirements.

Negligible: First aid or minor medical treatment only, non-serious equipment or facility damage, environmental release requiring routine cleanup without reporting.

Insignificant: Inconsequential with respect to injuries or illnesses, system loss or down time, or environmental release.

Incident or Exposure Probability Descriptions

Unlikely: Improbable, unrealistically perceivable.
Seldom: Could occur but hardly ever.
Occasional: Could occur intermittently.
Likely: Probably will occur several times.
Frequent: Likely to occur repeatedly.

Risk Levels

Combining the Severity and Occurrence Probability values yields a risk score in the matrix. The risk levels and the action levels are categorized below.

Risk Categories, Scoring, and Action Levels

Category	Risk Score	Action Level
Low risk	1 to 5	Remedial action discretionary.
Moderate risk	6 to 9	Remedial action to be taken at appropriate time.
Serious risk	10 to 14	Remedial action to be given high priority.
High risk	15 or greater	Immediate action necessary. Operation not permissible except in an unusual circumstance and as a closely monitored and limited exception with approval of the person having authority to accept the risk.

FIGURE 11.1   Risk assessment matrix: numerical gradings.

in relation to the extent of risk reduction to be obtained and performance requirements. In Chapter 2, "Achieving Acceptable Risk Levels: The Operational Goal," it is said that as every element in Z10 is applied, the outcome is to achieve acceptable risk levels so that the risk of harm is as low as reasonably practicable. It is also said that the risk assessment matrices in this chapter and the discussions of risk categories will help in determining acceptable risk levels.

Applying the concept of ALARP—as low as reasonably practicable—is recognized as a valuable tool to assist in determining acceptable risk levels. But a caution was offered indicating that, on occasion, achieving risk levels as low as reasonably practicable will not achieve acceptable risk levels. Prior to presenting the following definition, it was made clear that a workable and sound definition of acceptable risk must encompass hazards, risks, probability, severity, and economics.

> Acceptable risk is that risk for which the probability of an incident or exposure occurring and the severity of harm or damage that may result are as low as reasonably practicable (ALARP) in the setting being considered.

Thus far in this chapter we have dealt with hazards, risks, probability, and severity. In applying the ALARP concept, economics is brought into the decision making. ALARP is defined as follows: ALARP is that level of risk which can be further lowered only by an increment in resource expenditure that is disproportionate in relation to a resulting decrement of risk.

MANAGEMENT DECISION LEVELS

Remedial action or acceptance levels must be attached to risk categories to permit intelligent management decision making. Examples are given in Tables 11.10 and 11.11. It must be understood that they are examples only and that remedial action and acceptance levels should be developed to suit the needs and exposures in an individual operation. They must be agreed upon, understood, and supported by senior management.

TABLE 11.11 Management Decision Levels

Risk Category	Remedial Action or Acceptance
Low	Risk is acceptable; remedial action discretionary
Medium	Remedial action to be taken within appropriate time
Serious	Remedial action to have high priority
High	Immediate action necessary. Operator not permissible except in an unusual circumstance and as a closely monitored and limited exception with approval of the person having the authority to accept the risk

Going through the exercise of creating and reaching agreement on a risk assessment matrix and the management decision levels adds to a safety professional's effectiveness in communicating about risks and obtaining consideration of the remedial actions recommended.

For the discussion that follows of acceptable risk levels and the management actions to be taken to achieve them, the contents of Tables 11.10 and 11.11 are used as base data. Keep in mind that:

- An acceptable risk level must be tolerable in the situation being considered
- While economic considerations are a part of decision making, the risk level is to be as low as reasonably practicable, and acceptable

- Special consideration should be given to preventing incidents resulting in serious injuries and fatalities
- What follows represents my opinion; others may have different views.

If the risk category for worker injury or illness or environmental damage or other property damage is high, operation is not permissible except in an unusual circumstance and as a closely monitored and limited exception, only with approval of the person having the authority to accept the risk. If it is determined that the cost to reduce the risk to an acceptable level is excessive in relation to the risk reduction benefit to be achieved, the operation should cease in all but rare situations (e.g., society accepts the risks of deep-sea fishing, a high-hazard occupation).

If the risk category is serious, the risk is not acceptable and action should be undertaken on a high-priority basis, meaning very soon, to lower the risk to an acceptable level. While arrangements are made to reduce the risk, an extra-heavy application of the lower levels in the hierarchy of controls (warning systems, blocking off work areas, administrative controls, training, personal protective equipment) is in order. If it is determined that the cost to reduce the risk to an acceptable lower level is excessive in relation to the risk reduction benefit to be achieved, the operation should cease in all but rare situations.

Where the risk category is medium, even though the probability ratings for severe injury or illness are "improbable" or "remote," and the probability rating for minor injury is "occasional," and the probability ratings for negligible injury are "frequent" or "probable," remedial action should be taken, in good time, to reduce the risk in accord with good economics. This is the risk category where the lower levels in the hierarchy of controls, if applied more extensively and effectively, may be sufficient to achieve acceptable risk levels.

Where the risk category is low, the risk is considered acceptable. Nevertheless, there will be times when it is good business management and employee relations to reduce low risks further if they are perceived by employees to be more serious than they actually are. Remember, employee perceptions are their reality.

Some of the risk assessment matrices shown in this chapter combine elements pertaining to personal injury with the financial impact of an incident represented by the amount of property damage, business interruption time, and time to recover from an environmental incident. Safety professionals who have made such combinations in their risk assessment matrices say that they get better management response to their proposals for risk reduction if they tie the severity of injury potential to avoiding operational property damage, downtime, business interruption, and environmental damage. That has also been my experience.

DESCRIPTIONS OF HAZARDS ANALYSES AND RISK ASSESSMENT TECHNIQUES

Over the past 50 years, a large and unwieldy number of hazard analysis and risk assessment techniques have been developed. For example, Pat Clemens gave brief descriptions of 25 techniques in "A Compendium of Hazard Identification & Evaluation

Techniques for System Safety Applications." In the *System Safety Analysis Handbook*, 101 methods are described. In ANSI/ASSE Z690.3-2011, *Risk Assessment Techniques*, comparisons and descriptions are given of 31 risk assessment techniques. (Z690.3 is the American National Standard adoption of IEC/ISO 31010:2009.)

Brief descriptions are given here of purposely selected hazard analysis techniques. If a safety professional understands all of them and is capable of bringing them to bear in resolving hazard and risk situations, she or he will be exceptionally well qualified to meet the risk assessment requirements in Z10.

As a practical matter, having knowledge of three risk assessment concepts will be sufficient to address most occupational safety and health risk situations: preliminary hazard analysis, the what-if checklist analysis methods, and failure mode and effects and analysis.

It is important to understand that each of those techniques complement, rather than supplant, the others. Selecting a technique or a combination of techniques to be used to analyze a hazardous situation requires good judgment based on knowledge and experience. Qualitative rather than quantitative judgments will prevail. For all but the complex risks, qualitative judgments will be sufficient.

Sound quantitative data on incident probabilities are seldom available. Colleagues who are skilled in system safety, a field in which making quantitative risk assessments is ordinarily done, are not overly pleased when I say that most quantitative risk assessments are really qualitative risk assessments because so many judgments have to be made in the process to decide on the occurrence probability and the severity of outcome .

PRELIMINARY HAZARD ANALYSIS: HAZARD ANALYSIS AND RISK ASSESSMENT

Originally, the preliminary hazards analysis (PHA) technique was used to identify and evaluate hazards in the early stages of the design process. However, in actual practice the technique has attained much broader use. Principles on which preliminary hazards analyses are based have been adopted for use not only in the initial design process, but also in assessing the risks of existing products or operations.

For example, a standard adopted by the International Organization for Standardization is employed in the European Union, and the standard requires that risk assessments be made for all machinery that is to go into a workplace in member countries. That standard is EN ISO 12100–2010. *Safety of Machinery—General principles for design. Risk assessment and risk reduction.* Those risk assessment requirements have been met in some companies by applying an adaptation of the PHA technique.

In reality, the PHA technique needs a new name, reflecting its broader usage. In one organization, the process is called hazard analysis and risk assessment, a designation that is coming into greater us since it is more descriptive of its purpose.

(Also, take note to avoid confusion: In *OSHA's Rule for Process Safety Management of Highly Hazardous Chemicals* and *EPA's Risk Management*

Program for Chemical Accidental Release Prevention, "PHA" stands for process hazard analysis.)

Headings on preliminary hazard analysis forms include the typical identification data: date, names of evaluators, department, and location. The following information is usually included in a preliminary hazard analysis form.

- A hazard description, sometimes called a hazard scenario.
- A description of the task, operation, system, or product being analyzed.
- The exposures that are to be analyzed: people (employees, the public); facility, product, or equipment loss; operation downtime; environmental damage.
- The probability interval to be considered: unit of time or activity; events; units produced; life cycle.
- A numerical or alphabetical indicator for the severity of harm or damage that might result if the hazard's potential is realized.
- A numerical or alphabetical indicator for the occurrence probability.
- A risk assessment code, using the agreed-upon risk assessment matrix.
- Remedial action to be taken if risk reduction is required.

A communication accompanies the analysis, explaining the assumptions made and the rationale for them. Comment would be made on the assignment of responsibilities for the remedial actions to be taken, and when. A hazard analysis and risk assessment worksheet (formerly called a preliminary hazard analysis worksheet) appears as Addendum A in this chapter. That form, and similar forms, require entry of severity, probability, and risk codes before and after countermeasures are taken. Figure 11.1 is an add-on to Addendum A.

On Developing a Coding System

Assume that there are to be four severity categories, for which numerical codes are to be used: catastrophic, 1; critical, 2; marginal, 3; negligible, 4. Assume that there are to be five categories of occurrence probability, having alphabetical codes: frequent, A; probable, B; occasional, C; remote, D; improbable, E. Table 11.12 displays combinations of numerical and alphabetical indicators and the risk codes that derive from them.

TABLE 11.12   Risk Assessment Matrix, Including Probability and Severity Codes

Occurrence Probability	Severity Categories			
	Catastrophic (1)	Critical (2)	Marginal (3)	Negligible (4)
Frequent (A)	High	High	Serious	Medium
Probable (B)	High	High	Serious	Medium
Occasional (C)	High	Serious	Medium	Low
Remote (D)	Serious	Medium	Medium	Low
Improbable (E)	Medium	Medium	Low	Low

Risk codes would then be as follows, taking into account the combinations of the severity and probability codes:

Combinations	Risk Category	Risk Code
A–1, A–2, B–1, B–2, C–1	High	H
A–3, B–3, C–2, D–1	Serious	S
A–4, B–4, C–3, D–2, D–3, E–1, E–2, E–3	Medium	M
C–4, D–4, E–4	Low	L

The foregoing is intended as an example from which suitable adaptations can be made by safety professionals.

WHAT-IF ANALYSIS

For a what-if analysis, a group of people (as few as two, but often several more) use a brainstorming approach to identify hazards, hazard scenarios, how incidents can occur, and what their probable consequences might be. Questions posed during the brainstorming session may begin with "what-if," as in "What if the air-conditioning system fails in the computer room?" or may be expressions of more general concern, as in "I worry about the possibly of spillage and chemical contamination during truck offloading."

All of the questions are recorded and assigned for investigation. Each subject of concern is then addressed by one or more team members. They would consider the potential of the hazardous situation and the adequacy or inadequacy of risk controls in effect, suggesting additional risk reduction measures if appropriate.

CHECKLIST ANALYSIS

Checklists are primarily adaptations from published standards, codes, and industry practices. There are many such checklists. They consist of lists of questions pertaining to the applicable standards and practices, usually with a "yes" "no" or "not applicable" response. Their purpose is to identify deviations from the expected, and thereby possible hazards. A checklist analysis requires a walk-through of the area to be surveyed.

Checklists are easy to use and provide a cost-effective way to identify hazards that are customarily recognized. Nevertheless, the quality of checklists is dependent on the experience of the people who develop them. Further, they must be crafted to suit particular needs. If a checklist is not complete, the analysis may not identify some hazardous situations. An example of a checklist for machinery design is provided as Addendum C at the end of this chapter. A checklist for general design purposes appears in Chapter 16. These can serve as resources for those who choose to build their own checklists.

WHAT-IF/CHECKLIST ANALYSIS

A what-if/checklist hazard analysis technique combines the creative, brainstorming aspects of the what-if method with the systematic approach of a checklist. Combining the techniques can compensate for the weaknesses of each. The what-if part of the process, using a brainstorming method, can help the team identify hazards that have the potential to be the causal factors for incidents, even though no such incidents have yet occurred.

The checklist provides a systematic approach for review that can serve as an idea generator during the brainstorming process. Usually, a team experienced in the design, operation, and maintenance of the operation performs the analysis. The number of people required depends, of course, on the operation's complexity.

FAILURE MODES AND EFFECTS ANALYSIS

In several industries, failure mode and effects analysis (FMEA) has been the technique of choice by design engineers for reliability and safety considerations. Such techniques are used to evaluate the ways in which equipment fails and the response of the system to those failures. Although an FMEA is typically made early in the design process, the technique can also serve well as an analysis tool throughout the life of equipment or a process.

An FMEA produces qualitative, systematic lists that include failure modes, the effects of each failure, safeguards that exist, and additional actions that may be necessary. For example, for a pump, the failure modes would include: fails to stop when required, stops when required to run; seal leaks or ruptures; and pump case leaks or ruptures.

Both the immediate effects and the impact on other equipment would be recorded. Generally, when analyzing impacts, the probable worst case is assumed and analysts would conclude that existing safeguards do or do not work.

Although an FMEA can be made by one person, it's typical for a team to be appointed when there is complexity. In either case, the traditional process is similar.

1. Identify the item or function to be analyzed.
2. Define the failure modes.
3. Record the causes of failure.
4. Determine the effects of failure.
5. Enter a severity code and a probability code for each effect.
6. Enter a risk code.
7. Record the actions required to reduce a risk to an acceptable level.

Note that the FMEA process described here requires entry of probability, severity, and risk codes. The risk assessment matrix shown in Table 11.12 and the risk category codes that follow it fulfill the need for traditional FMEA purposes. A failure modes and effects analysis form on which those codes would be entered is provided as Addendum B.

HAZARD AND OPERABILITY ANALYSIS

The hazard and operability analysis (HAZOP) technique was developed to identify both hazards and operability problems in chemical process plants. An interdisciplinary team and an experienced team leader are required. In a HAZOP application, a process or operation is reviewed systematically to identify deviations from desired practices that could lead to adverse consequences. HAZOPs can be used at any stage in the life of a process.

HAZOPs usually require a series of meetings in which the team, using process drawings, systematically evaluates the impact of deviations from the desired practices. The team leader uses a set of guide words to develop discussions. As the team reviews each step in a process, they record any deviations, along with their causes, consequences, safeguards, and required actions, or the need for more information to evaluate the deviation.

FAULT TREE ANALYSIS

A fault tree analysis (FTA) is a top-down, deductive logic model that traces the failure pathways for a predetermined undesirable condition or event called a TOP event. An FTA can be carried out either quantitatively or subjectively.

An FTA generates a fault tree (a symbolic logic model) that enter's failure probabilities for combinations of equipment failures and human errors that can result in accidents. Each immediate causal factor is examined to determine its subordinate causal factors until the root causal factors are identified.

The strength of an FTA is its ability to identify combinations of basic equipment and human failures that can lead to an accident, allowing the analyst to focus preventive measures on significant basic causes. An FTA has a particularly high value when analyzing highly-redundant systems and high-energy systems in which high severity events can occur.

For systems vulnerable to single failures that can lead to accidents, the FMEA and HAZOP techniques are better suited. FTA is often used when another technique has identified a hazardous situation that requires more detailed analysis. Making a fault tree analysis of other than the simplest systems requires the talent of experienced analysts.

MANAGEMENT OVERSIGHT AND RISK TREE

As Pat Clemens wrote in a paper cited previously, the management oversight and risk tree (MORT) technique applies "a pre-designed, systematized logic tree to the identification of total system risks, both those inherent in physical equipment and processes and those which arise from operational/management inadequacies." MORT is an incident investigation and analysis technique. It is discussed here for a particular purpose. There are four major stages in operational risk management:

1. Pre-operational stage: in the initial planning, design, specification, prototyping, and construction processes, where the opportunities are greatest and the costs are lowest for hazard and risk avoidance, elimination, reduction, or control.

2. Operational stage: where hazards and risks are identified and evaluated and mitigation actions are taken through redesign initiatives or changes in work methods before incidents or exposures occur.

3. Post-incident stage: where investigations are made of incidents and exposures to determine the causal factors that will lead to appropriate interventions and acceptable risk levels.

4. Post-operational stage: where demolition, decommissioning, or reusing/rebuilding operations are undertaken.

All of the hazard analysis and risk assessment techniques discussed previously relate *principally* to the design process or achieving risk reduction in the operational mode *before hazards-related incidents occur*. MORT was developed for use *principally* when incident investigations are made—the post-incident stage. In the Introduction to the *NRI MORT User's Manual*, the following comments are made.

> The Management Oversight and Risk Tree (MORT) method is an analytical procedure for inquiring into causes and contributing factors of accidents and incidents. The MORT method is a logical expression of the functions needed by an organization to manage its risks effectively. MORT reflects a philosophy which holds that the most effective way of managing safety is to make it an integral part of business management and operational control. (p. viii)

MORT is a comprehensive analytical procedure that provides a disciplined method for determining the systemic causes and contributing factors of accidents. MORT directs the user to the hazards and risks deriving from both system design and procedural shortcomings. When used properly in the post-incident stage of the practice of safety, MORT provides an excellent resource from which decisions can be made to redesign technical and procedural aspects of operations.

ADDITIONAL RESOURCES

A Risk Assessment Tool made available by the European Agency for Safety and Health at Work can be found at http://hwi.osha.europa.eu/ra_tools_generic/. Parts I and II provide basic information on risk assessment. The subject matter consists of only eight pages. Nevertheless, it takes a reader through a basic hazard identification and risk assessment process. Part III provides checklists for hazard identification and selection of preventive measures for 10 subjects (e.g., moving machinery electrical installations and equipment, fire, etc.). Part IV consists of checklists for seven occupation settings (e.g., office work, food processing, small-scale surface mining, etc.)

In the UK, the Health and Safety Executive recently updated "Five steps to risk assessment." It can be accessed at http://www.hse.gov.uk/risk/fivesteps.htm. It is also a basic, uncomplicated system.

If a safety professional undertakes to inform supervisors and workers on the fundamentals of hazard identification and risk assessment, say at safety meetings, the two foregoing resources will serve well in developing the presentations and the written material.

A particularly valuable reference is the *Guidelines for Hazard Evaluation Procedures, Second Edition With Worked Examples*, issued by the Center for Chemical Process Safety. Although this text is issued by a chemical industry organization, it is largely generic.

The Basics of FMEA, a 75-page 5×7-inch paperback FMEA by McDermott, Mikulak, and Beauregard, is a primer on the FMEA process.

In Chapter 13, "Three and Four Dimensional Numerical Risk Scoring Systems," we comment on an FMEA publication issued for the semiconductor industry by International SEMATECH. It is well done. Its uniqueness is that environmental, safety, and health considerations are vital in the process described.

OSHA has issued a paper on Job Hazards Analysis which has enough guidance points to warrant inclusion of a modified version of it as Addendum C to this chapter. Its definition of a hazard is the same as mine: A hazard is the potential for harm. The paper also says that "Ideally, after you identify uncontrolled hazards, you will take steps to eliminate or reduce them to an acceptable risk level." But the data on eliminating or reducing hazards are a bit shallow.

Should a safety professional want to acquire extensive and valuable texts devoted entirely to applications in risk assessment, it is suggested that he or she consider two books written by Bruce Main, president of design safety engineering.

1. *Risk Assessment: basics and benchmarks* is a 485-page treatise published in 2004.
2. *Risk Assessment: Challenges and Opportunities* is a 364-page text published in 2012.

With respect to risk assessment, Main has been a researcher, writer, consultant, software developer, instructor, and a leader in standards development.

Addendum D provides additional information on types of hazards, hazardous situations, and hazardous events.

AVOIDING UNREALISTIC EXPECTATIONS

Making hazards analyses and risk assessments is both an art and a science. Whatever the methodology—the simplest or the most complex—many judgments will be made in determining the severity potentials of hazards and the probably of occurrence of incidents and exposures.

Even though appliers of the risk assessment methodologies make informed judgments, they may disagree on which hazards are most important because of their severity potential and which risks deserve the highest priority. One way to resolve those differences is to have qualified teams participate when the hazards and risks are considered significant, the intent being to reach consensus.

Some who oppose the use of qualitative risk assessment techniques do so because the outcomes are not stated in absolutely assured, precise numbers. Such accuracy is not attainable because incident probability data are lacking, and the severity of event outcomes is a best estimate. Expecting such results is unrealistic. Fortunately, recognition continues to grow that hazard analysis and risk assessment methods, although largely qualitative, add value to operational risk management decision making.

CONCLUSION

This chapter is simply a primer on hazard analysis and risk assessment. Its purpose is to provide a foundation for those who perceive that having additional knowledge in this aspect of safety and health risk management provides an opportunity for professional growth, accomplishment, and recognition. Having that knowledge adds to one's ability to evaluate hazardous situations and make more convincing presentations to management for resource allocation to accomplish the risk reduction measures proposed.

Looking to the future, safety professionals can expect that having knowledge of hazard analysis and risk assessment techniques will be required for job retention and career enhancement. Fortunately, it is not difficult to acquire the knowledge and skill required to fulfill almost all of their needs.

As safety and health professionals become more involved in risk assessments, they will come to understand that professional safety practice requires attention to the two distinct aspects of risk:

- Avoiding, eliminating, or reducing the probability of occurrence of a hazard-related incident or exposure
- Having the severity of the potential for harm or damage be as low as reasonably practicable

REFERENCES

"A Risk Assessment Tool." European Agency for Safety and Health at Work. At http://hwi.osha.europa.eu/ra-tools-generic/.

ANSI/AIHA Z10-2012. *Occupational Health and Safety Management Systems.* Fairfax, VA: American Industrial Hygiene Association, 2012. The American Society of Safety Engineers is now the secretariat. Available at https://www.asse.org/cartpage.php?link=z10-2005

ANSI/ASSE Z590.3-2011. *Prevention through Design: Guidelines for Addressing Occupational Hazards and Risks in Design and Redesign Processes.* Des Plaines, IL: American Society of Safety Engineers, 2011.

ANSI/ASSE Z690.3 *Risk Assessment Techniques.* Des Plaines, IL: American Society of Safety Engineers, 2011.

ANSI/PMMI B155.1-2011. *Safety Requirements for Packaging Machinery and Packaging-Related Converting Machinery.* Arlington, VA: Packaging Machinery Manufacturers Institute, 2011.

Clemens, Pat. "A Compendium Of Hazard Identification And Evaluation Techniques for System Safety Application." *Hazard Prevention*, Mar./Apr., 1982.

CSA Z1002-12. *Occupational Health and Safety—Hazard Identification and Elimination and Risk Assessment and Control.* Toronto, Canada: Canadian Standards Association, 2012.

EN ISO 12100–2010. *Safety of Machinery—General principles for Design. Risk Assessment and Risk Reduction.* Geneva, Switzerland: International Organization for Standardization, 2010.

EPA's Risk Management Program for Chemical Accidental Release Prevention. Washington, DC: U.S. Environmental Agency, 1996. Preview at http://www.epa.gov/emergencies/content/lawsregs/rmpover.htm.

Failure Mode and Effects Analysis (FMEA): A Guide for Continuous Improvement for the Semiconductor Equipment Industry. Technology Transfer No.92020963A-ENG. Austin, TX: International SEMATECH, 1992a. Also available at www.sematech.org.

"Five Steps to Risk Assessment." London: Health and Safety Executive, 2008.

Framework for Environmental Health Risk Management, Final Report, Vol 1. Washington, DC: Presidential/Congressional Commission on Risk Assessment and Risk Management, 1997.

Guidelines for Hazard Evaluation Procedures, Second Edition With Worked Examples. New York: Center for Chemical Process Safety of the American Institute of Chemical Engineers, 1992.

Job Hazard Analysis. U.S. Department of Labor, Occupational Safety and Health Administration, OSHA 3071, 2002 (Revised). Available at http://www.osha.gov/Publications/osha3071.html.

Johnson, William G. *MORT Safety Assurance Systems.* New York: Marcel Dekker, 1980.

Main, Bruce W. *Risk Assessment: basics and benchmarks* . Ann Arbor, MI: Design Safety Engineering, Inc., 2004. Information at www.designsafe.com.

Main, Bruce W. *Risk Assessment: Challenges and Opportunities.* Ann Arbor, MI: Design Safety Engineering, Inc., 2012. Information at www.designsafe.com.

McDermott, Robin E., Raymond J. Mikulak, and Michael R. Beauregard. *The Basics of FMEA.* New York: Productivity Press, 1996.

Mil-STD-882E. *Standard Practice for System Safety.* Washington, DC: Department of Defense, 2000. Also at http://www.systemsafetyskeptic.com/yahoo_site_admin/assets/docs/MIL-STD-882E_final.135152939.pdf.

NRI MORT User's Manual. Delft, The Netherlands: Noordwijk Risk Initiative Foundation, 2009. Enter "NRI MORT User's Manual" into a search engine for a free download.

OSHA's Rule for Process Safety Management of Highly Hazardous Chemicals, 29 CFR 1910.119. Washington, DC: Department of Labor, Occupational Safety and Health Administration, 1992.

Stephans, R. A. and W. W. Talso. *System Safety Analysis Handbook*, 2nd ed. Albuquerque, NM: New Mexico Chapter of the System Safety Society, 1997.

ADDITIONAL READING

Grose, Vernon L. *Managing Risk, Systematic Loss Prevention for Executives*. Originally published by Prentice-Hall, 1987. Now available through Omega Systems Group, Inc., Arlington, VA.

Manuele, Fred A. *On the Practice of Safety*, 4th ed. Hoboken, NJ: Wiley, 2014.

Roland, Harold E. and Brian Moriarty. *System Safety Engineering and Management*, 2nd ed. New York: Wiley, 1990.

Stevens, Richard A. *System Safety for the 21st Century*. Hoboken, NJ: Wiley, 2004. (This is an updated version of *System Safety 2000* by Joe Stephenson.)

ADDENDUM A–1

Preliminary Hazard Analysis and Risk Assessment
Codes are in Addendum A–2

Prepared by _____ Date _____ Location _____ Project No. _____ Page No. ____ of Pages ____

Description of task/operation/process _____

_____

_____

| Hazards Identified | E | Risk Before | | Risk Code | Risk Reduction Measures | Risk After | | Risk Code |
		S	P			S	P	

Advanced Safety Management: Focusing on Z10 and Serious Injury Prevention,
Second Edition. Fred A. Manuele.
© 2014 John Wiley & Sons, Inc. Published 2014 by John Wiley & Sons, Inc.

ADDENDUM A–2

Risk Assessment Matrix

Severity Levels and Values	Occurrence Probabilities and Values				
	Unlikely (1)	Seldom (2)	Occasional (3)	Likely (4)	Frequent (5)
Catastrophic (5)	5	10	15	20	25
Critical (4)	4	8	12	16	20
Marginal (3)	3	6	9	12	15
Negligible (2)	2	4	6	8	10
Insignificant (1)	1	2	3	4	5

Numbers were intuitively derived. They are qualitative, not quantitative and have meaning only in relation to each other.

Exposure Codes: P – personnel; E – environment; D – damage – facility, equipment, business interruption; M – material, product

Severity Descriptions

C – Catastrophic: One or more fatalities, total system loss and major business down time, environmental release with lasting impact on others with respect to health, property damage or business interruption.

Cr – Critical: Disabling injury or illness, major property damage and business down time, environmental release with temporary impact on others with respect to health, property damage or business interruption.

M – Marginal: Medical treatment or restricted work, minor subsystem loss or property damage, environmental release triggering external reporting requirements.

N – Negligible: First aid or minor medical treatment only, non-serious equipment or facility damage, environmental release requiring routine cleanup without reporting.

I – Insignificant: Inconsequential with respect to injuries or illnesses, system loss or down time, or environmental release.

Probability Descriptions

U – Unlikely: Could occur but hardly ever.
O – Occasional: Could occur intermittently.
L – Likely: Probably will occur several times.
F – Frequent: Likely to occur repeatedly.

Risk Levels

Combining the Severity and Occurrence Probability values yields a risk score in the matrix. The risks and the action levels are categorized below.

Risk Categories, Scoring, and Action Levels

Category	Risk Score	Action Level
L – Low risk	1 to 5	Remedial action discretionary.
M – Moderate risk	6 to 9	Remedial action to be taken at appropriate time.
S – Serious risk	10 to 14	Remedial action to be given high priority.
H – High risk	15 or greater	Immediate action necessary. Operation not permissible except in an unusual circumstance and as a closely monitored and limited exception only with approval of the person having authority to accept the risk.

ADDENDUM B

FMEA FORM
SYSTEM: DATE:
SUBSYSTEM: SHEET:
REFERENCE DRAWING: PREPARED BY:

FAILURE MODE AND EFFECTS ANALYSIS (FMEA)
FAULT CODE#

Subsystem/Module &Function	Potential Failure Mode	Potential Local Effect(s) of Failure	Potential End Effect(s) of Failure	SEV	Cr	Potential Cause(s) of Failure	OCC	Current Controls/Fault Detection	DET	RPN	Recommended Action(s)	Area/Individual Responsible & Completion Date(s)	Action Taken	SEV	OCC	DET	RPN
Subsystem name and function	"How can this subsystem fail to perform its function?" "What will an operator see?"	The local effect of the subsystem or end user	Downtime # of hours at the system level. 2. Safety 3. Environmental 4. Scrap loss	7	1	"How can this failure occur?" Describe in terms of something that can be corrected or controlled. 2. Refer to specific errors or malfunctions	5	"What mechanisms are in place that could detect, prevent, or minimize the impact of this cause?"	6	210	"How can we change the design to eliminate the problem?" "How can we detect (fault isolate) this cause for this failure?" "How should we test to ensure the failure has been eliminated?" "What PM procedures should we recommend?" 1. Begin with highest RPN. 2. Could say "no action" or "Further study is required." 3. An idea written here does not imply corrective action.	"Who is going to take responsibility?" "When will it be done?"	"What was done to correct the problem?" Examples: Engineering Change, Software revision, no recommended action at this time due to obsolescence, etc.	6	1	1	6

Critical Failure symbol (Cr). Used to identify critical failures that must be addressed (i.e., whenever safety is an issue).

Severity Ranking (1–10) (see Severity Table)

Occurrence Ranking (1–10) (see Occurrence Table)

Detection Ranking (1–10) (see Detection Table)

Risk Priority Number RPN = Severity*Occurrence *Detection

How did the "Action Taken" change the RPN?

ADDENDUM C

This presentation is an adaption from *Job Hazard Analysis*, U.S. Department of Labor, Occupational Safety and Health Administration, OSHA 3071, 2002 (revised). Available at http://www.osha.gov/Publications/osha3071.html.

What is a Hazard?

A hazard is the potential for harm. In practical terms, a hazard often is associated with a condition or activity that, if left uncontrolled, can result in an injury or illness. Identifying hazards and eliminating or controlling them as early as possible will help prevent injuries and illnesses.

What is a job hazard analysis?

A job hazard analysis is a technique that focuses on job tasks as a way to identify hazards before they occur. It focuses on the relationship between the worker, the task, the tools, and the work environment. Ideally, after you identify uncontrolled hazards, you will take steps to eliminate or reduce them to an acceptable risk level.

Why is Job Hazard Analysis Important?

Many workers are injured and killed at the workplace every day in the United States. Safety and health can add value to your business, your job, and your life. You can help prevent workplace injuries and illnesses by looking at your workplace operations, establishing proper job procedures, and ensuring that all employees are trained properly.

One of the best ways to determine and establish proper work procedures is to conduct a job hazard analysis. A job hazard analysis is one component of the larger commitment of a safety and health management system.

Advanced Safety Management: Focusing on Z10 and Serious Injury Prevention,
Second Edition. Fred A. Manuele.
© 2014 John Wiley & Sons, Inc. Published 2014 by John Wiley & Sons, Inc.

What is the Value of a Job Hazard Analysis?

Supervisors can use the findings of a job hazard analysis to eliminate and prevent hazards in their workplaces. This is likely to result in fewer worker injuries and illnesses; safer, more effective work methods; reduced workers' compensation costs; and increased worker productivity. The analysis also can be a valuable tool for training new employees in the steps required to perform their jobs safely.

For a job hazard analysis to be effective, management must demonstrate its commitment to safety and health and follow through to correct any uncontrolled hazards identified. Otherwise, management will lose credibility and employees may hesitate to go to management when dangerous conditions threaten them.

What Jobs are Appropriate for a Job Hazard Analysis?

A job hazard analysis can be conducted on many jobs in your workplace. Priority should go to the following types of jobs:

* Jobs with the highest injury or illness rates;
* Jobs with the potential to cause severe or disabling injuries or illness, even if there is no history of previous accidents;
* Jobs in which one simple human error could lead to a severe accident or injury;
* Jobs that are new to your operation or have undergone changes in processes and procedures; and
* Jobs complex enough to require written instructions.

Where do I Begin?

Involve your employees. It is very important to involve your employees in the hazard analysis process. They have a unique understanding of the job, and this knowledge is invaluable for finding hazards. Involving employees will help minimize oversights, ensure a quality analysis, and get workers to "buy in" to the solutions because they will share ownership in their safety and health program.

Review your accident history. Review with your employees your worksite's history of accidents and occupational illnesses that needed treatment, losses that required repair or replacement, and any "near misses" — events in which an accident or loss did not occur, but could have. These events are indicators that the existing hazard controls (if any) may not be adequate and deserve more scrutiny.

Conduct a preliminary job review. Discuss with your employees the hazards they know exist in their current work and surroundings. Brainstorm with them for ideas to eliminate or control those hazards.

If any hazards exist that pose an immediate danger to an employee's life or health, take immediate action to protect the worker. Any problems that can be corrected easily should be corrected as soon as possible. Do not wait to complete your job hazard analysis. This will demonstrate your commitment to safety and health and enable you to focus on the hazards and jobs that need more study because of their complexity. For those hazards determined to present unacceptable risks, evaluate types of hazard controls.

List, rank, and set priorities for hazardous jobs. List jobs with hazards that present unacceptable risks, based on those most likely to occur and with the most severe consequences. These jobs should be your first priority for analysis.

Outline the steps or tasks. Nearly every job can be broken down into job tasks or steps. When beginning a job hazard analysis, watch the employee perform the job and list each step as the worker takes it. Be sure to record enough information to describe each job action without getting overly detailed. Avoid making the breakdown of steps so detailed that it becomes unnecessarily long or so broad that it does not include basic steps. You may find it valuable to get input from other workers who have performed the same job.

Later, review the job steps with the employee to make sure that you have not omitted something. Point out that you are evaluating the job itself, not the employee's job performance. Include the employee in all phases of the analysis—from reviewing the job steps and procedures to discussing uncontrolled hazards and recommended solutions.

Sometimes, in conducting a job hazard analysis, it may be helpful to photograph or videotape the worker performing the job. These visual records can be handy references when doing a more detailed analysis of the work.

How do I Identify Workplace Hazards?

A job hazard analysis is an exercise in detective work. Your goal is to discover the following:

- What can go wrong?
- What are the consequences?
- How could it arise?
- What are other contributing factors?
- How likely is it that the hazard will occur?

To make your job hazard analysis useful, document the answers to these questions in a consistent manner. Describing a hazard in this way helps to ensure that your efforts to eliminate the hazard and implement hazard controls help target the most important contributors to the hazard. Good hazard scenarios describe:

- Where it is happening (environment),
- Who or what it is happening to (exposure),
- What precipitates the hazard (trigger),
- The outcome that would occur should it happen (consequence), and
- Any other contributing factors.

Rarely is a hazard a simple case of one singular cause resulting in one singular effect. More frequently, many contributing factors tend to line up in a certain way to create the hazard. Here is an example of a hazard scenario:

In the metal shop (environment), while clearing a snag (trigger), a worker's hand (exposure) comes into contact with a rotating pulley. It pulls his hand into the machine and severs his fingers (consequences) quickly.

To perform a job hazard analysis, you would ask:

- What can go wrong? The worker's hand could come into contact with a rotating object that "catches" it and pulls it into the machine.
- What are the consequences? The worker could receive a severe injury and lose fingers and hands.
- How could it happen? The accident could happen as a result of the worker trying to clear a snag during operations or as part of a maintenance activity while the pulley is operating. Obviously, this hazard scenario could not occur if the pulley is not rotating.
- What are other contributing factors? This hazard occurs very quickly. It does not give the worker much opportunity to recover or prevent it once his hand comes into contact with the pulley. This is an important factor, because it helps you determine the severity and likelihood of an accident when selecting appropriate hazard controls.

    Unfortunately, experience has shown that training is not very effective in hazard control when triggering events happen quickly because humans can react only so quickly.
- How likely is it that the hazard will occur? This determination requires some judgment. If there have been "near-misses" or actual cases, then the likelihood of a recurrence would be considered high. If the pulley is exposed and easily accessible, that also is a consideration.

In the example, the likelihood that the hazard will occur is high because there is no guard preventing contact, and the operation is performed while the machine is running. By following the steps in this example, you can organize your hazard analysis activities.

HOW DO I CORRECT OR PREVENT HAZARDS?

After reviewing your list of hazards with the employee, consider what control methods will eliminate or reduce them. The most effective controls are engineering controls that physically change a machine or work environment to prevent employee exposure to the hazard. The more reliable or less likely a hazard control can be circumvented, the better. If this is not feasible, administrative controls may be appropriate. This may involve changing how employees do their jobs.

Discuss your recommendations with all employees who perform the job and consider their responses carefully. If you plan to introduce new or modified job procedures, be sure they understand what they are required to do and the reasons for the changes.

ADDENDUM D

EXAMPLES OF HAZARDS, HAZARDOUS SITUATIONS, AND HAZARDOUS EVENTS

This checklist is an adaptation from a standard issued by the International Organization for Standardization titled *Safety of machinery. General Principles for Design. Risk assessment and risk reduction*, EN ISO 12100-2012. The checklist is a guide for companies located throughout the world that design and manufacture machinery and equipment that would go into workplaces. Although the checklist pertains to a broad range of equipment, those who use it as a reference must understand that it could not possibly include all hazards and all hazardous situations.

Mechanical Hazards

Due to machine parts or work pieces: e.g.,

- Shape
- Relative motion
- Mass and stability (potential energy of elements which may move under the effect of gravity)
- Mass and velocity (kinetic energy of elements in controlled and uncontrolled motion)
- Inadequacy of mechanical strength

Advanced Safety Management: Focusing on Z10 and Serious Injury Prevention,
Second Edition. Fred A. Manuele.
© 2014 John Wiley & Sons, Inc. Published 2014 by John Wiley & Sons, Inc.

Due to accumulation of energy inside the machinery: e.g.,

- Elastic elements (springs)
- Liquids and gases under pressure
- The effect of vacuum

Mechanical Hazards Due to the Potential for

- Crushing
- Shearing
- Cutting or severing
- Entanglement
- Drawing-in or trapping
- Impact
- Stabbing or puncture
- Friction or abrasion
- High-pressure fluid injection or ejection

Electrical Hazards Due to

- Contact of persons with live parts (direct contact)
- Contact of persons with parts which have become live under faulty conditions (indirect contact)
- Approach to live parts under high voltage
- Electrostatic phenomena
- Thermal radiation or other phenomena, such as the projection of molten particles and chemical effects from short circuits, overloads, etc.

Thermal Hazards, Resulting in

- Burns, scalds, and other injuries by a possible contact of persons with objects or materials with an extreme high or low temperature, by flames or explosions, and also by the radiation of heat sources
- Damage to health by hot or cold working environment

Hazards Generated by Noise, Resulting in

- Hearing loss (deafness), other physiological disorders (e.g., loss of balance, loss of awareness)
- Interference with speech communication, acoustic signals, etc.

Hazards Generated by Vibration

- Use of handheld machines resulting in a variety of neurological and vascular disorders
- Whole body vibration, particularly when combined with poor postures

Hazards Generated By Radiation

- Low-frequency, radio-frequency radiation; microwaves
- Infrared, visible, and ultraviolet light
- X- and gamma rays
- Alpha and beta rays, electron or ion beams, neutrons
- Lasers

Hazards Generated by Materials and Substances (and their Constituent Elements) Processed or Used by the Machinery

- Hazards from contact with or inhalation of harmful fluids, gases, mists, fumes, and dusts
- Fire or explosion hazards
- Biological or microbiological (viral or bacterial) hazards

Hazards Generated By Neglecting Ergonomic Principles In Machinery Design; A E.G.,

- Unhealthy postures or excessive effort
- Hazardous situations due to lifting
- Inadequate consideration of hand–arm or foot–leg anatomy
- Neglected use of personal protection equipment
- Inadequate local lighting
- Mental overload and underload, stress
- Human error, human behavior
- Inadequate design, location, or identification of manual controls
- Inadequate design or location of visual display units

Hazards Deriving from Unexpected Startup, Unexpected Overrun/Overspeed (or any Similar Malfunction) from

- Failure/disorder of the control system
- Restoration of energy supply after an interruption
- External influences on electrical equipment
- Other external influences (gravity, wind, etc.)
- Errors in the software
- Errors made by the operator (due to mismatch of machinery with human characteristics and abilities)
- Impossibility of stopping a machine under the best possible conditions
- Variations in the rotational speed of tools
- Failure of the power supply

- Failure of the control circuit
- Errors of fitting
- Breakup during operation
- Falling or ejected objects or fluids
- Loss of stability/overturning machinery
- Slip, trip, and fall of persons, related to machinery

Hazards, Hazardous Situation and Hazardous Events Due to Mobility:

Relating to the traveling function

- Movement when starting an engine
- Movement without a driver in the driving position
- Movement without all parts in a safe position
- Excessive speed of pedestrian-controlled machinery
- Excessive oscillation when moving

Linked to the work position (including driving station) on a machine

- Fall of persons during access (or at/from) the work position
- Exhaust gases/lack of oxygen at the work position
- Fire (flammability of the cab, lack of means extinguish to
- Mechanical hazards at the work position:
  1. Contact with the wheels
  2. Rollover
  3. Fall of objects, penetration by objects
  4. Breakup of parts
  5. Contact of persons with machine parts or tools (pedestrian-controlled machines)
- Insufficient visibility from the work position
- Inadequate lighting
- Inadequate seating
- Noise at the work position
- Vibration at the work position
- Insufficient means for evacuation/emergency

Due to the power source and to the transmission of power

- Hazards from an engine and batteries
- Hazards from transmission of power between machines
- Hazards from coupling and towing

From/to third persons

- Unauthorized startup/use
- Drift of a part away from its stopping position
- Lack of or inadequacy of means of visual or acoustic warning of

Hazards, Hazardous Situations, and Hazardous Events Due to Lifting

Mechanical hazards and hazardous events

- From load falls, collisions, machine tipping caused by:
  1. Lack of stability
  2. Uncontrolled loading, overloading, overturning moments exceeded
  3. Uncontrolled amplitude of movements
  4. Unexpected/unintended movement of loads
  5. Inadequate holding devices/accessories
  6. Collision of more than one machine
- From access of persons to load support
- From insufficient mechanical strength of parts
- From inadequate design of pulleys, drums

CHAPTER 12

PROVISIONS FOR RISK ASSESSMENTS IN STANDARDS AND GUIDELINES: SECTIONS 4.2 AND 5.1.1 OF Z10

As stated in Chapter 11, trends indicate that having the ability to make risk assessments will be expected of safety professionals. That premise has acquired considerable weight because of the more frequent inclusion of provisions in safety standards and guidelines requiring or recommending that risk assessments be made. This trend will have an impact on the knowledge and skills that safety professionals are expected to have. It will also provide career opportunities for them.

Addendum A in this chapter is a partial list of standards, guidelines, and initiatives that require or promote making risk assessments. To avoid having the list become overly lengthy, 2005 was selected as the year to begin recordings in the list. That is the year that the first version of Z10 was adopted as a national standard. Although there are 35 items in the list, it is probably not complete.

To provide guidance for safety professionals on trends throughout the world regarding requirements for risk assessments, we comment in this chapter on selected entries in the list to demonstrate:

- The variations in content for risk assessments in the standards and guidelines
- Specificity or lack thereof in their content
- The pace and importance of recent activity

There are similarities and differences in the approaches taken by the drafters of these standards and guidelines. Some are industry specific whereas others apply across

Advanced Safety Management: Focusing on Z10 and Serious Injury Prevention,
Second Edition. Fred A. Manuele.
© 2014 John Wiley & Sons, Inc. Published 2014 by John Wiley & Sons, Inc.

all industries. The message they give is clear: Safety professionals will be expected to have knowledge of a variety of hazard analysis and risk assessment methods and how to apply them.

EN ISO 12100–2010: *SAFETY OF MACHINERY — GENERAL PRINCIPLES FOR DESIGN. RISK ASSESSMENT AND RISK REDUCTION.*

This standard, issued in 2010 by the International Organization for Standardization (ISO), has had an interesting history. The standard combines and replaces three previously issued ISO standards. Note that "Risk assessment and risk reduction" are included in the title. That's significant, as it displays the significance that risk assessment has attained in designing for the safety of machinery. The impact of this standard worldwide has been substantial.

ISO 12100–1, Safety of machinery—Basic Concepts, General Principles for Design—Part 1, presented general design guidelines and required that risk assessments be made of machinery going into a workplace. ISO 12100–2, Safety of Machinery—Basic concepts, general principles for design—Part 2: Technical Principles, gave extensive details on design specifications for the "safety of machinery." ISO 14121, Safety of machinery—Principles of risk assessment, set forth the risk assessment concepts to be applied. EN ISO 12100–2010 combines these three standards and retains their content.

EN ISO 12100–2010 is truly an international standard and has had considerable influence worldwide. Its existence implies that a huge majority of countries agree on the principle that hazards should be identified and analyzed and their accompanying risks should be assessed in the design processes for machinery.

The "EN" that precedes "ISO" in the title indicates that the origins of the standard were in the European Community (later, the European Union). Several standards that were applicable in the European Union and had titles beginning with the EN designation became ISO standards. Some of the relative EN standards were written in the 1990s.

The European Union standards have had considerable influence on manufacturers throughout the world. An example follows. Suppliers of products that are to go into a country that is a member of the European Union are required to place a "CE" mark on products to indicate that all operable European Union directives have been met. Risk assessment provisions in EN ISO 12100–2010 are among those requirements.

Additional European Influence

Other developments originating in Europe have also had a noteworthy impact throughout the world. Comments on two of them follow.

BS OHSAS 18001:2007 is the designation for a guideline titled *Occupational health and safety management systems—requirements*, a British Standards Institution publication. In some contract situations, particularly in Asian countries, the bidder is required to establish that its safety management system has been "certified." Among other things, the British Standards Institution has attained prominence as a certifying entity, and 18001 is the base upon which certification is granted or withheld. In a 2007 revision, requirements for risk assessments became more explicit.

The guidelines now say in Section 4.3.1: "The organization shall establish, implement and maintain a procedure(s) for the ongoing hazard identification, risk assessment, and determination of necessary controls." As an indication of how broadly this guideline is known and used, Singapore adopted it fully as law in 2009.

In August 2008, the European Union (EU) launched a two-year health and safety campaign focusing on risk assessment. Their bulletin states:

Risk assessment is the cornerstone of the European approach to prevent occupational accidents and ill health. If the risk assessment process—the start of the health and safety management approach—is not done well or not at all, the appropriate preventive measures are unlikely to be identified or put in place.

That statement of the EU is seminal. It states boldly: "Risk assessment is the start of the health and safety management approach; if it is not done well or not at all, safety efforts will be misdirected." It is noteworthy that several European countries have continued with innovation on risk assessment since 2010, the year in which the campaign ended.

B11.TR3-2000: *RISK ASSESSMENT AND REDUCTION—A GUIDELINE TO ESTIMATE, EVALUATE AND REDUCE RISKS ASSOCIATED WITH MACHINE TOOLS*

"TR" stands for "Technical Report." "TR3" is the acronym for a report issued by the B11. TR3 Subcommittee formed by the Machine Tool Safety Standards Committee (B11) of the American National Standards Institute. The secretariat for this work is the Association for Manufacturing Technology.

TR3 became a registered document at ANSI in November 2000, the year immediately following the approval of the robotic standard. Its historic value and its influence are recognized here. Principles set by the TR3 subcommittee follow.

- A simple, practical, and generic hazard analysis and risk assessment process is to be developed that has the potential to be incorporated in all B11 standards.
- The process must apply to both suppliers and users.
- To the extent possible, the technical report is to be harmonized with European standards.

The document produced matches the principles. TR3 is a guideline, not a standard. However, the content of the guideline has been incorporated into almost all of the B11 ANSI standards, of which there are 24.

Over 90% of the guideline is generic. Thus, it is a basic document on hazard analysis and risk assessment that provides guidance on reducing risks according to a prioritized procedure and on the selection of appropriate design and protective measures. When the process is complete, an acceptable risk level is to be achieved.

As an indication of its influence, almost all of its provisions are now included in the 2010 version of B11.0, the American National Standard entitled *Safety of Machinery—General Requirements and Risk Assessment*.

A Standard of Major Consequence

Because of the breadth of its coverage, ANSI B11.0-2010, *Safety of Machinery— General Safety Requirements and Risk Assessment*, has major consequences. This is its stated purpose: "This standard describes procedures for identifying hazards, assessing risks, and reducing risks to an acceptable level over the life cycle of machinery."

Note that its Scope, as follows, has only one exclusion—portable hand tools: "This standard applies to new, modified or rebuilt power driven machines, not portable by hand, used to shape and/or form metal or other materials by cutting, impact, pressure, electrical or other processing techniques, or a combination of these processes."

The standard includes an explicit requirement that machinery suppliers, reconstructors, modifiers, and users achieve acceptable risk levels. ANSI B11.0 is the most comprehensive standard outlining the risk assessment process currently available for all of the operational categories just mentioned.

The ANSI B11.0 standard is built on standards that preceded it, including EN 1050, ANSI B11 TR3, ANSI/RIA R15.06 (robotics), ANSI B155.1 (packaging machinery), ISO 12100 (safety of machinery), SEMI S10 (semiconductors), and several others. As with most standard development efforts, the writing committee started, with the work completed by others, and then advanced the work to improve the content as part of continuous improvement.

The Foreword from B11.0 includes the following: "The concepts and principles contained in this standard can be applied very broadly to a wide variety of systems and applications." Documented risk assessments were first introduced to:

- The machine tool industry in 2000 with the publication of ANSI B11.TR3, *Risk Assessment and Risk Reduction—A Guide to Estimate, Evaluate and Reduce Risks Associated with Machine Tools*
- The robot industry in 1999 with the publication of ANSI/RIA R15.06, *Requirements for Industrial Robots and Robot Systems*
- The packaging and related machinery industry in 2006 with the publication of ANSI/PMMI B155.1, *Safety Requirements for Packaging Machinery and Packaging-Related Converting Machinery*

The Foreword also states: "Prevention through Design or PtD is a recent term in the industry; the objectives of risk assessment, risk reduction and elimination of hazards as early as possible are integral to and not new to this standard." This objective is also taken from the Foreword:

The objective of the B11 standards is to eliminate injuries to personnel from machinery or machinery systems by establishing requirements for the design, construction, reconstruction, modification, installation, set–up, operation and maintenance of machinery or machine systems. This standard should be used by suppliers and users, as well as by the appropriate authority having jurisdiction. Responsibilities have been assigned to the supplier (i.e., manufacturer, the reconstructor, and the modifier), the user, and the user personnel to implement this standard. This standard is not intended to replace good judgment and personal responsibility. Personnel skill, attitude, training and experience are safety factors that must be considered by the user.

Machines typically have long life spans and ANSI B11.0 addresses the life cycle of machinery. Once a supplier sells a machine, it usually has limited control over it. The user owns the machine and determines where it is to be installed, how it is used, and if, how, and who maintains it. Additionally, machinery is often modified—sometimes for the better or less so.

Comments made here on ANSI B11.0 were written, largely, by Bruce Main, president of design safety engineering. He was the chair of the committee that wrote the standard.

ANSI/AIHA Z10-2012: THE STANDARD FOR *OCCUPATIONAL HEALTH AND SAFETY MANAGEMENT SYSTEMS*

The first version of Z10, approved in 2005, required that processes be in place "to identify and take appropriate steps to prevent or otherwise control hazards and reduce risks associated with new processes or operations at the design stage." Its Annex E was captioned "Assessment and Prioritization". That annex commented on assessing the level of risk and included a hazard analysis and risk assessment guide. But a specific requirement that risk assessments be made was not included in the standard.

Thinking changed. The 2012 version of Z10 has a "shall" provision on risk assessment in Section 5.1.1. It says: "The organization shall establish and implement a risk assessment process(es) appropriate to the nature of hazards and level of risk." Annex F, "Risk Assessment." provides extended guidance.

ANSI/PMMI B155.1-2011: *SAFETY REQUIREMENTS FOR PACKAGING MACHINERY AND PACKAGING-RELATED CONVERTING MACHINERY*

The history of B155.1 dates back to 1972. A revision of the standard was approved by ANSI in 2006, one of several revisions in the intervening 34 years. The secretariat for the standard is the Packaging Machinery Manufacturers Institute.

The major revisions in the 2006 version are indicative of the progress in the 1990s toward acceptance of the premise that hazard analysis and risk assessment provisions should be included in ANSI safety standards.

In the following list of the major steps in the risk assessment process, all were included in the 2006 version except Section 6.8, which was added in the 2011 version. These are the key elements in Section 6:

Section 6.0: The Risk Assessment Process
     6.1 General
     6.2 Prepare for/set limits of the assessment
     6.3 Identify hazards
     6.4 Assess initial risk
     6.5 Reduce risk
     6.6 Assess residual risk
     6.7 Achieve acceptable risk
     6.8 Validate risk reduction measures
     6.9 Document the results

A review made of Section 6, which covers nearly eight pages in small print, indicates that the guidelines are largely generic. Note that in the risk assessment process, risk reduction measures are to be taken, if necessary, after the initial risk assessment and that the resulting residual risk is also to be assessed. The goal of all this is to attain acceptable risk levels through a continual application of the process. That's a sound methodology.

In Section 6.4.1, this requirement is stated: "Risks shall be assessed using a risk scoring system." The example risk scoring system shown in the standard's Table 1 is taken from TR3.

Annex D of the standard is a risk assessment matrix. Several scoring systems are shown, one of which, taken from MIL-STD-882, is given in Table 12.1.

TABLE 12.1   Risk Assessment Matrix

Occurrence Probability	Severity of Harm			
	Catastrophic	Serious	Moderate	Minor
Very likely	High	High	High	Medium
Likely	High	High	Medium	Medium
Unlikely	Medium	Medium	Low	Negligible
Remote	Low	Low	Negligible	Negligible

A MAJOR CONCEPT CHANGE

The following sentence appears in the Foreword for both the 2006 and 2011 versions of B155.1 (and in ANSI B11.0-2010):

This version of the standard has been harmonized with the international (ISO) and European (EN) standards by the introduction of hazard identification and risk assessment as the principal method for analyzing hazards to personnel and achieving a level of acceptable risk.

That language presents an interesting and weighty concept. It is a good match with the statement in an EU publication indicating that risk assessment is the cornerstone of the European approach to prevent occupational accidents and ill health.

If all safety professionals accept that hazard identification and risk assessment are the first steps in preventing injuries to personnel, a major concept change in the practice of safety will have been achieved. Adopting that premise takes the focus away from what have been called the unsafe acts of workers and redirects it to work system causal factors. This is sound thinking.

CERTAIN GOVERNMENTAL VIEWS

In a July 19, 2010 letter to the OSHA staff, the assistant secretary David Michaels wrote on several subjects, one of which follows: "Ensuring that American workplaces are safe will require a paradigm shift, with employers going beyond simply attempting to meet OSHA standards, to implementing risk-based workplace injury and illness prevention programs."

If elements in injury and illness prevention programs are to be risk-based, activity will be necessary to identify and assess the risks. That starts with hazard identification and analysis and, then, taking the next step to establish the risk level.

OSHA has not shown that it is adopting the concept of risk-based decision making. This statement by Michaels is noteworthy because it demonstrates that the head of a major governmental entity involved in occupational safety and health has recognized that injury and illness prevention programs should be risk-based. As will be seen, heads of other governmental agencies have reached similar conclusions.

In the December 8, 2010 *Federal Register*, the Federal Railroad Administration issued an advance notice of proposed rulemaking for certain railroads to have a risk reduction program. The *Federal Registry* entry stated: "It is proposed that the Risk Reduction Program be supported by a risk analysis and a Risk Reduction Plan." Enter "Federal Railroad Administration Risk Reduction Program" into a search engine and the following appears.

Risk Reduction Program

The primary mission of the Risk Reduction Program Division is ensuring the safety of the nation's railroads by evaluating safety risks and managing those risks in order to reduce the numbers and rates of accidents, incidents, injuries and fatalities.

Our mission is accomplished by:

- Identifying, collecting and analyzing precursor accident data to identify risks
- Developing voluntary pilot programs in cooperation with stakeholders that are designed to mitigate identified and potential risks

- Propagating and institutionalizing best practices and lessons learned to the entire rail industry
- Providing analytical support, data, and recommendations needed by stakeholders to develop strategies, plans and processes to improve safety and promote positive organizational change
- Developing and enforcing regulations promulgated in response to the Rail Safety Improvement Act of 2008

On March 11, 2011, the Pipeline and Hazardous Materials Safety Administration announced that hazardous materials regulations are to be modified to require that risk assessments be made of loading and unloading operations.

If http://primis.phmsa.dot.gov/meetings/MtgHome.mtg?mtg=70 is entered into an address bar, a report will be found on a meeting held in July 2011 to provide an opportunity for stakeholders to comment. Data-gathering activity continues. The report says, among other things:

This event is to provide an open forum for exchanging information on identifying threats, improving risk assessments and record keeping for onshore pipelines. Specifically it will:

- Provide a U.S. and International Regulatory perspective on pipeline integrity risk assessments.
- Provide an operator overview of the challenging factors with conducting risk assessments, canvassing effective approaches, and case studies.
- Identify options with addressing interactive threats, legacy pipelines and approaches for dealing with recordkeeping gaps.

On October 15, 2010, the Bureau of Ocean Energy Management Regulation and Enforcement (BOEMRE) published the Final Rule for 30 CFR Part 250 Subpart S, Safety and Environmental Management Systems, in the *Federal Register* (75 FR 63610). The Final Rule incorporates by reference, and makes mandatory, the American Petroleum Institute's Recommended Practice for Development of a Safety and Environmental Management Program for Offshore Operations and Facilities (API RP 75), 3rd edition, May 2004, reaffirmed May 2008. This recommended practice, including its appendices, constitutes a complete Safety and Environmental Management System (SEMS).

BOEMRE mandated that by November 15, 2011, all operators and lessees working in the Gulf of Mexico had to submit a comprehensive SEMS plan to the regulator that was required to address the following 13 elements of API RP 75:

1. General Management Program Principles
2. Safety & Environmental Information
3. Hazards Analysis
4. Management of Change
5. Operating Procedures

6. Safe Work Practice
7. Training
8. Quality Assurance/Mechanical Integrity
9. Pre-Startup Review
10. Emergency Response & Control
11. Incident Investigation
12. SEMS Element Audit
13. Documentation and Recordkeeping

This development is of particular interest for two reasons. Operators and lessees affected are required by regulation to make hazards analyses (the first step in making a risk assessment). Also, the plan required is a combination that includes occupational safety, public safety, and environmental safety in one instrument. That combination is a development that needs continual observation. I polled safety directors to determine what proportion of the safety professionals at their locations have responsibilities for both occupational safety and environmental concerns. The range was from 50 to 90%.

Risk assessments have been made for many years in the branches of the military; the National Aeronautics and Space Administration; some chemical operations; the atomic energy field; pharmaceutical companies operating under the rules of the Food and Drug Administration; research activities pertaining to public health; traffic control studies; and other fields.

That additional federal governmental entities have become risk conscious and are requiring that risk assessments be made is an indication of the trend—the subject of this chapter.

ANSI-ASSE Z590.3: *PREVENTION THROUGH DESIGN—GUIDELINES FOR ADDRESSING OCCUPATIONAL HAZARDS AND RISKS IN DESIGN AND REDESIGN PROCESSES*

This standard was approved by the American National Standards Institute on September 1, 2011. Extensive comments are made on the standard in Chapter 16, "Prevention through Design". It is mentioned in this chapter because it is another indication of the trend to have provisions for risk assessments in standards and guidelines. The core of Z590.3 is risk assessment.

MIL-STD-882E-2012: THE U.S. *DEPARTMENT OF DEFENSE STANDARD PRACTICE FOR SYSTEM SAFETY*

As is said in its Foreword: "This Standard is approved for use by all Military Departments and Defense Agencies within the Department of Defense." Certain contractors engaged by those departments and agencies are required to meet the requirements of the standard. This version was approved May 11, 2012. It is available at http://www.system-safety.org/. Scroll down and click on MIL-STD-882E in the right-hand column for a free copy. I strongly recommend that safety professionals obtain a copy of this standard for informative purposes.

The base document for the *Standard Practice for System Safety, MIL-STD-882*, was issued in 1969. It was a seminal document at that time and has continued to be an important reference. Four revisions of this standard have been issued over a span of 43 years. MIL-STD-882 has had considerable influence on the development of hazard identification and analysis, risk assessment, risk elimination, and risk control concepts and methods throughout the world. Much of the wording on risk assessments and hierarchies of control in safety standards and guidelines issued throughout the world relate to that in the several versions of 882.

MIL-STD-882E extends the previous issue—882D—considerably. For example, the 882D version, including addenda, had 26 numbered pages; the 882E version has 98 numbered pages. It replaces some of what was in 882C that was not included in 882D.

In 882E, achieving and maintaining acceptable risk levels dominates; revisions are made in the system safety process that give additional emphasis to hazard analysis and risk assessment; the use of a risk assessment matrix is required; noteworthy revisions are made in the design order of preference; appropriate emphasis is given to managing High and Serious risk levels; and a major section is devoted to software and software assessments. Excerpts follow, some of which are modified to avoid governmental terminology.

Section 4 in 882E, "General Requirements," sets forth the "requirements for an acceptable system safety effort." Section 4.3 and its subsections, outline and comment on the eight elements in the system safety process, as follows.

Element 1: Document the System Safety Approach. Describe the risk management effort and how the program is integrated into the overall business process.

Element 2: Identify and Document the Hazards. Hazards are identified through a systematic analysis process that includes the system hardware and software, system interfaces (to include human interfaces) and the intended use or application and operational environment.

Element 3: Assess and Document Risk. For each identified hazard, across all system modes, the mishap severity and probability are established in accord with the definitions given. A mishap risk assessment matrix is used to assess and display the risks.

Element 4: Identify and Document Risk Mitigation Measures. Potential risk mitigation(s) shall be identified, and the expected risk reduction(s) of the alternative(s) shall be estimated and documented. The goal should always be to eliminate the hazard if practicable. When a hazard cannot be eliminated, the associated risk should always be reduced to the lowest practicable acceptable risk level within the constraints of cost, schedule, and performance by applying the following system safety design order of precedence in their order of effectiveness.

a. Eliminate hazards through design selection. Ideally, the hazard should be eliminated by selecting a design or material alternative that removes the hazard altogether. (This is comparable to the "Avoidance" element in Chapter 16, "Prevention through Design".)

b. Reduce risk through design alteration. If adopting an alternative design change or material to eliminate the hazard is not feasible, consider design changes that reduce the severity and/or the probability of the mishap potential caused by the hazard(s).

c. Incorporate engineered features or devices. If mitigation of the risk through design alteration is not feasible, reduce the severity or the probability of the mishap potential caused by the hazard(s) using engineered features or devices. In general, engineered features actively interrupt the mishap sequence and devices reduce the risk of a mishap.

d. Provide warning devices. If engineered features and devices are not feasible or do not adequately lower the severity or probability of the mishap potential caused by the hazard, include detection and warning systems to alert personnel to the presence of a hazardous condition or occurrence of a hazardous event.

e. Incorporate signage, procedures, training, and personal protective equipment (PPE). Where design alternatives, design changes, and engineered features and devices are not feasible and warning devices cannot adequately mitigate the severity or probability of the mishap potential caused by the hazard, incorporate signage, procedures, training, and PPE. Signage includes placards, labels, signs, and other visual graphics. Procedures and training should include appropriate warnings and cautions. Procedures may prescribe the use of PPE. For hazards assigned Catastrophic or Critical mishap severity categories, the use of signage, procedures, training, and PPE as the only risk reduction method should be avoided.

Element 5: Reduce Risk. Mitigation measures are selected and implemented to achieve an acceptable risk level. Consider and evaluate the cost, feasibility, and effectiveness of candidate mitigation methods as a part of the overall operation process.

Element 6: Verify, Validate and Document Risk Reduction. Verify the implementation and validate the effectiveness of all selected risk mitigation measures through appropriate analysis, testing, demonstration, or inspection. Document the verification and validation.

Element 7: Accept risk and Document. Before exposing people, equipment, or the environment to known system-related hazards, the risks shall be accepted by the appropriate authority in accord with established acceptance authority levels. Definitions (in Tables and Matrices in this standard) shall be used to define the risks at the time of the acceptance decision, unless tailored alternative definitions and/or a tailored matrix are formally approved. The user representative shall be a part of this process and shall provide formal concurrence before all Serious and High risk acceptance decisions are made.

TABLE 12.2   Risk Assessment Matrix

Occurrence Probability	Severity of Concequence			
	Catastrophic(1)	Critical (2)	Marginal (3)	Negligible (4)
Frequent (A)	High	High	Serious	Medium
Probable (B)	High	High	Serious	Medium
Occasional (C)	High	Serious	Medium	Low
Remote (D)	Serious	Medium	Medium	Low
Improbable (E)	Medium	Medium	Medium	Low
Eliminated (F)	This category is used only for identified hazards that are totally removed.			

Element 8: Manage Life-Cycle Risk. After the system is fielded, the system program office uses the system safety process to identify hazards, assess the risks and maintain acceptable risk levels throughout the system's life cycle.

An instruction in Element 7 says that "Definitions (in Tables and Matrices in this standard) shall be used to define the risks at the time of the acceptance decision, unless tailored alternative definitions and/or a tailored matrix are formally approved."

Table I presents severity categories, Table II contains probability levels, and Table III in MIL-STD-882, shown here as Table 12.2, is a risk assessment matrix that combines the severity and probability categories and includes numerical and alphabetical indicators.

Numerical and alphabetical indicators are the base for expressing assessed risks in a risk assessment code (RAC), which is a combination of one severity category and one probability level. For example, a RAC of 1A is the combination of a catastrophic severity category and a frequent probability level.

For emphasis: MIL-STD-882E is an excellent resource document. Its base is hazard identification and analysis and risk assessment.

THE CANADIANS

CSA Standard Z1000-2006, *Occupational health and safety management*, was issued in the year following the first edition of Z10 and has a close relationship with respect to the content and order in the American standard. Section 4.3.4 reads as follows: "The organization shall establish and maintain a process to identify and assess hazards and risks on an ongoing basis. The results of this process shall be used to set objectives and targets and to develop preventive and protective methods." (CSA is the designation for the Canadian Standards Association.) The excerpt above is all that is said in the standard about hazard analysis and risk assessment. The subject is dealt with further in Annex A, which is informative. But the intent of the hazard analysis and risk assessment provision is amplified in the "shall" provision in Section 4.4.7, "Management of Change."

The organization shall establish and maintain procedures to identify, assess, and eliminate or control occupational health and safety hazards and risks associated with

(a) new processes or operations at the design stage
(b) significant changes to its work procedures, equipment, or organizational structure et al.

In September 2012, CSA Z1002-12, *Occupational health and safety—Hazard identification and elimination and risk Assessment and control* was issued. This was a major undertaking. It supports the purpose of Section 4.3.4 in Z1000-2006. The content relates entirely to hazards and risks in the workplace. Its issuance is another indication of the trend throughout the world whereby organizations are encouraged to have processes in place to identify and analyze hazards, to assess their accompanying risks, and to achieve acceptable risk levels.

ADVANCES IN FIRE PROTECTION

There are four entries in Addendum A of this chapter pertaining to activities of the National Fire Protection Association (NFPA) and the Society of Fire Protection Engineers (SFPE). In 2007, NFPA issued "Guidance Document for Incorporating Risk Concepts into NFPA Codes and Standards." This is an impressive, thought-provoking risk assessment–related document that will have a long-term affect in the fire protection field. It is available at http://www.nfpa.org/assets/files/PDF/Research/Risk-Based_Codes_and_Stds.pdf.

As an example of how risk concepts are being incorporated into NFPA standards, the 2112 edition of NFPA 70E, the *Standard for Electrical Safety in the Workplace*, has a new section on risk assessment.

SFPE developed an interesting course entitled "Introduction to Fire Risk Assessment," which is available on the Internet. (No publication date is shown, but it probably was 2006.) A paraphrased and brief version of what is said about the course on the Internet follows:

This five hour equivalent course is presented free of charge by the Society of Fire Protection Engineers. Although the course was developed primarily for fire service and fire prevention officers, it may be of value to engineers and students who would like to understand fire risk assessment. The full course consists of 19 lecture sessions each of which can be viewed in about 15 minutes.

This course is largely generic and deserves a look. Additional information, including the titles of the lecture sessions and how to access them, can be found at: http://www.sfpe.org/SharpenYourExpertise/Education/SFPEOnlineLearning/FireRiskAssessment.aspx.

In 2006, SFPE also issued the *Engineering Guide to Fire Risk Assessment*. This is a highly technical book that would be of particular interest to engineers. Nevertheless, its issuance demonstrates leadership by SFPE with respect to risk assessment.

DEVELOPMENTS IN AVIATION GROUND SAFETY

One of the most interesting innovations regarding hazard analysis and risk assessment can be found in the *Safety Handbook: Aviation Ground Operation* developed by the International Air Transport Section of the National Safety Council. The Air Transport Section is truly international, having representation from all the populated continents.

A sixth edition was published in July 2007. The following text is taken from Chapter 2, "Risk Management."

> Risk management takes aviation safety to the next level. It is a six-step logic-based approach to making calculated decisions on human, material, and environmental factors before, during and after operations.
>
> Risk management enables senior leaders, functional managers, supervisors and others to maximize opportunities for success while minimizing risks. Failure to successfully implement a risk management process will have a financial, legal and social impact. (p. 9)

The air transport group has outlined a way of thinking about and dealing with hazards and risks, applying a logical and sequential methodology. They have developed a "process to detect, assess, and control risk."

The captions in their six-step logic-based commonsense approach are shown in the *Handbook*'s Table 1. (p. 11) The process is shown here as Table 12.3.

Discussions of each step in the text are extensive. Comments will be made here on the first two only. The remaining steps are addressed in Chapter 14, "Hierarchy of Controls". For the first step—identify the hazards—the hazard analysis and risk assessment methodologies listed in Table 12.4, as in the *Handbook*'s Table 2, are discussed. (p. 10)

For Step 2—Assess the Risks—the text says: "The assessment is the application of quantitative or qualitative measures to determine the level of risk associated with a specific hazard. This process defines the probability and severity of an undesirable event that could result from the hazard. The Risk Assessment Matrix is a very useful tool in categorizing the effects of probability and severity as they relate to risk levels." (p. 12)

TABLE 12.3 The Risk Management Process

1. Identify the hazard.
2. Assess the risk.
3. Analyze risk control measures.
4. Make control decisions.
5. Implement risk controls.
6. Supervise and review.

TABLE 12.4   Hazard Analysis and Risk Assessment Methodologies

1. Operations analysis: Purpose—To understand the flow of events.
2. Hazard analysis: Purpose—To get a quick survey of all phases of an operation. In low-hazard situations, the preliminary hazard analysis may be the final hazard identification tool.
3. "What-if" analysis: Purpose—To capture the input of personnel in a brainstorming-like environment.
4. Scenario process tool: Purpose—To use imagination and visualizations to capture unusual hazards.
5. Change analysis: Purpose—To detect the hazard implications of both planned and unplanned change.

TABLE 12.5   Risk Assessment Matrix: Alphabetical Risk-Level Indicators[a]

| | Probability That Something Will Go Wrong | | | | |
Severity Categories	Frequent: Likely to occur immediately or soon: often	Likely: Quite likely to occur in time	Occasional: May occur in time	Seldom: Not likely to occur, but possible	Unlikely: Unlikely to occur
Catastrophic: death, multiple injuries, severe property or environmental damage	E	E	H	H	M
Critical: serious injuries, significant property or environmental damage	E	H	H	M	L
Marginal: may cause minor injuries, financial loss, negative publicity	H	M	M	L	L
Negligible: minimum threat to persons or damage to property	M	L	L	L	L

[a] **E, extremely high risk; H, high risk; M, moderate risk; L, low risk.**

A Risk Assessment Matrix is provided. Its configuration is unusual and it does not duplicate well. The terminology used in the matrix for probability and severity and for the risk gradings are identical with those in Table 11.9, "A Primer on Hazard Analysis and Risk Assessment," with one exception. In the *Handbook*, "M" is the designation for medium risk rather than for moderate risk. The version in Table 11.9 is shown here as Table 12.5.

The National Safety Council's *Safety Handbook: Aviation Ground Operation* is a good, thought-provoking, not overly complex resource document. It is an example of what trade groups can do as a service to their members.

SEMI 02-0712A AND SEMI S10-307E

Safety-related guidelines issued by the semiconductor industry are another indication of recognition by a trade group of the value of using hazard analysis and risk assessment techniques to eliminate or control hazards and to attain acceptable risk levels.

Manufacturers of machinery and equipment used in making semiconductors, and their customers (Intel, IBM, et al.), realized that they had mutual interests that would be better served if that equipment was designed to meet agreed-upon safety guidelines. Their trade association, SEMI (Semiconductor Equipment and Materials International), which has global participation, has issued several safety-related guidelines. Two of them are of interest here: SEMI S2-0712a and SEMI S10-307E.

SEMI S2-0712a, *Environmental, Health, and Safety Guideline for Semiconductor Manufacturing Equipment*, issued in July 2006, updated a 2003 version. The *Guideline* sets forth provisions for manufacturers of equipment to be used in the semiconductor industry. Certain aspects of the *Guideline's* Safety Philosophy (Section 6) are pertinent to this chapter.

6.2 The assumption is made that operators, maintenance personnel, and service personnel are trained in the tasks that they are intended to perform.

6.4 This guideline should be applied during the design, construction, and evaluation of semiconductor equipment, in order to reduce the expense and disruptive effects of redesign and retrofit.

6.8 A hazard analysis should be performed to identify and evaluate hazards. The hazard analysis should be initiated early in the design phase, and updated as the design matures.

6.8.1 The hazard analysis should include consideration of:
- the application or process
- the hazards associated with each task
- anticipated failure modes
- the probability of occurrence and severity of harm
- the level of expertise of exposed personnel and the frequency of exposure
- the frequency and complexity of operating, servicing and maintenance tasks
- safety critical parts

If the equipment is designed in accord with the *Guideline* and the hazard analyses prescribed are conducted, the responsibilities of users (employers) to meet the design review provision in Z10 are accomplished more easily. The hazard analysis is really a risk assessment since both occurrence probability and severity of harm are to be identified.

This *Guideline* also gives employers assistance in meeting the procurement provisions in Z10 that require including safety specifications in purchasing documents. This is item 7.1 in the General Provisions: "This guideline should be incorporated by reference in equipment purchase specifications."

Section 6.8.2 states: "The risks associated with hazards should be characterized using SEMI S10-307E, the title of which is *Safety Guideline for Risk Assessment and Risk Evaluation Process*. This is the purpose of S10.

> The purpose of this guideline is to establish general principles for risk assessment to enable identification of hazards, risk estimation and risk evaluation in a consistent and practical manner. The document provides a framework for carrying out risk assessments on equipment in the semiconductor and similar industries and is intended for use by supplier and purchaser as a reference for EHS considerations.

The hazard identification and analysis processes shown in SEMI S10 duplicate those in SEMI S2. In the risk assessment process, severity of outcome and likelihood of occurrence are to be identified and categorized. In appendices, recommended categories for likelihood and severity are given as well as matrices showing risk categories. The exhibits are comparable to those shown in Chapter 11, "A Primer on Hazard Analysis and Risk Assessment".

ANSI/ASSE Z244.1-2009

In July 2003, approval was given by ANSI for the reissuance of a standard entitled *Control of Hazardous Energy—Lockout/Tagout and Alternative Methods*. It was reaffirmed without change in 2009. This standard will have a broad impact because it affects a huge number of locations. Alternative methods of control are discussed in Section 5.4, which can be paraphrased as follows:

> When lockout/tagout is not used for tasks that are routine, repetitive, and integral to the production process, or traditional lockout/tagout prohibits the completion of those tasks, an alternative method of control shall be used. Selection of an alternative control method by the user shall be based on a risk assessment of the machine, equipment, or process.

The foregoing is significant because a risk assessment is required prior to selecting an alternative risk control method.

THE CHEMICAL INDUSTRY: OSHA REQUIREMENTS

OSHA's Rule for Process Safety Management of Highly Hazardous Chemicals 29 CFR 1910.119, issued in 1992, applies to employers at about 50,000 locations, many of which are not considered chemical companies. With respect to requirements for hazard analyses being included in standards, this OSHA standard merits a review by safety practitioners. The standard requires that:

> The employer shall perform an initial hazard analysis (hazard evaluation) on processes covered by this standard. The process hazard analysis shall be appropriate to the complexity of the process and shall identify, evaluate, and control the hazards involved in the process. The employer shall use one or more of the following methodologies that are appropriate to determine and evaluate the hazards of the process being analyzed:

> - What-If;
> - Checklist;
> - What-If/Checklist;
> - Hazard and Operability Study (HAZOP);
> - Failure Mode and Effects Analysis (FMEA);
> - Fault Tree Analysis; or
> - An appropriate equivalent methodology.

Also, the hazard analysis shall address:

> - The hazards of the process;
> - The identification of any previous incident which had a likely potential for catastrophic consequences in the workplace;
> - Engineering and administrative controls applicable to the hazards and their interrelationships;
> - Consequences of failure of engineering and administrative controls;
> - Facility citing;
> - Human factors; and
> - A qualitative evaluation of a range of the possible safety and health effects of failure of controls on employees in the workplace.

Under the requirements for pre-startup safety review for new facilities and for significant modifications, the employer is required to provide a process hazard analysis—among other considerations. In no place in the standard is there mention of occurrence probability. This appears in the preamble to the standard:

> OSHA has modified the paragraph [editorial note: paragraph on consequence analysis] to indicate that it did not intend employers to conduct probabilistic risk assessments to satisfy the requirement to perform a consequence analysis.

However, all risks are not equal. And managements do consider incident probability in their decision making when determining the priority levels that individual projects are to have when allocating resources.

THE CHEMICAL INDUSTRY: EPA REQUIREMENTS

The U.S. Environmental Protection Agency (EPA) and OSHA have different legal authority with respect to accidental releases of harmful substances. The concerns at EPA center on off-site consequences: that is, harm to the public and the environment. At OSHA, the legal authority pertains to on-site consequences.

On August 19, 1996, EPA issued rule 40 CFR Part 68, *Risk Management Programs for Chemical Accidental Release Prevention*. Risk Management Plans required of location managements by the rule were due by June 21, 1999. Although the provisions of the rule are extensive, only the specifications for hazards analyses are addressed here.

Processes subject to this rule are divided into three groups, labeled by the EPA as programs 1, 2, and 3. Program levels relate to the quantities and extent of exposure to toxic and flammable chemicals. For locations qualifying for program levels 1 and 2, those with less exposure, EPA will accept hazard reviews done by qualified personnel using suitable checklists.

Hazard reviews must be documented and show that problems have been addressed. In its literature, EPA comments on the desirability of using the "what-if" hazard identification and analysis process. EPA also proposes the use of more involved analytical techniques if findings suggest that to be desirable.

Hazard review requirement for program level 3 locations are more specific and extensive. But those locations that are compliant with the OSHA rule for process safety management of highly hazardous chemicals will need to do little that is new, although they do need to extend their hazard analyses to consider the probability of harm to the public or to the environment. As with OSHA, a team must complete the process hazard analyses required by EPA. One member of the team, at least, is to have experience with the process.

As would be expected at locations with more significant exposures, the process hazard analysis requirements are more extensive. They must be documented and include:

- Hazards of the process
- Identification of previous, potentially catastrophic incidents
- Engineering and administrative controls applicable to the hazards
- Siting
- Human factors
- Qualitative evaluation of health and safety impacts of control failure

For American industry, EPA has obviously extended knowledge and skill requirements regarding hazard analysis techniques.

THE CHEMICAL INDUSTRY: THE EXTENSIVE BODY OF INFORMATION

Completing hazard analyses was a common practice in the chemical industry many years before requirements for them were established by OSHA and EPA. Although that practice is not of recent origin, it is mentioned here because of its extensive knowledge requirements. The body of information in the chemical industry on hazard analysis is extensive. But reference will be made here to only one publication because of its particular significance.

The Center for Chemical Process Safety is a part of the American Institute of Chemical Engineers. One of its several books is entitled *Guidelines For Hazard Evaluation Procedures, Second Edition With Worked Examples*. Publication of the text by a chemically oriented group should not dissuade those who want an education in the following evaluation techniques. Their descriptions are generic.

- Safety Review
- Checklist Analysis
- Relative Ranking
- Preliminary Hazard Analysis
- What-If analysis
- What-If Checklist Analysis
- Hazard and Operability Analysis
- Fault Tree Analysis
- Event Tree Analysis
- Cause-Consequence Analysis
- Human Reliability Analysis

These techniques are dealt with broadly in the *Guidelines* in chapters entitled "Overview of Hazard Evaluation Techniques" and "Using Hazard Evaluation Techniques." Brief descriptions of some of those techniques were given in Chapter 11.

CONCLUSION

The message is clear. Including provisions requiring hazard analyses and risk assessments in safety standards and guidelines is becoming ordinary. It is logical to assume that this trend will continue and that safety professionals will be expected to have the knowledge and skill necessary to give counsel on applying those provisions.

REFERENCES

ANSI/AIHA Z10-2012. *Occupational Health and Safety Management Systems*. Fairfax, VA: American Industrial Hygiene Association, 2012. ASSE is now the secretariat. Available at https://www.asse.org/cartpage.php?link=z10_2005.

ANSI/ASSE Z241.1-2009. *Control of Hazardous Energy: Lockout/Tagout and Alternative Methods*. Des Plaines, IL: American Society of Safety Engineers, 2009.

ANSI/ASSE Z690.3, *Risk Assessment Techniques*. Des Plaines, IL: American Society of Safety Engineers, 2011.

ANSI B11.0. *Safety of Machinery—General Safety Requirements and Risk Assessments.* Leesburg, VA: B11 Standards, Inc., 2010.

ANSI/PMMI B155.1-2011. *Safety Requirements for Packaging Machinery and Packaging-Related Converting Machinery.* Arlington, VA: Packaging Machinery Manufacturers Institute, 2011.

API RP 75 (R2008). *Recommended Practice for Development of a Safety and Environmental Management Program for Offshore Operations and Facilities,* 3rd ed. American Petroleum Institute. Go to http://www.techstreet.com/standards/api/rp_75_r2008_?product_id=1157045.

B11.TR3. *Risk Assessment and Reduction—A Guideline to Estimate, Evaluate and Reduce Risks Associated with Machine Tools.* McLean, VA: Association for Manufacturing Technology, 2000.

BOEMRE (Bureau of Ocean Energy Management Regulation and Enforcement). Requirements for SEMS: Safety and Environmental Management Systems, at www.boemre.gov/semp.

BS OHSAS 18001:2007. *Occupational Health and Safety Management Systems—Requirements.* London: British Standards Institution, 2007.

CSA Standard Z1000-06. *Occupational Health and Safety Management.* Mississauga, Ontario, Canada: Canadian Standards Association, 2006.

CSA Standard Z1002-12. *Occupational Health and Safety—Hazard Identification and Elimination and Risk Assessment and Control.* Mississauga, Ontario, Canada: Canadian Standards Association, 2012.

EN ISO 12100–2010. *Safety of Machinery—General Principles for Design. Risk Assessment and Risk Reduction.* Geneva, Switzerland: International Organization for Standardization, 2010.

EPA. *Risk Management Programs for Chemical Accidental Release Prevention.* http://www.epa.gov/emergencies/content/lawsregs/rmpover.htm.

European Union,. Risk Assessment 2008. http://osha.europa.eu/en/topics/riskassessment.

Federal Railroad Administration Risk Reduction Program. Enter the title into a search engine to bring up information; or go to http://www.fra.dot.gov/Page/P0049.

Guidance Document for Incorporating Risk Concepts into NFPA Codes and Standards. National Fire Protection Association, 2007. Available at http://www.nfpa.org/assets/files/PDF/Research/Risk-Based_Codes_and_Stds.pdf.

Guidelines for Hazard Evaluation Procedures, Second Edition with Worked Examples. New York: Center for Chemical Process Safety of the American Institute of Chemical Engineers, 1992.

ISO 10218:1992. *Manipulating Industrial Robots—Safety.* Geneva, Switzerland: International Organization for Standardization, 1992. http://www.iso.ch/iso/en/aboutiso/introduction/index.html.

ISO 12100–1. *Safety of Machinery—Basic concepts, general principles for design; Part 1. Basic terminology, methodology.* Geneva, Switzerland: International Organization for Standardization, 2003.

ISO 12100–2. *Safety of machinery – Basic concepts, general principles for design; Part 2. Technical principles and specifications.* Geneva, Switzerland: International Organization for Standardization, 2003.

ISO 14121. *Safety of Machinery—Principles for risk assessment.* Geneva, Switzerland: International Organization for Standardization, 1999.

Michaels, David. Assistant Secretary at OSHA, July 9, 2010 letter to the OSHA. At http://orc-dc.com/?q=node/3649.

MIL-STD-882E. *Standard Practice for System Safety*. Washington, DC: Department of Defense, 2012 Available at http://www.system-safety.org/. Scroll down and click on MIL-STD-882E in the right-hand column for a free copy.

NFPA 70E, Standard for Electrical Safety in the Workplace, 2112 ed. Begin inquiry at http://us.yhs4.search.yahoo.com/yhs/search;_ylt=A0oG7psAQy1R2CoAgLVXNyoA? p=32.%09NFPA%2070E%2C%20Standard%20for%20Electrical%20Safety%20 in%20the%20Workplace%2C%202112%20Edition&fr2=sb-top&hspart=att&hsimp =yhs-att_001& type=att_lego_portal_home&type_param=att_lego_portal_home.

OSHA's Rule for Process Safety Management of Highly Hazardous Chemicals, 29 CFR 1910.119. Washington, DC: Department of Labor, Occupational Safety and Health Administration, 1992.

Pipeline and Hazardous Materials Safety Administration announced their hazardous materials Regulations. See comments at http://primis.phmsa.dot.gov/meetings/MtgHome.mtg?mtg=70.

Risk Management Programs for Chemical Accidental Release Prevention, 40 CFR Part 68. Washington, DC: Environmental Protection Agency, 1996.

Safety Handbook: Aviation Ground Operation, 6th ed. Itasca, IL: National Safety Council, 2007.

SEMI S2-0712a. *Environmental, Health, and Safety Guideline for Semiconductor Manufacturing Equipment*. San Jose, CA: SEMI (Semiconductor Equipment and Materials International), 2006.

SEMI S10-307E. *Safety Guideline for Risk Assessment and Risk Evaluation Process*. San Jose, CA: SEMI (Semiconductor Equipment and Materials International), 2007.

Society of Fire Protection Engineers. *Engineering Guide to Fire Assessment*, 2006. Available at http://findpdf.net/documents/SFPE-engineering-guide-to-fire-risk-assessment.html.

Society of Fire Protection Engineers. *Introduction to Fire Risk Assessment*, 2006b. Enter the title in a search engine for information on the course. Or go to http://www.sfpe.org/ SharpenYourExpertise/Education/SFPEOnlineLearning/FireRiskAssessment.aspx.

ADDENDUM A

PARTIAL LIST OF STANDARDS, GUIDELINES, AND INITIATIVES THAT REQUIRE OR PROMOTE MAKING RISK ASSESSMENTS BEGINNING IN 2005

1. ANSI/AIHA Z10-2005, *Occupational Health and Safety Management Systems* standard.

   Z10 sets a benchmark provision requiring that processes be in place: To identify and take appropriate steps to prevent or otherwise control hazards and reduce risks associated with new processes or operations at the design stage.

2. *Guidance On The Principles Of Safe Design For Work*. Australian Safety and Compensation Council, Australian government, 2006.

3. In 2006, NIOSH announced a major national initiative on Prevention through Design.

4. SFPE, *Engineering Guide to Fire Assessment*, 2006. This is a technical book that would be of particular interest to engineers.

5. SFPE, *Introduction to Fire Risk Assessment* [Believe release date was 2006.]

   Enter the title in a search engine for course modules on fire risk assessment.

6. CSA Z1000-2006, the *Occupational Health and Safety Management Standard*, issued by the Canadian Standards Association.

Advanced Safety Management: Focusing on Z10 and Serious Injury Prevention, Second Edition. Fred A. Manuele.
© 2014 John Wiley & Sons, Inc. Published 2014 by John Wiley & Sons, Inc.

7. The Industrial Safety and Health Act of Japan was revised: effective in April 2006. It stipulates—without penalty—that employers should make efforts to implement risk assessment.

8. ISO 14121–1, *Safety of Machinery—Principles for risk assessment*. 2007.

9. In 2007, the OSHA Alliance Construction Roundtable developed a video training program entitled "Design for Construction Safety."

10. NFPA, *Guidance Document for Incorporating Risk Concepts into NFPA Codes and Standards*, 2007.

11. BS OHSAS 18001:2007, *Occupational health and safety management systems— requirements*, a British Standards Institution publication.

    In the 2007 revision, requirements for risk assessments are more explicit. The guidelines now say: "The organization shall establish, implement and maintain a procedure(s) for the ongoing hazard identification, risk assessment, and determination of necessary controls."

12. The Nano Risk Framework, issued in June 2007 through the combined efforts of the Environmental Defense Fund and DuPont, includes a six-step guidance framework for "the responsible development of nanoscale materials."

    They are: 1. Describe the material and its application; 2. Profile life cycle(s); 3. Evaluate risks; 4. Assess risk management; 5. Decide, document, and act; 6. Review and adapt.

13. ANSI B11.TR7 2007: *ANSI Technical Report for Machines – A Guide on Integrating Safety and Lean Manufacturing principles in the use of Machinery*.

14. China's State Administration of Work Safety published provisional regulations on risk assessment in 2008.

15. The Health and Safety Executive in the UK issued "Five steps to risk Assessment" in 2008.

16. All employers in the UK must conduct a risk assessment. An HSE bulletin says: "The law does not expect you to eliminate all risk, but you are required to protect people as far as is 'reasonably practicable'."

17. In August 2008, the European Union launched a two-year health and safety campaign focusing on risk assessment. Their bulletin states:

    Risk assessment is the cornerstone of the European approach to prevent occupational accidents and ill health. If the risk assessment process—the start of the health and safety management approach—is not done well or not at all, the appropriate preventive measures are unlikely to be identified or put in place.

18. *Machine Safety: Prevention of mechanical hazards*. Issued by The Institute for research for safety and security at work and The Commission for safety and security at work in Quebec, 2009.

19. ASSE Technical Report Z790.001: *Prevention Through Design: Guidelines for Addressing Occupational Risks in the Design and Redesign Processes*, 2009.

20. Singapore Standard SS 506: *Occupational Safety and Health (OSH) Management Systems; Part 1: Requirements*, 2009.

21. ANSI-ITAA GEIA-STD-0010-2009: *Standard Best Practices for System Safety Program Development and Execution.*

> Foreword: Coupled with use of the system safety risk mitigation order of precedence, functional hazard analysis lets a program identify early in the life cycle those risks which can be eliminated by design, and those which must undergo mitigation by other controls in order to reduce risk to an acceptable level.

22. ExxonMobil issued its *Operations Integrity Management System* in July 2009. It pertains to safety, health, the environment and product safety. The first four of 11 elements in this management system are:
    1. Management leadership, commitment, and accountability
    2. Risk assessment and management
    3. Facilities design and construction
    4. Information and documentation

23. ISO/IEC 31000–2009: *Risk Management—Principles and guidelines* and ISO/IEC 31010–2009: *Risk assessment techniques.*

24. EN ISO 12100–2010: *Safety of Machinery—General principles for design. Risk assessment and risk reduction.*

> This standard combines three previously issued ISO standards (including item 8 in this list) and replaces them. Risk assessments are explicitly required.

25. In a July 19, 2010 letter to the OSHA staff, assistant secretary David Michaels wrote on several subjects, one of which follows.

> Ensuring that American workplaces are safe will require a paradigm shift, with employers going beyond simply attempting to meet OSHA standards, to implementing risk-based workplace injury and illness prevention programs.

26. ANSI B11.0: *Safety of Machinery—General Safety Requirements and Risk Assessments*, December 2010.

> Purpose: This standard describes procedures for identifying hazards, assessing risks, and reducing risks to an acceptable level over the life cycle of machinery.

27. In the December 8, 2010 *Federal Register*, the Federal Railroad Administration issued an advance notice of proposed rulemaking for certain railroads to have a risk reduction program.

It is proposed that the Risk Reduction Program be supported by a risk analysis and a Risk Reduction Plan.

28. ANSI/PMMI B155.1—March 2, 2011: *Safety Requirements for Packaging Machinery and Packaging-Related Converting Machinery.*

   Foreword: This version of the standard has been harmonized with international (ISO) and European (EN) standards by the introduction of hazard identification and risk assessment as the principal method for analyzing hazards to personnel and achieving a level of acceptable risk.

29. Pipeline and Hazardous Materials Safety Administration, March 11, 2011.

   Hazardous materials regulations are to be modified to require that risk assessments be made of loading and unloading operations.

30. *OSH Management System: A tool for continual improvement* – issued by the International Labour Organization, Geneva. April 28, 2011.

   Hazard and risk assessments have to be carried out to identify what could cause harm to workers as well as property so that appropriate preventive and protective measures can be developed and implemented.

31. ANSI-ASSE Z590.3—September 1, 2011: *Prevention through Design: Guidelines for Addressing Occupational Hazards and Risks in Design and Redesign Processes.*

   This is an American National Standard. The core of Z590.3 is risk assessment, to be performed as a continuum in the design and redesign processes.

32. NFPA 70E: *Standard for Electrical Safety in the Workplace*, 2112 ed., has a new section on risk assessment.

33. MIL-STD-882E. *Department of Defense Standard Practice for System Sa*fety, approved May 11, 2012. It is available at http://www.system-safety.org/links/; click on Home. Click on 882E for a free download.

34. ANSI/AIHA Z10-2012: *Occupational Health and Safety Management Systems* standard. The second version of Z10, approved in June 2012, now contains a specific requirement for a risk assessment process to be in place.

   5.1.1 Risk Assessment

   The organization shall establish and implement a risk assessment process(es) appropriate to the nature of hazards and level of risk.

35. CSA Z1002-12: *Occupational Health and Safety—Hazard Identification and Elimination and Risk Assessment and Control.* Canadian Standards Association, 2012.

CHAPTER 13

THREE- AND FOUR-DIMENSIONAL RISK SCORING SYSTEMS: SECTIONS 5.1.1 AND 5.1.2 OF Z10

In all of the risk assessment matrices discussed in Chapter 11, "A Primer on Hazard Analysis and Risk Assessment", risk levels are based on the *probability* of an event occurring and the *severity* of harm or damage that could result. Therefore, they are two-dimensional risk scoring systems. For the hazards and risks that most safety professionals encounter, two-dimensional systems will be sufficient.

Some systems now in use are three- or four-dimensional. Safety professionals can expect that variations of numerical risk scoring systems will proliferate. For the knowledge that safety professionals should have, in this chapter we inform on:

- Transitions in numerical risk assessment methods
- Precautions to be considered in using three- and four-dimensional systems, particularly those that diminish the significance of the severity potential
- Three- and four-dimensional numerical risk scoring systems
- An extended three-dimensional numerical risk scoring system that includes a method to justify the risk amelioration costs in relation to the amount of risk reduction to be attained
- A numerical risk scoring system developed by this author

Advanced Safety Management: Focusing on Z10 and Serious Injury Prevention,
Second Edition. Fred A. Manuele.
© 2014 John Wiley & Sons, Inc. Published 2014 by John Wiley & Sons, Inc.

TRANSITIONS IN RISK ASSESSMENT

It is typical for engineers to have a passion for quantification and the application of numbers. They become comfortable with statistical measurements. Quantification is stressed in their education. Engineering texts still quote Lord Kelvin, who wrote the following over 150 hundred years ago:

> When you can measure what you are speaking about, and express it in numbers, you know something about it: but when you cannot measure it, when you cannot express it in numbers, your knowledge is of a meager and unsatisfactory kind. (http://www.cromwell-intl.com/3d/Index.html)

Some practitioners in risk assessment such as Vernon L. Gorse, disagree with Lord Kelvin's absolutist statement. The quote above from Lord Kelvin can also be found in Grose's book, *Managing Risk: Systematic Loss Prevention For Executives*. (p. 259) Gorse expressed concern over the use of "numerology" in making risk assessments, and cautions about the:

> Increasingly common abuse of mathematical statistics by "numericalizing without data" to obtain risk probabilities, and the deception-by-numbers that lulls executives into a false sense of security. (p. 259)

Nevertheless, the influence that engineers have had in developing risk assessment methods is obvious. Their passion for numerical precision encourages the use of quantitative methods.

As an example, in at least two industries, quantitative Failure Modes and Effects Analyses (FMEAs), rather than qualitative FMEAs, are now required to meet quality assurance requirements. Generally, engineers make those FMEAs.

Similarly, some engineers are using numerical risk scoring systems to meet the risk assessment requirements placed on manufacturers who sell machinery that is to go into workplaces in European Union countries. One such system in use for that purpose is four-dimensional. Its negative aspects will be discussed later.

Also, a three-dimensional numerical risk scoring system is in use in a segment of the heavy machinery manufacturing industry. It was introduced by engineering personnel to meet the demands for product safety.

A NEED FOR CAUTION AND PERCEPTIVE EVALUATION

Although three- and four-dimensional risk scoring systems are now more commonly used, there are several reasons to be inquisitive and cautious concerning their content and results in their application, the meanings of the terms used in them, the numerical values applied to the gradations within elements to be scored, and how they are applied in determining risk levels. My observations follow.

- Creating numerical risk scorings begins with subjective judgments on the values to be given to the elements in the system, and those judgments are then translated into numbers. Thus, judgmental observations become finite numbers, which leads to an image of preciseness. Further, those numbers are multiplied or totaled to produce a risk score, giving the risk assessment process a scientific appearance. The fact is that the risk assessment process is as much art as science.

- There are no universally applied rules to assign value numbers to elements to be scored. Value numbers in all numerical risk scoring systems are judgmental and reflect the experience and views of those who create a system.

- Historically, frequency of exposure has been one of the elements considered to determine event probability. However, giving frequency of exposure its own multiplier, separate from and in addition to a probability multiplier, diminishes the necessary emphasis on the severity of harm or damage that could result from an event.

- Meanings of terms such as *probability, likelihood, frequency of exposure or endangerment,* and *severity* used in risk scoring systems are not consistent. Also, it may be that within a system, a term may appear more than once and have different uses. Thus, consistency in the meanings of the terms used in a system may not be maintained.

- If risk scoring systems are applied rigidly whereby a higher score indicates a greater risk than a lower score, and the scores are not considered in light of employee, community, social, and financial concerns, the best interests of an organization may not be well served. This applies especially to low-probability incidents that may have serious consequences, for which risk scores may be low.

THREE- AND FOUR-DIMENSIONAL NUMERICAL RISK SCORING SYSTEMS

There are substantial variations in the elements to be scored in the three- and four-dimensional numerical risk scoring systems reviewed in this chapter. To begin the discussion, excerpts are taken from a National Safety Council publication, *Accident Prevention, Manual: Administration & Programs,* 13th edition, and appear here with permission.

NATIONAL SAFETY COUNCIL

In the *Accident Prevention Manual,* the Council outlines a system for "Ranking Hazards by Risk (Severity, Probability, and Exposure)." The text says that "Ranking provides a consistent guide for corrective action, specifying which hazards warrant immediate action, which have secondary priority, and which can be addressed in the future." (p. 155)

In the process, for each hazard identified, numerical scorings are given to Severity, Exposure and Probability. It is significant that they are totaled to arrive at a final rating, not multiplied in sequence as is the case in other three- and four-dimensional risk scoring systems. Thus, the status of the severity rating is not diminished as much.

Severity: Consider the potential losses or destructive and disruptive consequences that are most likely to occur if the job/task is performed improperly. Assumptions would be made as to the worst credible outcome of an incident in selecting a numerical rating for severity. Use the point values in Table 13.1.

Probability: Consider the probability of loss that occurs each time the job is performed. The key question is: How likely is it that things will go wrong when this job is performed? Probability is influenced by a number of factors, such as the hazards associated with the task, difficulty in performing the job, the complexity of the job, and whether the work methods are error-provocative. Use the point values in Table13.2.

Exposure: Consider the number of employees that perform the job, and how often. Use the point values in Table 13.3.

After numerical point values are given to each job or task for severity, probability, and exposure, the point values are added to produce a total risk assessment code (RAC). A total score can be as low as 3 or as high as 10. So, each task is given a risk ranking from which judgments can be made in setting priorities. and, Management can use these values to select particular jobs for further analysis. The Council's text clearly states that "The RAC rating scale is a guideline and is not intended to be used as an absolute measurement system." (p. 156)

In the Council's system, there are four severity ratings, while probability and exposure each have three. Choosing to set up such a system and deciding on the numerical values to be assigned is entirely at the discretion of the creator of the system.

TABLE 13.1   NSC Severity Descriptions and Point Values

1. Negligible	Probably no injury or illness; no production loss; no lost days workdays
2. Marginal	Minor injury or illness; minor property damage
3. Critical	Severe injury or occupational illness with lost time; major property damage; no permanent disability or fatality
4. Catastrophic	Permanent disability; loss of life; loss of facility or major process

TABLE 13.2   NSC Probability Descriptions and Point Values

1. Low probability of loss occurrence
2. Moderate probability of loss occurrence
3. High probability of loss Occurrence

TABLE 13.3   NSC Exposure Descriptions and Point Values

1. A few employees perform the task up to a few times a day
2. A few employees perform the task frequently
3. Many employees perform the task frequently

Other choices, reflecting a person's experience, are equally valid. However, a variation of this system, or any other system, should keep severity, frequency of exposure, and probability considerations in balance so that disproportionate weightings are not given to any of the categories.

FAILURE MODE AND EFFECTS ANALYSIS

SEMATECH is a consortium of semiconductor equipment manufacturing companies. Its *Failure Mode and Effects Analysis (FMEA): A Guide for Continuous Improvement for the Semiconductor Equipment Industry* describes its recommended FMEA technique. This guide was chosen to illustrate a three-dimensional risk assessment technique because of its uniqueness and because of its special character:

- Consideration of environmental, safety, and health needs is prominent throughout the analysis.
- There is a separate table for severity levels that encompasses several forms of harm, damage, or loss—injury to personnel, exposure to harmful chemicals or radiation, fire, or a release of chemicals to the environment. (p. 14)
- The value of counsel available from safety engineering personnel is recognized by including that designation—safety engineering—among the titles of people who would form a team to conduct FMEAs. (p. 4)

With the permission of SEMATECH, excerpts from that publication appear here. In the SEMATECH FMEA, three criteria are to be ranked numerically: Severity, Occurrence, and Detection. This is how they are defined in the *Guide*. As stated earlier, definitions of the terms used in risk scoring systems vary, and the definitions in this system have their own distinct characteristics.

Severity ranking criteria. Customer satisfaction is key in determining the effect of a failure mode. Safety criticality is also determined at this time related to Environmental, Safety and Health (ES&H) levels. Based on this information, a severity ranking is used to determine criticality of the failure mode on the subassembly to the end effect. The end (global) effect of the failure mode is the one to be used for determining the severity ranking. Calculating the severity levels provides for a classification ranking that encompasses safety, production continuity, scrap loss, etc. (p. 8)

This system also provides for the use of an Environmental, Safety and Health Severity Code—which is a qualitative means of representing the worst case incident that could result from an equipment or process failure or for lack of a contingency plan for such an incident. (p. 14)

Occurrence ranking criteria. The probability that a failure will occur during the expected life of the system can be described in potential occurrences per unit of time. (p. 15)

Detection ranking criteria. This section provides a ranking based on an assessment of the probability that the failure mode will be detected given the controls in place. (p. 16)

Two scoring tables are given for the severity ranking (Tables 13.4 and 13.5). One has five possible scoring levels and relates to the effect that a failure may have on customer relations; the other pertains to environmental, safety, and health levels and has four possible scoring levels. Abbreviated versions of the severity criteria are provided in the tables.

Note that for most of the scoring levels for the Severity, Occurrence and Detection criteria, two numerical scoring possibilities are available for each narrative description.

A statement, as follows, defines the risk level to be achieved: "ES&H severity levels are patterned after the industry standard, SEMI S2-91, *Product Safety Guidelines.* All equipment should be designed to level IV severity. Types I, II, III are considered unacceptable risks." (p. 16)

TABLE 13.4 SEMATACH Scoring Table for Severity Ranking: Customer Related

Rank	Description
1–2	Failure is of such a minor nature that the customer (internal or external) will probably not detect the failure.
3–5	Failure will result in slight customer annoyance and/or slight deterioration of part or system performance.
6–7	Failure will result in customer dissatisfaction and/or slight deterioration of part or system performance.
8–9	Failure will result in a high degree of customer or cause nonfunctionality of system.
10	Failure will result in major customer dissatisfaction and cause nonsystem operation or noncompliance with government regulations. (p. 14)

TABLE 13.5 SEMATACH Scoring Table for Severity Ranking: ES&H Definitions

Rank	Severity Level	Description
10	Catastrophic (I)	A failure results in major injury or death of personnel.
7–9	Critical (II)	A failure results in minor injury to personnel, personnel exposure to harmful chemicals or radiation, fire, or a release of chemicals to the environment.
4–6	Major (III)	A failure results in a low-level exposure to personnel, or activities facility alarm system.
1–3	Minor (IV)	A failure results in minor system damage but does not cause injury to personnel, allow any kind of exposure to operational or service personnel, or allow any release of chemicals into environment. (p. 14)

TABLE 13.6   SEMATECT Scoring Table for Occurrence Ranking Criteria

Rank	Description
1	An unlikely probability of occurrence during the item operating time interval. *Unlikely* is defined as a single failure mode (FM) probability < 0.001 of the overall probability of failure during the time operating interval.
2–3	A remote probability of occurrence during the item operating time interval (i.e., once every two months), defined as a single FM probability > 0.001 but < 0.01.
4–6	An occasional probability of occurrence during the item operating time interval (i.e., once a month), defined as a single FM probability > 0.01 but < 0.10.
7–9	A moderate probability of occurrence during the item time operating interval (i.e., once every two weeks), defined as a single FM probability > 0.10 but < 0.20.
10	A high probability of occurrence during the item time operating interval (i.e., once a week), defined as a single FM probability > 0.20. (p. 15)

TABLE 13.7   SEMATECH Scoring Table for Detection Ranking Criteria

Rank	Probability That The Defect Will Be Detected
1–2	Very high—almost certain
3–4	High—good chance
5–7	Moderate—likely
8–9	Low—not likely
10	Very low (or zero)

Thus, the guide clearly establishes that Type IV severity is the acceptable risk level. It is well known that this industry has established high environmental, safety, and health standards for itself. A slightly reduced display of the Occurrence Ranking Criteria. Interestingly, it relates probability to a time operating interval and expresses probability as a failure rate.

A condensed version of the Detection ranking scale is provided in Tables 13.7. It pertains to what the verification system and/or controls in place are expected to accomplish.

At this point, a Risk Priority Number (RPN) is calculated:

$$RPN = Severity \times Occurence \times Detection$$
$$RPN = S \times O \times D$$

For each failure mode, codes for severity, occurrence probability and detectability, and the risk priority number are entered into the FMEA form. Recommended actions are listed to resolve situations that need attention. After the recommended actions are taken, revised codes and a priority number are entered, reflecting the effect of the work done to reduce the risk.

There is a major difference in this system that should be of particular interest to safety professionals. In the semiconductor industry FMEA form, "Cr" is at the top of a column. This is a Critical Failure Symbol. A "Y" is to be entered for "yes" if the failure potential is considered safety-critical.

That gives importance to the Environmental, Safety, and Health Severity Level Definitions shown previously. On the form, the instruction is that critical failures must be addressed when safety is an issue.

Although the SEMATECH FMEA publication was drafted for the semiconductor equipment industry, it is highly recommended as a reference if three-dimensional risk scoring systems are to be considered. Its inclusion of environmental, safety, and health considerations makes it unique and exceptionally valuable. It is downloadable, at no cost, at http://www.sematech.org/docubase/document/0963beng.pdf.

Addendum A is an example of an FMEA form that relates to the SEMATECH provisions. It includes instructions as to its use.

THREE-DIMENSIONAL RISK ASSESSMENT SYSTEM: HEAVY EQUIPMENT BUILDERS

Hazard analysis and risk assessment systems have been used for many years by certain heavy equipment manufacturers in their design processes to build inherently safer products. In the use of those systems, the principal concern is product safety and the avoidance of injury to equipment users or to bystanders.

There are variations in the terms used in the several versions of this industry's hazard analysis system. In one version, the risk assessment variables are the severity of the injury, the frequency of exposure, and the probability that injury is not avoided and injury occurs.

In another version, the parameters to be scored are severity, frequency, and vulnerability. The latter version and its definitions are discussed here. Definitions of terms follow.

- *Severity:* the most probable injury which would be expected from an accident.
- *Frequency:* an estimate of how often a product user or bystander may be exposed to a hazard. The hazard may result from a machine part or system failure or from a man–machine interface failure.
- *Vulnerability:* the degree of user or bystander susceptibility to injury when exposed to a hazard. The likelihood that personal injury will occur once exposure to a hazard has occurred, taking into consideration the detectability of a hazard, risk assumptions, the presence of environmental or stress conditions, skills and attitudes, and so on.

For each of these three variables, there are five possible ratings, all using identical numerics: 1, 3, 5, 7, and 9. Other numerical selections are equally valid as long as

TABLE 13.8   Severity Scale (S)

1.	Minor first aid. Immediate return to work/activity.
3.	Doctor's office or emergency room treatment. Up to one week lost time.
5.	Hospitalization. Up to one month lost time. No loss in work capacity.
7.	Permanent partial loss in work capacity. Increased work difficulty.
9.	Death or complete disability.

TABLE 13.9   Frequency Scale (F)

1.	Theoretically can occur, but highly unlikely during the life of the machine population.
3.	Only once in the life of a small percentage (10%) of the product.
5.	Once per use season or once annually.
7.	Once daily.
9.	Continuous exposure.

TABLE 13.10   Vulnerability Scale (V)

1.	Practically impossible to complete injury sequence.
3.	Remotely possible, but unlikely.
5.	Some conditions favorable to completing the injury sequence.
7.	Very possible, but not assured.
9.	Almost certain to complete injury sequence.

they are proportional and giving excessive weight to one of the factors being scored is avoided. The ratings are described in Tables 13.8 to 13.10.

The literature on this system indicates the following: hazards analyses and risk assessments are done by a team; consensus is to be reached on risk scores; risk scores are listed and ranked; high and low values would be examined; and actions for improvement are to be developed when required. This is the risk scoring system:

$$\text{Risk Score} = \text{Severity} \times \text{Frequency} \times \text{Vulnerability}$$
$$\text{Risk Score} = S \times F \times V$$

This risk scoring system gives equal weight to all variables. Thus, the necessary emphasis on severity of injury is diminished when the risk score is produced. In a discussion with a user of this system, it was acknowledged that with all variables being given equal weight, the significance of severity was subordinated. However, it was also said that, in practice, the team gives the potential for severe injury due consideration when deciding on product improvement recommendations.

THE WILLIAM T. FINE SYSTEM: THREE-DIMENSIONAL NUMERICAL RISK SCORING MODEL

While at the Naval Ordnance Laboratory in White Oak, Maryland in 1971, William T. Fine submitted a report entitled "Mathematical Evaluations For Controlling Hazards." Fine was an early proponent of determining whether the expenditure needed to reduce risk could be justified after considering the amount of reduction to be attained.

In a sense, Fine's work was a precursor of the concept on which ALARP is based. ALARP is defined as that level of risk which can be lowered further by an increment in resource expenditure that cannot be justified by the resulting decrement in risk. Fine made this comment about his work in the paper's abstract:

> A formula has been devised which weighs the controlling factors and "calculates the risk" of a hazardous situation, giving a numerical evaluation to the urgency for remedial attention to the hazard. Calculated Risk Scores are then used to establish priorities for corrective action. An additional formula weighs the estimated cost and effectiveness of any contemplated corrective action against the Risk Score and gives an indication on whether the cost is justified.

Fine's paper is thought provoking, valuable, a good resource, and somewhat difficult to locate. For those who would like to explore a three-dimensional risk scoring system to which an additional formula is applied to determine whether the risk reduction cost can be justified, a summary version of Fine's paper is provided in Addendum B to this chapter. Although condensed, the substance of the paper has been maintained.

FOUR-DIMENSIONAL NUMERICAL RISK SCORING SYSTEM

Although reference is made here to a four-dimensional risk scoring system as it appears in *Pilz: Guide to Machinery Safety*, 6th edition, it should be understood that this system has appeared elsewhere and is in the public domain. For example, it has been used with minor modifications by at least one U.S. company to meet the EU risk assessment requirements as set forth in the International Organization for Standardization standard EN ISO 12100–2010, *Safety of Machinery—General principles for design. Risk assessment and risk reduction.* As will be illustrated later, this risk scoring system has major shortcomings.

In the Pilz text, the following statement is made in Chapter 4: Risk Assessment under the caption 4.0: Background to risk assessment.

In simple terms, there are only two real factors to consider:

- The severity of foreseeable injuries (ranging from a bruise to a fatality)
- The probability of their occurrence. (p. 75)

As the text proceeds, however, the system becomes more complex. There are four elements to be scored:

- Likelihood of occurrence/contact with hazard (*LO*)
- Frequency of exposure to the hazard (*FE*)
- Degree of possible harm (*DPH*), taking into account the worst possible case
- Number of persons exposed to the hazard (*NP*) (p. 85)

This system has a particular focus. It applies only to personal injury, particularly to employees, that could derive from machinery operation. All of the listings for "degree of possible harm" involve personal injuries. There are no entries for possible damage to property or the environment. Terms used to establish gradations and scores in the likelihood of occurrence and frequency of exposure categories are comparable to those in risk assessment systems cited previously. They follow in Tables 13.11 and 13.12. Note the position within the listings of "Even chance—could happen" in the Likelihood category, and "Annually" in the Frequency of exposure category. A computation appears later in which these ratings are used.

TABLE 13.11 Likelihood of Occurrence (*LO*) and the Scores

Category	Score
Almost impossible—possible only under extreme circumstances	0.033
Highly unlikely—though conceivable	1
Unlikely——but could occur	1.5
Possible——but unusual	2
Even chance—could happen	5
Probable—not surprising	8
Likely—only to be expected	10
Certain—no doubt	15

TABLE 13.12 Frequency of Exposures to the Hazards (*FE*) and the Scores

Frequency of Exposure	Scores
Annually	0.5
Monthly	1
Weekly	1.5
Daily	2.5
Hourly	4
Constantly	5

The third dimension in this system is the degree of possible harm (Table 13.13), and the fourth dimension is the number of persons exposed to the hazard (Table 13.14).

TABLE 13.13 Degree of Possible Harm (*DPH*) and the Scores

Degree of Harm	Score
Scratch/bruise	0.1
Laceration/mild ill effect	0.5
Break minor bone or minor illness (temporary)	2.0
Break major bone or major illness (temporary)	4.0
Loss of one limb, eye, hearing loss (permanent)	6.0
Loss of two limbs, eyes (permanent)	10.0
Fatality	15.0

TABLE 13.14 Number of Persons Exposed (*NP*) and the Scores

Number of Persons	Score
1 or 2	1
3–7	2
8–15	4
16–50	8
50+	12

Using the numerical ratings given to each element, the formula produces a Risk level by simple multiplication. The formula and the Risk levels (Table 13.15) follow.

$$LO \times FE \times DPH \times NP = \text{Risk level}$$

TABLE 13.15 Risk Level

Risk Level	Scoring Range
Negligible—presenting very little risk to health and safety	0–5
Low but significant—containing hazards that require control measures	5–50
High—having potentially dangerous hazards, which require control measures to be implemented urgently	50–500
Unacceptable—continued operation in this state is unacceptable	500+

Assume that during a company's annual shutdown for retooling and maintenance, a task is to be performed for which the likelihood of occurrence of a hazardous event was rated as "Even chance — could happen" and the outcome would be Fatalities. In the illustration in Table 13.16, the ratings are purposely kept the same for likelihood of occurrence, frequency of exposure, and degree of possible harm. Ratings vary for the number of people exposed.

For such a hazard scenario, if there is an even chance that 2 people could be killed, the risk level is Low. That is not good risk management. Computations never fall within the Unacceptable risk level (500+) even when there is an even chance that 51 people will be killed. Score levels in this system are ill conceived. They greatly diminish the value of life. Nevertheless, promotion of this risk scoring system continues.

Using the number of persons exposed as a category in risk assessment requires careful consideration. In most risk assessment matrices, a single death or a permanent

TABLE 13.16 Use of a Four-Dimensional Risk Scoring System

Rating	Number of Persons Exposed				
	2	7	15	50	51
Likelihood: even chance	5	5	5	5	5
Frequency of exposure: annual	0.5	0.5	0.5	0.5	0.5
Degree of possible harm: fatality	15	15	15	15	15
For: number of persons exposed	1	2	4	8	12
Risk levels	37.5	75	150	300	450
	Low But Significant	High	High	High	High

total disability is in the Catastrophic category. For example, in the FMEA system published by SEMATECH, a single fatality gets a Catastrophe rating. In the Fine system, a fatality that has an even chance of occurring and an annual exposure falls in the unacceptable risk category.

In the scenario just described, an occurrence resulting in 51 fatalities does not receive the severity grading it deserves. This risk scoring system is not acceptable.

MODEL THREE-DIMENSIONAL NUMERICAL RISK SCORING SYSTEM

One of the aims in my study of multidimensional numerical risk scoring systems was to determine whether a model could be proposed that:

1. Serves the needs of those who are more comfortable with statistics
2. Addresses the strong beliefs of those who want frequency of exposure given separate consideration in the risk assessment process, as is the case in Z10
3. Maintains credibility and efficacy

As the work proceeded in crafting the numerical risk scoring model presented here, the following guidelines emerged and were adopted.

- In a statistical risk scoring system, all scores are to meet a plausibility test. High- risk scores must be produced for high risks: a lower-level risk is not to fall in a high-risk category.
- To create a focal point for the relative placing of other risk scores, it was assumed that for an incident having a severity outcome of one or more fatalities with an even chance of occurring (50/50) and an annual frequency of exposure, the risk score must be high.
- The frequency of exposure can be evaluated separately without negatively affecting the validity of risk determinations, provided that adequate weighting is given to severity of outcome in the scoring system.
- Adaptations can be made of the risk assessment matrices shown in Chapter 11, "A Primer on Hazard Analysis and Risk Assessment," to develop a single risk scoring model that addresses injury to people (employees and the public); facilities, product, or equipment loss; operations downtime; and chemical releases and environmental damage.

- .If the number of gradations for probability, frequency of exposure, or severity is excessive, distinctions between them are difficult to make: Five gradations were chosen for probability and for frequency of exposure; for severity, there are four gradations.
- Although an attempt was made not to use a descriptive word more than once in the elements to be scored, it was not successful.

DEFINITIONS

Definitions follow for the terms relevant to the numerical risk scoring system presented here.

- *Hazard*: the potential for harm to people, property, and the environment. The dual nature of hazards must be understood. Hazards encompass all aspects of technology or activity that produce risk. Hazards include the characteristics of things and the actions or inactions of people.
- *Risk*: an estimate of the probability (likelihood) of a hazards-related incident or exposure occurring and the severity of harm or damage that could result.
- *Probability*: an estimate of the likelihood of an incident or exposure occurring that could result in harm or damage for the selected unit of time, events, population, items, or activity being considered.
- *Frequency of exposure*: the frequency and duration of exposure to the hazard over time.
- *Severity*: an estimate of the magnitude of the harm or damage that could result from a hazard-related incident or exposure, taking into consideration the number of people exposed, occupational health and environmental exposures, and the potential for property damage and production loss.

THE RISK SCORE FORMULA

Having decided to include frequency of exposure as a separate element to be scored, the question became how to assure that scoring computations, particularly for severity, are not skewed adversely. Thus, for this Risk Score Formula, the rating for frequency of exposure is not an equal multiplier; rather, it is added to the rating for occurrence probability to produce a midlevel score that is then multiplied by the severity rating.

$$\text{Risk Score} = (\text{Probability Rating} + \text{Frequency of Exposure Rating})$$
$$\times \text{Severity Rating}$$
$$RS = (PR + FER) \times SR$$

GRADATION AND SCORING DEVELOPMENT

To achieve plausibility, a variety of scorings were tested until credible results were obtained. The gradations and ratings selected are shown in Table 13.17.

TABLE 13.17   Descriptive Words and Ratings

Probability		Frequency of Exposure		Severity	
Frequent (Fre)	15	Often (Of)	13	Catastrophic (Cat)	50
Likely (Lik)	9	Occasional (Oc)	10	Critical (Cri)	40
Occasional (Occ)	4	Infrequent (In)	7	Medium (Med)	25
Remote (Rem)	1	Seldom (Se)	4	Minimal (Min)	10
Improbable (Imp)	0.5				

TABLE 13.18   Incident Probability

Category: Descriptive Word	Definition: Applies for the Selected Unit of Time, Events, Population, Item Unit of Time, or Activity
Frequent	Likely to occur repeatedly, to even chance
Likely	Likely to occur several times
Occasional	Occurs sporadically, likely to occur sometime
Remote	Not likely to occur, but could possibly occur
Improbable	So unlikely, can assume occurrence will not be experienced

TABLE 13.19   Severity of Consequences

Category: Descriptive Word	People: Employees, Public	Facilities, Product, or Equipment Loss	Operations Downtime	Environmental Damage
Catastrophic	Fatality	Exceeds 2 million	Exceeds 6 months	Major event, requiring several years for recovery
Critical	Disabling injury or illness	500,000 to 2 million	4 Weeks to 6 months	Event requires 1 to 2 years for recovery
Marginal	Minor injury or illness	50,000 to 500,000	2 days to 4 weeks	Recovery time is less than 1 year
Negligible	No injury or illness	Less than 50,000	Less than 2 days	Minor damage, easily repaired

WHAT THE DESCRIPTIVE WORDS MEAN

To give substance to the words used to establish gradations within the probability, severity, and frequency of exposure categories, their meanings as they are used in this risk scoring system are presented in Tables 13.18 to 13.20.

Four risk categories were considered adequate: Low, Moderate, Serious and High. Risk scores for each category were assigned as well as management decision indicators with respect to risk reduction actions to be taken or for risk acceptance. They are shown in Table 13.21.

It must be understood that the score levels provided in Table 13.22 were established through subjective judgments. They should be considered as advisory indicators and

TABLE 13.20   Frequency of Exposure

Category: Descriptive Word	Definition
Often	Continues to daily
Occasional	Daily to monthly
Infrequent	Monthly to yearly
Seldom	Less than yearly

TABLE 13.21   Risk Categories, Score Levels, and Action or Risk Acceptance Levels

Risk Category	Score Levels	Remedial Action, or Acceptance
Low	199 and below	Risk is acceptable; remedial action discretionary
Moderate	200 to 499	Remedial action to be taken at appropriate time
Serious	500 to 799	Remedial action to be given high priority
High	800 and above	Immediate action necessary. Operation not permissible except in an unusual circumstance and as a closely monitored and limited exception with approval of the person having authority to accept the risk

TABLE 13.22   The Risk Scoring System

Probabiltiy Category and Rating		Frequency of Exposure and Rating		Mid-Score	Cata-strophic	Final Score	Critical	Final Score	Medium	Final Score	Minimal	Final score
Fre	15	Of	13	28	50	1,400	40	1,120	25	700	10	280
	15	Oc	10	25	50	1,250	40	1,000	25	625	10	250
	15	In	7	22	50	1,100	40	880	25	550	10	220
	15	Se	4	19	50	950	40	760	25	475	10	190
Lik	9	Of	13	22	50	1,100	40	880	25	550	10	220
	9	Oc	10	19	50	950	40	760	25	475	10	190
	9	In	7	16	50	800	40	640	25	400	10	160
	9	Se	4	13	50	650	40	520	25	325	10	130
Occ	4	Of	13	17	50	850	40	680	25	425	10	170
	4	Oc	10	14	50	700	40	560	25	350	10	140
	4	In	7	11	50	550	40	440	25	275	10	110
	4	Se	4	8	50	400	40	320	25	200	10	80
Rem	1	Of	13	14	50	700	40	560	25	350	10	140
	1	Oc	10	11	50	550	40	440	25	275	10	110
	1	In	7	8	50	400	40	320	25	200	10	80
	1	Se	4	5	50	250	40	200	25	125	10	50
Imp	0.5	Of	13	13.5	50	675	40	540	25	338	10	135
	0.5	Oc	10	10.5	50	525	40	420	25	263	10	105
	0.5	In	7	7.5	50	375	40	300	25	188	10	75
	0.5	Se	4	4.5	50	225	40	180	25	113	10	45

not as absolutes in management decision making. For example, in a real-world case that has a high severity potential, remedial actions brought the risk score down to 225, which is very close to 199, the level where the risk would be considered acceptable and not requiring further remedial action. The work proceeded, with close observation.

Table 13.22, The Risk Scoring System, shows the various combinations with respect to incident probability, the frequency of exposure, and the severity of consequences; how probability and frequency ratings are totaled to become midscores; and the final score, arrived at by multiplying the severity score by the midscore.

The goal was to create a three-dimensional numerical risk scoring system that serves the needs of those who are more comfortable with statistics in their risk assessments; addresses the strong beliefs of those who want frequency of exposure given separate consideration in the risk assessment process; and maintains credibility and efficacy. The example in Table 13.23 demonstrates how the risk scoring system is applied.

TABLE 13.23 Example of How the Risk Scoring System Is Applied

Probability of Occurrence	Score	Frequency of Exposure	Score	Severity	Score	Risk Score
Frequent	15	Often	13	Critical	40	(15 + 13) x 40 = 1120
Likely	9	Occasional	10	Critical	40	(9 + 10) x 50 = 760
Occasional	4	Infrequent	7	Medium	25	(4 + 7) x 25 = 275
Remote	1	Seldom	4	Minimal	10	(1 + 4) x 10 = 50

CONCLUSION

Three-dimensional numerical risk scoring systems can be crafted and have credibility. But how are such systems to be used? Numerical risk scores carry an image of preciseness, and that can influence decision making and priority setting. In reality they should not be the sole or absolute determinant.

In the research for this chapter, one of the most interesting discussions of the real-world use of a numerical risk scoring system took place with a person whose principal interest is the safety of the products that her company produces. The risk determination system in that company requires that an independent facilitator serve as a discussion leader and that the risk review committee consist of at least five knowledgeable people. Consensus must be reached on risk scores and recommendations for any necessary risk reduction actions.

In these deliberations, it is common that severity scores for injuries to users or bystanders are moved up a level or two. For example, an injury requiring only a visit to a doctor and immediate return to work receives the lowest severity rating in the scoring system. However, when arriving at the recommendations to be made affecting product design or revising the operating procedures in instruction manuals, the review group regularly gives more weight to the injury than the scoring system indicates is necessary. Recommendations to achieve risk reduction are often influenced by the possible damage to corporate image, customer relations, and an assumed societal responsibility.

Numerical risk scoring systems can serve a real need. But it should be remembered that they consist of numerics arrived at through subjective judgments. Risk assessment is still as much an art as a science.

REFERENCES

Accident Prevention Manual, Administration and Programs, 13th ed. Itasca, IL: National Safety Council. 2009.

EN ISO 12100–2010: *Safety of Machinery—General principles for design. Risk assessment and risk reduction.* Geneva, Switzerland: International Organization for Standardization, 2010.

Failure Mode and Effects Analysis (FMEA): A Guide for Continuous Improvement for the Semiconductor Equipment Industry. Technology Transfer No.92020963A-ENG. Albany, NY: SEMATECH, 1992. Also available at http://www.sematech.org/docubase/document/0963beng.pdf.

Fine, William T. "Mathematical Evaluations for Controlling Hazards." White Oak, MD: Naval Ordinance Laboratory, 1971.

Grose, Vernon L. *Managing Risk: Systematic Loss Prevention for Executives.* Originally published by Prentice-Hall, 1987. Now available through Omega Systems Group, Arlington, VA.

Pilz: Guide to Machinery Safety, 6th ed. Farmington, MI: Pilz Automation Technology, 1999.

ADDENDUM A

FMEA FORM

FMEA FORM
SYSTEM:
SUBSYSTEM:
REFERENCE DRAWING:

FAILURE MODE AND EFFECTS ANALYSIS (FMEA)
FAULT CODE#

DATE:
SHEET:
PREPARED BY:

Subsystem/ Module &Function	Potential Failure Mode	Potential Local Effect(s) of Failure	Potential End Effect(s) of Failure	SEV	Cr	Potential Cause(s) of Failure	OCC	Current Controls/ Fault Detection	DET	RPN	Recommended Action(s)	Area/ Individual Responsible & Completion Date(s)	Action Taken	SEV	OCC	DET	RPN
Subsystem name and function	"How can this subsystem fail to perform its function?" "What will an operator see?"	The local effect of the subsystem or end user	Downtime # of hours at the system level. 2. Safety 3. Environ- mental 4. Scrap loss	7	1	"How can this failure occur?" Describe in terms of something that can be corrected or controlled. 2. Refer to specific errors or malfunctions	5	"What mechanisms are in place that could detect, prevent, or minimize the impact of this cause?"	6	210	"How can we change the design to eliminate the problem?" "How can we detect (fault isolate) this cause for this failure?" "How should we test to ensure the failure has been eliminated?" "What PM procedures should we recommend?" 1. Begin with highest RPN. 2. Could say "no action" or "Further study is required." 3. An idea written here does not imply corrective action.	"Who is going to take responsibility?" "When will it be done?"	"What was done to correct the problem?" Examples: Engineering Change, Software revision, no recommended action at this time due to obsolescence, etc.	6	1	1	6

Critical Failure symbol (Cr). Used to identify critical failures that must be addressed (i.e., whenever safety is an issue).

Severity Ranking (1–10) (see Severity Table)

Occurrence Ranking (1–10) (see Occurrence Table)

Detection Ranking (1–10) (see Detection Table)

Risk Priority Number
RPN = Severity*Occurrence *Detection

How did the "Action Taken" change the RPN?

ADDENDUM B

MATHEMATICAL EVALUATIONS FOR CONTROLLING HAZARDS

William T. Fine
Naval Ordinance Laboratory
White Oak, Silver Springs, MD
1971

Abstract: To facilitate expeditious control of hazards for accident prevention purposes, two great needs have been recognized. These are for:

1. a method to determine the relative seriousness of all hazards for guidance in assigning priorities for preventive effort; and
2. a method to give a definite determination as to whether the estimated cost of the contemplated corrective action to eliminate a hazard is justified.

To supply these needs, a formula has been devised which weighs the controlling factors and "calculates the risk" of a hazardous situation, giving a numerical evaluation to the urgency for remedial attention to the hazard. Calculated Risk Scores are then used to establish priorities for corrective effort. An additional formula weighs the estimated cost and effectiveness of any contemplated corrective action against the Risk Score and gives a determination as to whether the cost is justified.

Advanced Safety Management: Focusing on Z10 and Serious Injury Prevention,
Second Edition. Fred A. Manuele.

Chapter 1: Introduction

General. The purpose of this chapter is to illustrate the need for quantitative evaluations to aid in the control of hazards and to explain the general plan of this report.

A problem frequently facing the head of any (field type) safety organization is to determine just how serious each known hazard is, and to decide to what extent he should concentrate his resources and strive to get each situation corrected. Normal safety routines such as inspections and investigations usually produce varying lists of hazards which cannot all be corrected at once. Decisions must be made as to which ones are the most urgent. On costly projects, management often asks whether the risk due to the hazard justifies the cost of the work required to eliminate it. Since budgets are limited, there is a necessity to assign priorities for costly projects to eliminate hazards.

The question of whether a costly engineering project is justified is usually answered by a general opinion which may be little better than guesswork. Unfortunately in many cases, the decision to undertake any costly correction of a hazard depends to a great extent on the salesmanship of safety personnel. As a result, due to insufficient information, the cost of correcting a very serious hazard may be considered prohibitive by management, and the project postponed; or due to excellent selling jobs by Safety, highly expensive engineering or construction jobs may be approved when the risks involved really do not justify them.

In Chapter 2 of this report, a formula is presented to "calculate the risk" due to a hazard, or to quantitatively evaluate the potential severity of a hazardous situation. Use of this formula will provide a logical system for safety and management to determine priorities for attention to hazardous situations, and guidance for safety personnel in determining the areas where their efforts should be concentrated.

In Chapter 3 of this report, a formula is presented for determining whether or not the cost of eliminating a hazard is justified. Use of the formula will provide a solid foundation upon which safety personnel may base their recommendations for engineering-type corrective action. It will assure that projects which are not justified will not be recommended.

This report deals with justification of costs to eliminate hazards. This does not imply in any way that a cost, no matter how great, is not worthwhile if it will prevent an accident and save a human life. However, we must also consider accident prevention with reason and judgment. Budgets are not unlimited. Therefore, the maximum possible benefit for safety must be derived from any expenditure for safety. When an analysis results in a decision that the cost of certain measures to eliminate a hazard "is not justified," we do not say or suggest that the hazard is not serious and may be ignored.

We do say that, based on evaluation of the controlling factors, the return on the investment, or in other words, the amount of accident prevention benefit, is below the standards we have established. The amount of money involved will no doubt provide greater safety benefit if used to alleviate other higher-risk hazards which this system will identify. As for the hazard in question, less costly preventive measures should be sought.

Definitions: For the purpose of this presentation, three factors are defined as follows.

a. *Hazard*: Any unsafe condition or potential source of an accident. Examples are: an unguarded hole in the ground; defective brakes on a vehicle; a deteriorated wood ladder; a slippery road.

b. *Hazard-event*: An undesirable occurrence; the combination of a hazard with some activity or person which could start a sequence of events to end in an accident. Examples of hazard events are: a person walking through a field which contains a hazard such as an unguarded well opening; a person not wearing eye protection while in an eye hazardous area; a person driving a vehicle that has defective brakes; a man climbing up a defective ladder; a vehicle being driven on a slippery road.

c. *Accident- sequence:* This chain of events or occurrences which take place starting with a "hazard-event" and ending with the consequences of an accident.

d. Additional definitions will be provided in later pages as needed.

Chapter 2. Formula for Evaluating the Seriousness of the Risk Due to a Hazard

General. The purpose of this chapter is to present a complete explanation of the method for quantitatively evaluating the seriousness of hazards, and some of the benefits that may be derived from such analyses.

The expression "a calculated risk" is often used as a catchall for any case when work is to be done without proper safety measures being taken. But usually such work is done without any actual calculation. By means of this formula, the risk is calculated. The seriousness of the risk due to a hazard is evaluated by considering the potential consequences of an accident, the exposure or frequency of occurrence of the hazard-event that could lead to the accident, and the probability that the hazard-event will result in the accident and consequences.

The *formula* is as follows:

$$\text{Risk Score} = \text{Consequences} \times \text{Exposure} \times \text{Probability}$$
$$\text{Abbreviated}: R = C \times E \times P$$

Definitions of the elements of the formula and numerical ratings for the varying degrees of the elements are given below.

a. *Consequences C:* The most probable result of a potential accident, including injuries and property damage. This is based on an appraisal of the entire situation surrounding the hazard, and accident experience. Classifications and ratings are:

Description	Rating
(1) Catastrophic: numerous fatalities; extensive damage (over $1,000,000);major disruption of activities of national significance	100
(2) Multiple fatalities; damage $500,000 to $1,000,000	50
(3) Fatality; damage $100,000 to $500,000	25
(4) Extremely serious injury (amputation, permanent disability); damage $1,000 to $100,000	15
(5) Disabling injuries; damage up to $1,000	5
(6) Minor cuts, bruises, bumps, minor damage	1

b. *Exposure E: Frequency of occurrence of the hazard-event*—the undesired event which could star the accident-sequence. Classifications are below. Selection is based on observation, experience, and knowledge of the activity concerned.

Description	Rating
The hazard event occurs:	
(1) Continuously (or many times daily)	10
(2) Frequently (approximately once daily)	6
(3) Occasionally (from once per week to once per month)	3
(4) Unusually (from once per month to once per year)	2
(5) Rarely (it has been known to occur)	1
(6) Very rarely (not known to have occurred but considered remotely possible)	0.5

c. *Probability P:* This is the likelihood that, once the hazard-event occurs, the complete accident-sequence of events will follow with the necessary timing and coincidence to result in the accident and consequences. This is determined by careful consideration of each step in the accident sequence all the way to the consequences, and based upon experience and knowledge of the activity, plus personal observations. Classifications and ratings follow:

Description	Rating
The accident-sequence, including the consequences:	
(1) Is the most likely and expected result if the hazard-event takes place	10
(2) Is quite possible, would not be unusual, has an even 50/50 chance	6
(3) Would be an unusual sequence or coincidence	3
(4) Would be a remotely possible coincidence. (It has happened here.)	1
(5) Extremely remote but conceivably possible. (Has never happened after many years of exposure.)	0.5
(6) Practically impossible sequence or coincidence; a "one in a million" possibility. (Has never happened in spite of exposure of many years.)	0.1

Examples   The use of this formula is demonstrated by actual examples. Six widely different types of situations have been selected to illustrate the broad applicability of the formula. *[In this condensation of Fine's paper, three examples are given.]*

a. Example No. 1

1. *Problem.* There is a quarter-mile stretch of two-lane road used frequently by both vehicles and pedestrians departing or entering the grounds. There is no sidewalk, so pedestrians frequently walk in the road, especially when he grass is wet or snow covered. There is little hazard to pedestrians when all the traffic is going in one direction only; but when vehicles are going in both directions and passing by each other, the vehicles require the entire width of the road, and pedestrians must then walk on the grass alongside the road: It is considered that an accidental fatality could occur if a pedestrian steps into the road, or remains in the road at a point where two vehicles are passing.

2. *Steps to Use the Risk Score Formula*:

Step 1. List the accident-sequence of events that could result in the undesired consequence.

1. It is a wet or snowy day, making the grass along the road wet and uninviting to walk on.
2. At quitting time, a line of vehicles, and some pedestrians, are leaving the grounds, using this road.
3. One pedestrian walks on the right side of this road, and he has an attitude which makes him oblivious to the traffic. (This is the hazard-event.)
4. Although traffic is "one way" out at this time, one vehicle comes from the opposing direction, causing the outgoing traffic line to move to the right edge of the road.

5. The pedestrian on the right side of the road fails to observe the vehicles, and he remains in the road.

6. The driver of one vehicle fails to notice the pedestrian and strikes him from the rear.

7. The pedestrian is killed.

Step 2. Determine values for elements of formula: *Consequences*: A fatality. Therefore, $C = 25$.

- *Exposure*: The hazard-event is event 3 above, the pedestrian remaining in road and refusing to notice the line of traffic. It is considered that this type of individual appears or is "created" by conditions occasionally. Therefore, $E = 3$.

- *Probability* of all events of the accident sequence following the hazard-event is: "conceivably possible, although it has never happened in many years." Reasoning is as follows: events 4, 5, 6, and 7 are individually unlikely, so the combination of their occurring simultaneously is extremely remote.

Event 4 is unlikely because traffic is "one way" at quitting time.
Event 5 is unlikely because a number of drivers would undoubtedly sound their horns and force the pedestrian's attention.
Event 6 is unlikely because most drivers are not deliberately reckless.
Event 7, a fatality, is unlikely because vehicle speeds are not great on the road, and the most likely case would be a glancing blow and minor injury. Not even a minor injury has ever been reported here. In view of the above, Probability $P = 0.5$.

Step 3. Substitute into formula and determine the Risk Score.

$$R = C \times E \times P = 25 \times 3 \times 0.5 = 37.5$$

(*Note*: The Risk Score or one case alone is meaningless. Additional hazardous situations must also be calculated for comparative purposes and a definite pattern. Additional cases are similarly calculated below.)

b. Example No.2

1. *Problem.* A 12,000-gallon propane storage tank is subject to two hazards. One hazard is the fact that the tank is located alongside a well-traveled road. The road slopes, and is occasionally slippery due to rain, snow, or ice. It is considered possible that a vehicle (particularly a truck) could go out of control, leave the road, strike and rupture the tank, and cause a propane gas explosion and fire that could destroy several buildings, with consequences amounting to damage costing $200,000, plus a fatality.

   The second hazard is the tank's location close to ultrahigh-compressed air lines and equipment. A high-pressure pipeline explosion could result from a malfunctioning safety valve, a human error in operating the equipment, damage to a pipeline, or from other causes. Blast or flying debris could conceivably strike the propane tank, rupture it, and cause it to explode with the same consequences as for a runaway vehicle.

2. *Using the Risk Score Formula* (*Note*: In this case there are two hazards, so the evaluation is done in two parts, one for each of the hazards, and the total scores are added.)

Step 1. Consider—just the first hazard, that due to a vehicle. List the sequence of events that would result in an accident:

1. Many vehicles are driven down the hill alongside the storage tank
2. The road has suddenly become slippery due to an unexpected freezing rain.
3. One truck starts to slide on the slippery road as it goes down this hill. (*Note*: This is the "hazard-event" that starts the accident sequence.)
4. The driver loses his steering control at a point when he is uphill from and approaching the tank.
5. Brakes fail to stop the vehicle from sliding.
6. Vehicle heads out of control toward the tank.
7. Vehicle strikes the tank with enough force to rupture it and permit the propane gas to leak out.
8. A spark ignites the propane
9. Explosion and conflagration occur.
10. Building and equipment damage is $200,000, and one man is killed.

Step 2. Substitute numerical values into formula:

- *Consequences*: One fatality and damage loss of $200,000. Therefore, $C = 25$.
- *Exposure*: The hazard-event that would start the accident sequence is—the truck starting to slide on this road. This happened "rarely." Therefore, $E = 1$.
- *Probability*. To decide on the likelihood that the complete accident-sequence will follow the occurrence of the hazard-event, we consider the probability of each event:

a. Loss of steering control to occur at the precise point in the road approaching the tank is possible but would be a coincidence.
b. Once the vehicle started to slide, if the road was ice covered, it would be expected that the brakes would fail to stop the slide.
c. The vehicle heading toward the tank is highly unlikely. Momentum would cause the vehicle to continue straight down the road.
d. The vehicle striking the tank with great force is extremely unlikely.

If a vehicle were sliding on an ice-covered surface toward the tank, it would be easily diverted from its direction of travel by a number of obstructions between the road and the tank. When roads are slippery, travel is curtailed and drivers are cautioned to drive slowly. A slow rate of speed would be unlikely to produce enough force to damage the tank. The shape and position of the tank are such that a vehicle would tend to glance off it.

In summary, because of the highly unlikely nature of most of the events, this sequence has a one-in-a-million probability. It has never happened. But it is conceivable. Therefore, $P = 0.5$.

Step 3. Substitute in the formula.

$$\text{Risk Score} = 25 \times 1 \times 0.5 = 12.5$$

The entire process is to be repeated for the second hazard, which is the location near the high-pressure air lines and equipment.

Step 1. List the sequence of events.

1. Normal daily activities involve operation of equipment and pressuring of pipelines, some of which are in the vicinity of the propane storage tank.
2. A pipeline containing air compressed to 3,000 pounds per square inch, approximately 50 feet away from the storage tank has become deteriorated or damaged. (This is the hazard-event.)
3. The pipeline bursts.
4. Metal debris is thrown by the blast in all directions, several pieces flying and striking the propane tank with such force that the tank is ruptured.
5. Propane starts to leak out of the tank.
6. A spark ignites the propane fumes.
7. The propane and air mixture explodes.
8. Building damage is $200,000, and one man is killed.

Step 2. Determine values and substitute in the formula.

- *Consequences*: One fatality and damage loss of $200,000. $C = 25$.
- *Exposure*: High-pressure air lines have been known to have been neglected or damaged. Frequency of such occurrences is considered "unusual." Therefore, $E = 2$.
- *Probability*: Now we estimate the likelihood that a damaged pipeline will explode and the explosion will occur close enough and with enough blast to throw debris and strike the propane tank with such force as to complete the accident sequence. Several bursts have occurred in the past few years, but none have damaged the propane tank. Few of the pipelines are close enough to endanger the tank. After careful consideration, the accident sequence is considered "very remotely possible." $P = 0.5$.

Step 3. Substituting into the formula.

$$\text{Risk Score} = 25 \times 2 \times 0.5 = 25$$
$$\text{Totaling Risk Score} = 12.5 + 25 = 37.5$$

Chapter 3. Formula to Determine the Justification for Recommended Corrective Actions

General. The purpose of this chapter is to describe the method of determining whether the cost of corrective action to alleviate a hazard is justified. Once a hazard has been recognized, appropriate corrective action must be tentatively decided upon and its cost estimated. Now the "Justification" formula can be used to determine whether the estimated cost is justified.

The *formula* is as follows:

$$\text{Justification} = \frac{\text{Consequences} \times \text{Exposure} \times \text{Probability}}{\text{Cost Factor} \times \text{Degree of Correction}}$$

Elements are abbreviated:

$$J = \frac{C \times E \times P}{CF \times DC}$$

It should be noted that the elements of the numerator of this formula are the same as the Risk Score formula described in Chapter 2. We have simply added a denominator made up of two additional elements, which are as follows:

a. *Cost Factor CF*: A measure of the estimated dollar cost of the proposed corrective action. Classifications and ratings are:

Cost	Ratings
(1) Over $50,000	10
(2) $25,000 to #50,000	6
(3) $10,000 to $25,000	4
(4) $1,000 to $10,000	3
(5) $100 to $1,000	2
(6) $25 to $100	1
(7) Under $25	0.5

b. *Degree of Correction DC*: An estimate of the degree to which the proposed corrective action will eliminate or alleviate the hazard, forestall the hazard-event, or interrupt the accident sequence. This will be an opinion based on experience and knowledge of the activity concerned. Classifications and ratings are:

Description	Rating
(1) Hazard positively eliminated, 100%	1
(2) Hazard reduced at least 75%, but not completely	2
(3) Hazard reduced by 50 to 75%	3
(4) Hazard reduced by 25 to 50%	4
(5) Slight effect on hazard, less than 25%	6

Criteria for Justification   Values are substituted into the formula to determine the numerical value for Justification. The Critical Justification Rating is 10. For any rating over 10, the expenditure will be considered justified. For a score less than 10, the cost of the contemplated corrective action is not justified.

Note: The Critical Justification Rating has been arbitrarily set at 10, based on experience, judgment, and the current budgetary situation. After extended experience at an individual organization, based on accident experience, budgetary situations, and appraisals of the safety status, it may be found desirable to raise or lower the critical score.

Examples   The use of the Justification formula will be illustrated by the use of the same six examples discussed in Chapter 2. [Three are shown here.]

a. Example No. 1   The hazard of pedestrians and vehicles using the same road. To reduce this risk, the corrective action being considered is to construct a sidewalk alongside the road, at an estimated cost of $1,500. The "*J*" formula is now used to determine whether this contemplated expenditure is justified.

1. *Substitute values in the "J" formula*:

$$J = \frac{C \times E \times P}{CP \times DC}$$

    a. *C, E*, and *P*, for this situation were discussed as Example No. 1 in Chapter 2 of this report and determined to be 25, 3, and 0.5, respectively.

    b. *Cost Factor*: The estimated cost factor is $1,500. Therefore, *CF* is 3.

    c. *Degree of Correction*: The probability of the hazard-event occurring is considered to be reduced at least 75% but 100%, by the construction of the sidewalk: Therefore, *DC* = 2.

    d. *Justification Rating*

$$J = \frac{25 \times 3 \times 0.5}{3 \times 2} = \frac{37.5}{6} = 6.25$$

2. *Conclusion. J* is less than 10. Therefore, the cost of construction of the sidewalk is not justified.

Note: This lack of sufficient justification evaluates the situation from the safety viewpoint only. Management could feel there is added justification for morale or other purposes.

3. *Additional consideration.* Since the Risk Score is still a substantial 37.5, other less costly corrective measures should be sought. This includes improved administrative controls to enforce one-way traffic, reduce speed, and encourage pedestrians to use another exit gate. This will reduce the Risk Score by reducing both Exposure and Probability.

b. Example No. 2 The hazard due to compressed air being used in a shop without proper pressure reduction nozzles. The proposed corrective action is installation of proper pressure reducing nozzles on the 50 air hoses, at a cost of $8 each, or $400. To determine justification for the expenditure:

1. Determine values for the elements of the J formula:
   a. C, E, and P were discussed in Example No. 2 of Chapter 2 and evaluated at 5, 10, and 6, respectively.
   b. *Cost Factor.* The cost of the corrective action is $400, so $CF = 2$.
   c. *Degree of Correction.* The corrective action will reduce the hazard by at least 50 percent, so $DC = 3$.
   d. Substituting in the formula:

$$J = \frac{5 \times 10 \times 6}{2 \times 3} = \frac{300}{6} = 50$$

2. *Conclusion.* "J" is well above 10. The cost of installing pressure reduction nozzles is strongly justified.

c. Example No. 3 The hazardous location of the 12,000-gallon propane storage tank. The proposed corrective action is to relocate the tank to a place where it will be less likely to be damaged by any external source, at an estimated cost of $16,000.

1. Determine values for elements of the formula.
   a. C, E, and P were determined in Example No. 3 of Chapter 2 to be 25, 1, and 1.5 (the two hazards combined).
   b. *Cost Factor:* Cost of relocation is $16,000. $CF = 4$.
   c. *Degree of Correction.* In the very best location available, there still remains a remote possibility of damage to the tank, so $DC = 2$.
   d. Substituting in the formula:

$$J = \frac{25 \times 1 \times 1.5}{4 \times 2} = \frac{37.5}{8} = 4.7$$

2. *Conclusion.* Based on the established criteria, the cost of relocation of the tank is not justified.
3. It is emphasized that the conclusion in this case that the proposed corrective action is not justified does not mean that the hazard is of little or no significance. The Risk Score is still 37.5, and this remains of appreciable concern.

   Since the potential consequences of an accident are quite severe, effort should be expended to reduce the risk, by reducing either the Exposure of the Probability, or devising other less costly corrective action. In this case, it is considered that an additional steel plate barrier could be erected to protect the tank from the compressed air activities, and one or two strong posts in the ground could minimize danger from the road. Thus, the Probability of serious damage to the truck, and the Risk Score, would be considerably lessened at a very nominal cost.

Recommended Procedure for Using the "*J*" Formula

A convenient "*J*" Formula worksheet is furnished for undertaking a hazard analysis to determine the Justification Rating. Once a hazard has been recognized, the following procedure is recommended:

a. State the problem briefly.

b. Decide on the most likely consequences of an accident due to the hazard.

c. Review all factors carefully, on the scene. List the actual step-by-step sequence of events that is most likely to result in the consequences chosen. You must be specific.

d. Decide on the most appropriate corrective action and obtain or make a rough estimate of its cost.

e. Consider carefully the effect of the proposed corrective action on the hazard, and estimate roughly the degree to which the dangerous situation will be alleviated.

f. If alternative corrective measures are possible, repeat steps (d) and (e) for them.

g. Select the hazard-event: the first undesirable occurrence that could start the accident sequence.

h. Consider the existing situation carefully to determine the frequency of the occurrence of the hazard-event: by on the scene observation, and then decide on the Exposure Rating. If in doubt between two ratings, interpolate.

i. For the Probability Rating, consider the likelihood of the occurrence of each event of the accident sequence, including the resulting injury and/or damage, and form an opinion based on the descriptive words. For example, if two unusual coincidences are required, this could be considered "remotely possible"; two "remotely possible" occurrences could be "conceivably possible" etc. If in doubt between two ratings, interpolate.

Endeavor to be consistent. Consider the occurrence of only the same consequences which were decided on in step (b) above. For example, if you decided on consequences of a fatality, then in this step you may only consider the probability of a fatality. If you also wish to consider lesser injuries, a separate and additional computation must be made, since both the Consequences and Probability evaluations would be different. Scores should be added.

j. You have now obtained ratings for all the elements of the "*J*" formula. Substitute in the formula and compute the Justification Score.

k. If alternative corrective measures are being considered to alleviate the hazard, compute their Justification Scores also.

l. If there are alternative corrective measures which have acceptable Justification Scores, the most desirable from the Safety standpoint is the one which would make the greatest reduction in the Risk Score. Therefore, for each alternative, assume that the corrective measures are in effect and re-compute the Risk Score. Of course, this selection may also be affected by external (non-safety) considerations, such as the size of investment required, the relative effects on morale, esthetics, efficiency, convenience, ease of implementation, etc.

Exception to Reliance on the "*J*" Formula

A highly hazardous situation may exist for which no corrective action that can be devised will give an acceptable Justification Score. Obviously in such a case, whatever corrective action is necessary to reduce the Risk Score should be taken, regardless of the Justification Score.

"*J*" FORMULA WORKSHEET

Problem:

Sequence of events or factors necessary for accident:
1.
2.
3.
4.
5.
6.
7.
8.

Formula Factors	Rating

C   Consequences _____

E   Exposure _____

P   Probability _____

CF   Cost Factor _____

DC   Degree of Correction _____

J   Justification: $J = \dfrac{C \times E \times P}{CF \times DC} = \dfrac{\times \quad \times}{\times}$

CHAPTER 14

HIERARCHY OF CONTROLS: SECTION 5.1.2 OF Z10

Section 5.1.2 of ANSI/AIHA Z10-2012, the *Occupational Health and Safety Management Systems* standard, deals with the hierarchy of controls. The opening sentence in that section states: "The organization shall establish and implement a process for achieving feasible risk reduction based on the following order of controls." A prescribed hierarchy of controls immediately follows that provision.

The hierarchy in Z10 is the base for decision making when action is taken to resolve occupational health and safety issues. Those issues are "defined as hazards, risks, management system deficiencies, and opportunities for improvement," as in the Planning Section (4.0).

Application of a hierarchy of controls is of such importance that a separate chapter is devoted to it. Thus, in this chapter we:

- Comment on the evolution of hierarchies of control
- Discuss the hierarchy of controls in Z10
- Provide guidelines on the application of a hierarchy of controls
- Establish the logic of taking steps in the hierarchy of controls in the order given
- Place the hierarchy of controls within a good problem-solving techniques, as in The Safety Decision Hierarchy

Advanced Safety Management: Focusing on Z10 and Serious Injury Prevention,
Second Edition. Fred A. Manuele.
© 2014 John Wiley & Sons, Inc. Published 2014 by John Wiley & Sons, Inc.

EVOLUTION OF THE HIERARCHY OF CONTROLS

Z10's hierarchy of controls has six elements. Hierarchies in other published standards and guidelines have three to nine elements. Z10's hierarchy of controls is the outcome of the work of a large number of safety professionals over many years. All of the contributors cannot be recognized here. A limited review of the evolution of the hierarchy of controls follows.

AT THE NATIONAL SAFETY COUNCIL

The third edition of the National Safety Council's *Accident Prevention Manual* was published in 1955. Section 4, "Removing the Hazard from the Job," provides a three-step "order of effectiveness and preference," This is taken from the *Accident Prevention Manual*.

> The engineer should include in his planning and follow-through such measures as will attain one of the accident prevention goals listed as follows (in the order of effectiveness and preference):
>
> - Elimination of the hazard from the machine, method, material, or plant structure.
> - Guarding or otherwise minimizing the hazard at its source if the hazard cannot be eliminated.
> - Guarding the person of the operator through the use of personal protective equipment if the hazard cannot be eliminated or guarded at its source. (p. 4–1)
>
> Company policies should be such that safety can be designed and built into the job rather than added after the job has been put into operation.

Establishing the concept that risk reduction actions should be taken in an order of effectiveness and preference was an important step in the evolution of the practice of safety. It implies that some steps in the process are preferable since they achieve greater risk reduction than others. Declaring that safety policies should require that safety be designed and built into the job rather than be dealt with as an add-on is also a premise that influenced later versions of hierarchies of control.

THE COUNCIL OF THE EUROPEAN COMMUNITIES

Although the excerpts taken from the Directive issued in 1989 on workplace safety by the Council of the European Communities are not labeled a hierarchy of controls, that's what they are: a nine-element list of "principles of prevention." It is intriguing because it begins with risk avoidance, stresses adopting the work to the capabilities of individuals, gives prominence to the design of the workplace and the choice of work equipment; and advises employers "to take account of changing circumstances and aim to improve existing situations."

The full title of the bulletin is "Council Directive 89/391/EEC of 12 June 1989 on the introduction of measures to encourage improvements in the safety and health of workers at work." It is available at http://eur-lex.europa.eu/LexUriServ/LexUriServ.do?uri=CELEX:31989L0391:en:HTML.

The following general obligations on employers is from Article 6.

1. Within the context of his responsibilities, the employer shall take the measures necessary for the safety and health protection of workers, including prevention of occupational risks and provision of information and training, as well as provision of the necessary organization and means. The employer shall be alert to the need to adjust these measures to take account of changing circumstances and aim to improve existing situations.

2. The employer shall implement the measures referred to in the first subparagraph of paragraph 1 on the basis of the following general principles of prevention:

   a. avoiding risks;

   b. evaluating the risks which cannot be avoided:

   c. combating the risks at source;

   d. adapting the work to the individual, especially as regards the design of work places, the choice of work equipment and the choice of working and production methods, with a view, in particular, to alleviating monotonous work and work at a predetermined work-rate and to reducing their effect on health.

   e. adapting to technical progress;

   f. replacing the dangerous by the non-dangerous or the less dangerous;

   g. developing a coherent overall prevention policy which covers technology, organization of work, working conditions, social relationships and the influence of factors related to the working environment;

   h. giving collective protective measures priority over individual protective measures;

   i. giving appropriate instructions to the workers.

ANSI/PMMI B155.1-2011

The Packaging Machinery Manufacturers Institute is the secretariat for the standard *Safety Requirements for Packaging Machinery and Packaging-Related Converting Machinery*. A revision of B155.1 was approved by ANSI in 2011. It replaced a version issued in 2006. In effect, this is an eight-element process within three classifications. In part, this is the guidance given in section 6.5.1, "Use the Hazard Control Hierarchy."

In selecting the most appropriate risk reduction measures, apply the following principles in the order in which they appear.

Classification—Design Out
   Elimination or substitution
Classification—Engineering Controls
   Guards and safeguarding devices
Classification—Administrative Controls
   Awareness devices
   Training and procedures
   Personal protective equipment

This hierarchy of controls repeats the elements in the hierarchies in other standards. They are close to the provisions in MIL-STD-882E, comments for which follow.

MIL-STD-882E-2012

The Department of Defense *Standard Practice For System Safety, MIL-STD-882*, issued in 1969, was a seminal document at that time. Four revisions of 882 have been issued over a span of 43 years. This standard has had considerable influence on the development of risk assessment, elimination, and amelioration concepts and methods. Much of the wording on risk assessments and hierarchies of control in safety standards and guidelines issued throughout the world is comparable to that in the several versions of 882.

The fifth edition, issued in May 2012, is designated MIL-STD-882E. It is available on the Internet at http://www.systemsafetyskeptic.com/yahoo_site_admin/assets/docs/MIL-STD-882E_final.135152939.pdf. It can be downloaded at no cost. Scroll down and click on the 882E indicator on the right-hand side.

A "system safety design order of precedence" is outlined in 882E. *Precedence* means priority in order, rank, or importance. In 882D, the design order of preference had four elements. This latest version has five. (p. 13)

The system safety design order of precedence identifies alternative mitigation approaches and lists them in order of decreasing effectiveness.

a. Eliminate hazards through design selection.

b. Reduce risk through design alteration.

c. Incorporate engineered features or devices.

d. Provide detection and warning devices.

e. Incorporate signage, procedures, training, and PPE.

Instruction is given in 882E to assure that:

• Mitigation measures, if necessary, are selected and implemented to achieve an acceptable risk level.

• Implementation measures are evaluated to verify that they have achieved the risk reduction expected.

• Documentation comments on the validation steps taken and the logic in support of risk acceptance.

HIERARCHIES OF CONTROL: PREMISES AND GOALS

A *hierarchy* is a system of persons or things ranked one above the other. The hierarchy of controls in Z10 is to provide a systematic way of thinking, considering steps in a ranked and sequential order, to choose the most effective means of eliminating or reducing hazards and the risks that derive from them. Acknowledging that premise, that risk reduction measures should be considered and taken in a prescribed order, represents an important step in the evolution of the practice of safety.

A model of hierarchies of control may give examples of the types of actions to be taken for each of its elements, as does Appendix G in Z10. But little is written about the purpose and the goals to be achieved in applying a hierarchy of controls. An attempt to do so follows.

A major premise to be considered in applying a hierarchy of controls is that the outcome of the actions taken is to be an acceptable risk level, defined as follows.

Acceptable risk is that risk for which the probability of a hazard-related incident or exposure occurring and the severity of harm or damage that could result are as low as reasonably practicable in the situation being considered.

That definition requires taking into consideration each of the two distinct aspects of risk as risk reduction actions are decided upon:

- Avoiding, eliminating, or reducing the *probability* of a hazard-related incident or exposure occurring
- Reducing the *severity* of harm or damage that may result if an incident or exposure occurs

The definition of acceptable risk also requires reflection on the feasibility and effectiveness of the risk reduction measures to be taken and their costs in relation to the amount of risk reduction to be achieved. The following appears in the "shall" column as a part of section 5.1.2 of Z10.

Feasible applications of this hierarchy of controls shall take into account:

- The nature and extent of the risks being controlled
- The degree of risk reduction desired
- The requirements of applicable local, federal and state statutes, standards, and regulations
- Recognized best practices in industry
- Available technology
- Cost effectiveness
- International organization standards

Decision makers should understand that with respect to the six levels of action shown in the hierarchy of controls in Z10:

The ameliorating actions described in the first, second, and third action levels are more effective because they

- Are *preventive* actions that eliminate or reduce risk by design, substitution, and engineering measures
- Rely the least on personnel performance
- Are less defeatable by supervisors or other employees

Actions described in the fourth, fifth, and sixth levels are *contingent* actions and rely greatly on the performance of personnel.

What Kepner and Tregoe wrote in *The New Rational Manager* about taking preventive and contingent actions in the problem-solving process fits precisely with the risk elimination and amelioration concepts set forth here.

Two kinds of actions are available to anyone conducting a Potential Problem Analysis: preventive actions and contingent actions. The effect of preventive actions is to remove, partially or totally, the likely cause of a potential problem.

The effectiveness of a contingent action is to reduce the impact of a problem that cannot be prevented. Preventive actions, if they can be taken, are obviously more efficient than contingent actions. (p. 147)

As decisions are made in applying each step in the hierarchy of controls, the following should be considered as goals:

- Avoiding work methods that are overly stressful, taking into consideration worker capabilities and limitations
- Keeping the probability of human error as low as reasonably practicable by designing workplaces and work methods that are not error-provocative, meaning that they do not (as in Chapanis, p. 119):
  - Violate operator expectations
  - Require performance beyond what an operator can deliver
  - Induce fatigue
  - Provide adequate facilities or information for the operator
  - Present unnecessarily difficult or unpleasant requirements
  - Include unnecessarily dangerous methods
- Designing systems so that human interaction with equipment and processes is as low as reasonably practicable
- Designing systems so that use of personal protective equipment is as low as reasonably practicable

THE HIERARCHY OF CONTROLS IN Z10

We said in Chapter 1 that although Z10 is a management system standard and not a specification standard, the provisions pertaining to a hierarchy of controls are the exception. Rather than present a performance statement relating to the outcomes to be achieved through a risk reduction process, a specifically defined hierarchy of controls is outlined in section 5.1.2.

The organization shall establish a process for achieving feasible risk reduction based upon the following preferred order of controls:

A. Elimination

B. Substitution of less hazardous materials, processes, operations, or equipment

C. Engineering controls

D. Warnings

E. Administrative controls

F. Personal protective equipment

This hierarchy of controls contains six elements. The first step, Elimination, is separated from the Substitution element. In reality, undertaking separate processes is required to accomplish what is needed. Those subjects were combined in some hierarchies issued previously.

THE LOGIC OF TAKING ACTION IN THE DESCENDING ORDER GIVEN

Comments follow on each of the action elements listed in Z10's hierarchy of controls, including the rationale for the listing in the order given. Taking actions as *feasible and practicable* in the prescribed order is the most effective means to achieve risk reduction.

A. Elimination

Use of the term *elimination* as the first step in applying a hierarchy of controls is a bit simplistic. My experience requires that it be replaced with such as: "Eliminate or reduce hazards and risks through system design and redesign." (In Chapter 16, "Prevention through Design", Avoidance precedes Elimination.)

The theory is stated plainly. If the hazards are eliminated in the design and redesign processes, risks that derive from those hazards are also eliminated. But the complete elimination of hazards by modifying the design may not always be practicable. Then the goal is to modify the design, within practicable limits, so as to limit the:

• Probability of personnel making human errors because of design inadequacies

• Ability of personnel to defeat the work system and the work methods prescribed

Examples include designing to eliminate or reduce the risk from:

- Fall hazards
- Ergonomic hazards
- Confined-space hazards
- Noise hazards
- Chemical hazards.

Obviously, hazard elimination or reduction is the most effective way to remove or reduce risk. If a hazard is eliminated or reduced, the need to rely on worker behavior to avoid risk is diminished.

B. Substitution of Less Hazardous Materials, Processes, Operations, or Equipment

Methods that illustrate substituting less hazardous methods, materials, or processes for that which is more hazardous include:

- Using automated material-handling equipment rather than manual material handling
- Providing an automatic feed system to reduce machine hazards
- Using a less hazardous cleaning material
- Reducing speed, force, amperage
- Reduce pressure, temperature
- Replacing an ancient steam-heating system and its boiler explosion hazards with a hot-air system

Substitution of a less hazardous method or material may or may not result in equivalent risk reduction in relation to what might be the case if the hazards and risks were reduced to an acceptable level through system design or redesign.

Consider this example. Considerable manual material handling is often necessary in a mixing process for chemicals. A reaction takes place and an employee sustains serious chemical burns. There are identical operations at two of the company's locations. At one location, a decision is made to redesign the operation so that it is completely enclosed, fed automatically, and operated by computer from a control panel, thus greatly reducing operator exposure.

At the other location, funds for doing the same were not available. To reduce the risk, a substitution took place in this manner:

- It was arranged for the supplier to premix the chemicals before shipment.
- Some mechanical feed equipment for the chemicals was also installed.

The risk reduction achieved by substitution was not equivalent to that attained by redesigning the operation.

C. Engineering Controls

When safety devices are incorporated in a system in the form of engineering controls, substantial risk reduction can be achieved. Engineered safety devices are to prevent access to a hazard by workers. They are to separate hazardous energy from the worker and deter worker error. They include such devices as:

- Machine guards
- Interlock systems
- Circuit breakers
- Startup alarms
- Presence-sensing devices
- Safety nets
- Ventilation systems
- Sound enclosures
- Fall-prevention systems
- Lift tables, conveyors, and balancers

D. Warnings [Warning Systems]

Warning system effectiveness, and the effectiveness of instructions, signs, and warning labels, rely considerably on administrative controls, such as training, drills, the quality of maintenance and human reaction capabilities. Further, although vital in many situations, warning systems may be reactionary in that they alert persons only after a hazard's potential is in the process of being realized (e.g., a smoke alarm). Examples are:

- Smoke detectors
- Alarm systems
- Backup alarms
- Chemical detection systems
- Signs
- Alerts in operating procedures or manuals

A comment is necessary on my preferred use of the term *warning systems* rather than *warnings* or *warning signs*. The latter terms appear in some published hierarchies of control, as in Z10. The entirety of the needs of a warning system must be considered, for which warning signs or warning devices alone may be inadequate.

For example, the NFPA Life Safety Code 101 may require, among other things, detectors for smoke and products of combustion; automatic and manual audible and visible alarms; lighted exit signs; designated, alternate, properly lit exit paths; adequate spacing for personnel at the end of the exit path; proper hardware for doors; and emergency power systems. Obviously, much more than merely "Warnings" is needed.

E. Administrative Controls

Administrative controls rely on the methods chosen being appropriate in relation to the needs, the capabilities of people responsible for their delivery and application, the quality of supervision, and the expected performance of the workers. Some administrative controls are:

- Personnel selection
- Developing appropriate work methods and procedures
- Training
- Supervision
- Motivation, behavior modification
- Work scheduling
- Job rotation
- Scheduled rest periods
- Maintenance
- Management of change
- Investigations
- Inspections

Achieving a superior level of effectiveness in all of these administrative methods is difficult, and not often attained.

F. Personal Protective Equipment

The proper use of personal protective equipment relies on an extensive series of supervisory and personnel actions, such as the identification of the type of equipment needed, its selection, fitting, training, inspection, maintenance, and so on. Examples include:

- Safety glasses
- Face shields
- Respirators
- Welding screens
- Safety shoes
- Gloves
- Hearing protection.

Although the use of personal protective equipment is common and necessary in many occupational situations, it is the least effective method to deal with hazards and risks. Systems put in place for their use can easily be defeated. In the design process, one of the goals should be to reduce reliance on personal protective equipment to as low a level as practicable.

APPLICATION OF THE HIERARCHY

For many risk situations, a combination of the risk management methods shown in the hierarchy of controls is necessary to achieve acceptable risk levels. But the expectation is that consideration will be given to each of the steps in a descending order, and that reasonable attempts will be made to eliminate or reduce hazards and their associated risks through steps higher in the hierarchy before lower steps are considered. A lower step in the hierarchy of controls is not to be chosen until practical applications of the preceding level or levels are exhausted.

ATTACHING THE HIERARCHY OF CONTROLS TO PROBLEM-SOLVING TECHNIQUES

The following observations are a reflection of my experience: encompassing the design and engineering, the operational, the post-incident, and the post-operational aspects of the practice of safety:

- Safety practitioners often recommend solutions to resolve hazard/risk situations before they define the problem: that is, before they identify the specifics of the hazards and assess the associated risks
- Rarely are safety management systems in place to determine whether the preventive actions taken achieve the risk reduction intended

Those observations led to research into the feasibility of encompassing the hierarchy of controls within a sound problem-solving technique which:

- Begins with problem identification and analysis
- Requires measurement of results of the actions taken to determine their effectiveness
- Necessitates taking further preventive measures if the residual risk is not acceptable

The initial step in this research was to review what is proposed in several texts regarding the basics of problem solving. The problem-solving methods the authors propose have great similarity. A composite of those techniques is presented in Table 14.1.

In every problem-solving method reviewed, the first steps are to identify and analyze the problem. Also, they end with a provision requiring that evaluations be made of the effects of the actions taken. Figure 14.1, the Safety Decision Hierarchy, presents a logical sequence of actions that safety professionals should consider in resolving safety issues: identify and analyze the problem; consider the possible solutions; decide on and implement an action plan; and determine whether the actions taken achieved the intended risk reduction results.

TABLE 14.1   Problem-Solving Methodology

1. Identify the problem.
2. Analyze the problem.
3. Explore alternative solutions.
4. Select a plan and take action.
5. Examine the effects of the actions taken.

The Safety Decision Hierarchy

A. Problem identification and analysis:

> 1. Identify and analyze hazards.
>
> 2. Assess the risks.

B. Consider these actions in their order of effectiveness:

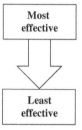

1. Avoid, eliminate, or reduce risks in the design processes.

2. Reduce risks by substituting less hazardous methods or materials.

3. Incorporate safety devices.

4. Provide warning systems.

5. Apply administrative controls (work methods, training, etc.).

6. Provide personal protective equipment.

C. Select risk-reduction measures and implement them.

D. Measure for effectiveness.

E. Accept the residual risk, or start over if it is unacceptable.

FIGURE 14.1   The Safety Decision Hierarchy

The safety decision hierarchy depicts a way of thinking about hazards and risks and establishes an effective order for risk avoidance, elimination, or amelioration. Why propose that safety professionals adopt a safety decision hierarchy? This quote from *The New Rational Manager*, reflecting the real-world observations of Kepner and Tregoe in dealing with many clients, makes the case:

> The most effective managers, from the announcement of a problem until its resolution, appeared to follow a clear formula in both the orderly sequence and the quality of their questions and actions. (p. vii)

It makes sense to apply a safety decision hierarchy encompassing methods in an orderly sequence of effectiveness to resolve safety issues.

ON PROBLEM IDENTIFICATION AND ANALYSIS

In utilizing the safety decision hierarchy, the goal in the problem identification and analysis phase is to identify and analyze the hazards and assess the risks. Hazard and risk situations cannot be dealt with intelligently until the hazards are analyzed and assessments are made of the probability of incidents or exposures occurring and the severity of their consequences are estimated.

Chapter 11, "A Primer on Hazard Analysis and Risk Assessment," is a resource for this problem identification and analysis phase.

EXPLORING ALTERNATIVE SOLUTIONS

The action steps shown in The Safety Decision Hierarchy under the caption "Consider These Actions, in Their Order of Effectiveness" provide a base for considering alternative risk avoidance, elimination, or reduction measures. They are similar to items A through F in the preferred order of controls outlined in section 5.1.2, the hierarchy of controls, in Z10. The logic in support of those steps and the order in which they are listed were given earlier in the chapter.

DECIDING AND TAKING ACTION

All facets of the safety decision hierarchy apply when considering the hazards and risks in a specific facility, process, system, piece of equipment, or tool in its simplest form. Also, they are broadly applicable in all four of the major aspects of the practice of safety:

- In the design processes, pre-operational: where the opportunities are greatest and the costs are lower for hazard avoidance, elimination, or control
- In the operational mode: where, integrated within a continual improvement process, hazards are eliminated or controlled before their potentials are realized and hazard-related incidents or exposures occur
- Post-incident: through investigation of hazard-related incidents and exposures to determine and eliminate or control their causal factors
- Post-operational: when demolition, decommissioning, or reusing/rebuilding operations are undertaken

MEASURING FOR EFFECTIVENESS

Provisions in the Management Review Process – Section 7.1 of Z10, require that systems be in place to measure the effectiveness of the risk reduction measures taken. Those provisions are relative to the measuring for effectiveness and re-analyzing steps in The Safety Decision Hierarchy.

Assuring that the actions taken accomplish what was intended is an integral part of the PDCA process. Follow-up activity would determine that:

- The problem was resolved, or only partially resolved, or not resolved.
- The actions taken did or did not create new hazards.

ACCEPT THE RESIDUAL RISK, OR START OVER IF IT IS UNACCEPTABLE

If the follow-up activity indicates that the residual risk is not acceptable, the thought process set forth in the safety decision hierarchy would be applied again, beginning with hazard identification and analysis.

CONCLUSION

The hierarchy of controls in Z10 derives from work that has evolved over many years. As management "implement(s) and maintain(s) a process for achieving feasible risk reduction," the hierarchy presents the actions to be considered in a logical order.

Encompassing a hierarchy of controls within a sound problem-solving technique furthers the ability of management and safety professionals to achieve effective risk avoidance, elimination, or reduction, and to meet the requirements of certain provisions in Z10. Adopting established problem-solving techniques to address hazard and risk situations is a fundamentally sound approach. That is the purpose of The Safety Decision Hierarchy.

REFERENCES

Accident Prevention Manual, 3rd ed. Itasca, IL: National Safety Council, 1955.

ANSI/AIHA Z10-2012. *Occupational Health and Safety Management Systems.* Fairfax, VA: American Industrial Hygiene Association, 2005. ASSE is now the secretariat. Available at https://www.asse.org/cartpage.php?link=z10_2005.

ANSI/PMMI B155.1-2011. *Safety Requirements for Packaging Machinery and Packaging-Related Converting Machinery.* Arlington, VA: Packaging Machinery Manufacturers Institute, 2011.

Chapanis, Alphonse. "The Error-Provocative Situation." In *The Measurement of Safety Performance*, William E. Tarrants, Ed. New York: Garland Publishing, 1980.

Kepner, Charles H. and Benjamin B. Tregoe. *The New Rational Manager.* Princeton, NJ: Princeton Research Press, 1981.

Manuele, Fred A. *On the Practice of Safety*, 4th ed. Hoboken, NJ: Wiley, 2013.

Manuele, Fred A. "Risk Assessment and Hierarchies of Control." *Professional Safety*, May 2005.

MIL-STD-882E. *Standard Practice for System Safety.* Washington, DC: Department of Defense, 2012. Available at http://www.system-safety.org/. Scroll down and click on MIL-STD-882E in the right-hand column for a free copy.

NFPA 101. Life Safety Code. Quincy, MA: National Fire Protection Association, 2012.

CHAPTER 15

SAFETY DESIGN REVIEWS: SECTION 5.1.3 OF Z10

Requirements for design review and management of change are addressed jointly in Section 5.1.3 of Z10. Although the subjects are interrelated, each has its own importance and uniqueness. Comments on the management of change concept and how to institute the management of change process are the focus of Chapter 19. Only the design review requirements of Z10 are addressed in this chapter.

Although the term *design review* appears in the title of Section 5.1.3, the "shall" requirement and the advisory comments indicate that much more than a design review is expected. The standard states that "The organization shall establish a process to identify, and take appropriate steps to prevent or otherwise control hazards at the design and redesign stages to reduce potential risks to an acceptable level."

A design review is one part of that process. For a design review to occur, there must be a design to review. As Bruce Main writes in *Risk Assessment: Challenges and Opportunities*: "A design review evaluates a design against the design criteria or requirements." (p. 27)

In the advisory column, E5.1.3 says: "The design review should consider all aspects including design, procurement, construction, operation, maintenance, and decommissioning." And E5.1.3.1 says: "The design process typically includes consideration of some or all of the following life cycle phases: Concept design stage; Preliminary design; Detailed design; Build or purchase process; Commissioning, installing and debugging processes; Production and maintenance operations; and Decommissioning activity.

Advanced Safety Management: Focusing on Z10 and Serious Injury Prevention,
Second Edition. Fred A. Manuele.
© 2014 John Wiley & Sons, Inc. Published 2014 by John Wiley & Sons, Inc.

To do as Z10 implies, systems must be in place to avoid, eliminate, reduce, or control hazards and the risks that derive from them:

- As early as possible and as often as needed in every aspect of the design and redesign processes
- In all phases of operations

What needs to be agreed upon in an organization is a well-understood concept, a way of thinking that is translated into a system that effectively addresses hazards and risks in the design and redesign processes, a way of thinking that is applied universally by all who are involved in the design processes.

I hold to the premise that over time, the level of safety achieved will relate directly to whether acceptable risk levels are achieved or not achieved in the design and redesign processes. This premise relates to the design of facilities, hardware, equipment, tooling, operations layout, the work environment, the work methods, and products. *Design*, as the term is used here, encompasses all processes applied in devising a system to achieve results.

Methods used to avoid bringing hazards and risks into the workplace will be discussed here broadly, with the hope that safety professionals can adopt from the materials presented to their advantage. To assist in applying the Z10 design provisions, this chapter includes:

- A review of safety through design and prevention through design concepts and procedures
- Comments on how some safety professionals are engaged in activities that lessen the probability of hazards and risks being brought into the workplace
- An edited composite of procedures in place in several companies to achieve hazard avoidance and control in the design process, and procedures to be followed before modified equipment is released for operation
- A general checklist as a reference from which a specifically tailored checklist can be developed for use in design reviews, and for equipment acceptance
- A company's equipment design philosophy, a model

SAFETY THROUGH DESIGN AND PREVENTION THROUGH DESIGN

If the design review requirements of the Z10 standard are to be met, "to reduce potential risks to an acceptable level," the focus in the design and redesign processes has to be on hazards and the risks that derive from them. In the advisory column following E5.1.3F, the instruction given is: "See Appendix F: Risk Assessment for more information on risk assessment methodologies to consider when assessing risks during design reviews and managing hazards and risks associated with operational change." Appendix F is a good reference, as is Chapter 11, "A Primer on Hazard Analysis and Risk Assessment" in this book. There is no appendix in Z10 on safety in the design and redesign processes.

Activities undertaken by the National Safety Council and the National Institute for Occupational Safety and Health that encourage addressing hazards and risks in the design and redesign processes are related directly to Z10's design requirements.

NATIONAL SAFETY COUNCIL

The book *Safety Through Design* was a creation of the Institute For Safety Through Design, an entity at the National Safety Council. The Institute's vision was to achieve a future in which "Safety, health, and environmental considerations are integrated into the design and development of systems meant for human use." (p. xi) An advisory committee, the members of which were drawn from industry, organized labor, academia, and others interested in the cause, agreed on the following definition.

Safety Through Design: The integration of hazard analysis and risk assessment methods early in the design and engineering stages and taking the actions necessary so that risks of injury or damage are at an acceptable level. (p. 3)

That definition serves well in understanding and developing the design requirements in Z10. The theme of safety through design is that if decisions affecting safety, health, and the environment are integrated into the early stages of the design and redesign processes:

- Productivity will be improved.
- Operating costs will be reduced.
- Expensive retrofitting to correct design shortcomings will be avoided.
- Significant reductions will be achieved in injuries, illnesses, and damage to the environment, and their attendant costs. (p. 3)

Effectiveness and economics were addressed in the following citation: "Hazards and the risks that derive from them are most effectively and economically avoided, eliminated, or controlled if they are considered early in the design process, and where necessary as the design progresses." (p. 11)

All of the foregoing derives from the following premises. Risks of injury derive from hazards. If hazards are avoided, eliminated, reduced, or controlled in the design process and acceptable risk levels are achieved, the potential for harm or damage and operational waste is diminished. These premises tie in neatly with the reference in Z10 to addressing occupational health and safety management system issues, which are defined "as hazards, risks, management system deficiencies, and opportunities for improvement". (Section 4.0 – Planning).

Item C in Z10's Section 3.1.3 – Responsibility and Authority – indicates that the occupational health and safety management system should be integrated into an organization's other business systems and processes. There are seven chapters in Part II of *Safety Through Design* that are good resources for those who attempt to integrate the Z10 design requirements into other business processes.

In accord with an initially established sundown provision, the operations of the Institute for Safety through Design were discontinued in 2005.

NATIONAL INSTITUTE FOR OCCUPATIONAL SAFETY AND HEALTH

In 2006, several of the participants in the activities of the Institute For Safety Through Design, and others, received an e-mail from an executive at the National Institute for Occupational Safety and Health (NIOSH) encouraging our participation in a major initiative promoting Prevention through Design. One aspect of the NIOSH initiative was to develop and approve a broad generic voluntary consensus standard on Prevention through Design that is aligned with international design activities and practice.

I volunteered to lead that endeavor. On September 1, 2011, the American National Standards Institute (ANSI) approved the standard ANSI/ASSE Z590.3-2011, *Prevention Through Design: Guidelines for Addressing Occupational Hazards and Risks in the Design and Redesign Processes.*

By its charter, activities at NIOSH are limited to occupational safety and health hazards and risks and are the focus of Z590.3. But, by intent, the terminology in Z590.3 was kept broad enough so that the guidelines could additionally be applicable to all hazard-based fields: product safety, environmental controls, property damage that could result in business interruption, and so on. In the standard, the definition of prevention through design is, of necessity, applicable to work-related hazards and risks only.

Prevention through design: Addressing occupational safety and health needs in the design and redesign process to prevent or minimize the work-related hazards and risks associated with the construction, manufacture, use, maintenance, retrofitting, and disposal of facilities, processes, materials, and equipment. (p. 10)

The content of the Z590.3 standard is closely related to Z10's design provisions. Prevention through design is addressed in Chapter 16.

THE SAFETY DESIGN REVIEW PROCESS

A safety design review process will not be effective until the participants on the review team have acquired knowledge of hazards and risks and have agreed on the risk assessment methods and risk assessment matrix to be used. Chapter 11, "A Primer on Hazard Analysis and Risk Assessment," serves as a base for the education of team members. Also, a safety design review process can be employed successfully only if senior management has been convinced of its value.

Paul Adams has had considerable experience in design reviews. He was asked to describe the process for integrating safety into the design process as it is implemented in a major manufacturing company with which he has been affiliated. His comments appear here with his permission.

The Safety Design Review Process

Formal safety design reviews may sound like tedious exercises, but they are effective processes for delivering inherently safer designs. Design reviews are systematic processes for carefully reviewing design attributes, applications, misapplications, energy control systems, and human interactions. Safety design reviews aim to identify hazards and hazardous conditions that are foreseeable throughout the lifecycle of a product, process or facility, and to develop mitigation strategies.

In most cases, a safety design review is best conducted by a team of stakeholders and at least one objective, disinterested engineer. Typical participants include representatives from Engineering, Production, Maintenance, and Health and Safety. Both the system designer(s) and the review team share responsibility for the safety of the final design.

Safety design reviews should be approached as important problem-solving events. A spirit of cooperation, and even fun, can be maintained by restricting criticism to constructive debate on specific design features. Although the focus should always be on safety, review teams frequently identify additional opportunities to reduce costs and improve productivity. Constructability and maintainability issues often surface, and these alone can pay huge dividends for the comparatively short time invested in reviewing a project. While it is the role of the facilitator to assure that these topics do not derail or compromise the safety focus, such opportunities should obviously be captured as value-added by-products. Review sessions also provide excellent learning opportunities for both novice and experienced engineers, as collective knowledge and experience are openly shared and questioned.

A common approach for conducting a formal design safety review is to methodically work through a Design Safety Checklist. Some organizations use a generic checklist, supplemented with additional checklists for specific disciplines, such as electrical or chemical systems. For each system element, reviewers address the various forms of energy present and the steps taken to control unwanted or hazardous release. A typical design safety review meeting proceeds as follows:

1. Project manager distributes drawings and copies of checklists, or they can be projected on large screens.
2. Review team chooses a facilitator, often a disinterested engineer; i.e., an engineer not assigned to or intimately familiar with the project.
3. Project engineer/manager describes the project and its scope, and answers general questions about major areas of concern.
4. Project engineer/manager keeps notes; a.k.a. design punch list.
5. Facilitator leads a methodical review using a generic checklist, with team members asking detailed questions to ensure thorough consideration of hazards and their control. Checklist items and sections that are not applicable are so noted. Both production and maintenance tasks should be considered.

6. Additional discipline checklists are reviewed as appropriate.

7. A marked-up copy of the checklists, along with signatures of the participants, is retained with project documents in accord with records retention policy.

In the event an organization does not have a design safety review checklist, a systematic review can still be performed. A suggested method is to carefully review layout and assembly drawings, considering each form of energy separately, including the potential hazard exposure, and possible scenarios and consequences of unintended release of this energy form.

For example, review the drawing considering all uses and possible exposures to electrical energy, then pneumatic systems, hydraulic systems, chemical processes, mechanical kinetic and potential energy, thermal and radiation energy, etc. For each form of energy, work systematically through the layout or assembly drawing using a geographical approach, such as starting at the beginning of a production line and moving through it to its end. Again, it is important to consider both production and maintenance activities.

The development of simulation software has added a new dimension to the design safety review process. Modern computer assisted design software systems convert two-dimensional drawings to three-dimensional images that can be moved and reoriented with ease to show different viewing perspectives. When used in design reviews, this provides a powerful means of illustrating designs for those who have difficulty visualizing workstations and processes from traditional two dimensional drawings. In addition, some software products enable engineers to place humanoid figures into scaled three-dimensional depictions of proposed workstations, and have figures of various body sizes move through simulated work tasks. It also allows reviewers to more effectively consider ergonomic issues and risk factors; i.e., potential for strains and cumulative trauma disorders resulting from the expenditure of "human energy" in awkward postures.

Specific deliverables from a design review typically include:

- A set of marked-up drawings and specifications.
- A list of design modifications requiring attention prior to release for bid or construction/fabrication.
- A list of specific items to be checked during installation and the final walk-down or post-fabrication inspection. (There are often items identified as potentially problematic that cannot be adequately assessed at the time of the review due to lack of detailed knowledge.)
- Assignments for resolving specific details or making design corrections.

Following the review, it is the responsibility of the project engineer/manager to ensure that all issues raised during the design safety review are resolved and appropriate revisions are completed. Verification through a post-construction/

fabrication inspection should be completed prior to release of the process, facility or system.

That's what a design safety review is all about. Done well, such reviews diminish the likelihood of bringing hazards and risks into the workplace. Having the right facilitator is important in the design review process.

In one company, it is standard practice to engage a consultant facilitator to lead safety design reviews to assure that the views of one or two people do not dominate and that all participants are given an opportunity to be heard.

EMPHASIZING CONSIDERATION OF THE WORK METHODS IN THE DESIGN PROCESS

Much has been made in this book of the need to design work methods and procedures so that they are not error-provocative or overly stressful. Considerations of hazards and risks in the design process should not be limited to the facility, equipment, and processes (i.e., to the hardware). They should also focus on the hazards and risks in the work methods prescribed, taking into consideration:

- The capabilities and limitations of the workers so that the risks of injury and damage are at as low as reasonably practicable
- The benefits of not creating work situations that are error-provocative

In a paper entitled "The Titanic and Risk Management," Roy Brander wrote this: "Safe design of the procedure is as important as design of the artifact." (Item 30) His point is important. It fits closely with my observation that, too often, inadequate attention is given to avoiding hazards in work methods, the result being that what workers are expected to do is inherently risky.

HOW SOME SAFETY PROFESSIONALS ARE ENGAGED IN THE DESIGN PROCESS

To obtain information on how safety professionals are involved in activities to avoid bringing hazards and risks into the workplace, a request for comments on the subject was made through an Internet safety server. These are some of the responses—the most unfortunate first. As they are reviewed, safety professionals may want to assess their place in the design process and look for hints on how they can improve their positions.

- From an industrial hygienist: If the engineers would only let me into the design process, I know that many of the health hazards I deal with could be better controlled, we wouldn't need to do so much testing, and our operations would be more productive because we would reduce the amount of time employees spend on testing and on the use of personal protective equipment.

- We don't have any forms or established procedures for our getting into what engineers are designing, and there is no formal method for engineering to notify us of a project. We do get copies of the engineering weekly reports, and we read those carefully. Then, we invite ourselves into the discussion.

- There was a lot of resistance by engineering to our getting into what they were designing because they had not recognized that we could be an asset. It took us a long time and quite a few money-saving successes to get it ingrained into their procedures that it was to their benefit to refer capital expenditure requests to us for our input. We don't have a written procedure, but it gets done.

- Almost all of our engineering is done by outside firms. I haven't been successful in convincing the few engineers we have left that I can help them write specifications that will save them money. But I have convinced the manager of my plant that I am to have sign-off authority for new installations.

  My sign-off reviews can be embarrassing to the people who laid out the specs and I have to be diplomatic in how I do what I need to do. But the plant manager gives me support now. He has been educated.

- This from a construction safety professional: I request to see all drawings at the 10% level while changes are still easily incorporated. I get a schedule of all construction plans and visit contractor lay down areas to inspect materials being used on the job. I attempt to educate quality assurance personnel on what to look for as safety indicators.

  I visit the engineering department and the contracting department on a daily basis when I'm in town. Yes, it is time consuming but it saves money and lives as well as equipment. It took maximum effort on my part to get where I am.

- In my company, it's in the capital expenditure procedure manual that all funding requests will receive a safety review. Managers have the same responsibility for safety as they do for productivity and quality, and when they approve the capital expenditure request, they are also signing off for safety. But, my name appears on the capital expenditure distribution list among the management people who have to sign off as approving the request. This isn't as burdensome as it was in the beginning because I have educated a fairly stable engineering and management staff on what it takes to get my approval.

- Through our successes in ergonomics, engineering recognized our contributions not only for safety, but also for productivity. They now invite us into the design process in the idea stage. They make it plain that they look to us to see that they don't mess up. We learned the hard way over a lot of years that it is expensive to correct hazards that are in the machines and equipment we buy. We had to do some costly equipment modifications for employee safety and health, and for environmental situations after the installations were completed and in operation. Our engineers aren't easy on us. But that's okay. We had to accept the criticism from them that our ergonomics checklists were so general that they weren't helpful.

- Our Facility Safety Manual includes extensive procedures for documented reviews for safety, health, environmental, and ergonomic standards before budget approval is obtained on a new project, and for equipment reviews before the equipment is released to normal production.

  As you will see from what I'm mailing to you, we are deep into specifications, and the signature of a safety specialist is required in the project and budget review procedures.

Involvement of safety professionals in the design processes, specification writing, purchasing, and sign-off processes varies from nothing at all to being required to be participants by written procedures. In all but one of the cases cited, there is some involvement by safety professionals—and that came about because safety professionals took the initiative and proved their value.

At the Professional Development Conference held by the American Society of Safety Engineers in 2012, I was among about 40 others attending a late afternoon session—a typical number for a presentation late in the day. A later session, for which the subject was risk assessments, began at 4:30 P.M. and extended into the happy hour. The presenters were long-time associates. I thought it appropriate to caution them not to expect a large audience. But surprise, surprise. Well before starting time it became aware that the attendance figure would be huge.

When the attendees were asked for a show of hands on how many were doing risk assessments in the design and redesign processes, over half responded favorably. That represents substantial progress in relation to the case some 20 years ago.

A GOOD PLACE TO START

Safety professionals who are not involved in the design processes should consider ergonomics as fertile ground in which to get started. It is well established that successfully applied ergonomics initiatives result not only in risk reduction, but also in improved productivity, lower costs, and waste reduction. Furthermore, musculoskeletal injuries are a large segment of the spectrum of injuries and illnesses in all organizations. Since they are costly, reducing their frequency and severity will show notable results.

Ergonomists know how to write specifications that result in designing the workplace and the work methods to fit the capabilities and limitations of workers. A company that established detailed ergonomics design criteria was DaimlerChrysler. How their design criteria were used demonstrates that there can be a close relationship between establishing safety design parameters and including safety specifications in purchasing documents. The version of the ergonomics design criteria available on the Internet was last updated in May 2006, and it is still presented as a DaimlerChrysler document. This is how the DaimlerChrysler ergonomic design criteria were introduced.

This document attempts to integrate new technology around the human infrastructure by providing uniform ergonomic design criteria for DaimlerChrysler's manufacturing, assembly, power train and components operations, as well as part distribution centers. These criteria supply distinct specifications for the Corporation, to be used by all DaimlerChrysler engineers, designers, builders, vendors, suppliers, contractors etc. providing new or refurbished/rebuilt materials, services, tools, processes, facilities, task designs, packaging and product components to DaimlerChrysler.

In effect, the Ergonomic Design Criteria used internally at DaimlerChrysler also became the ergonomic specifications that vendors and suppliers had to meet. In a section of the criteria entitled "Supplier Roles and Responsibilities," it is made clear that all suppliers were to "make all reasonable efforts to implement all of the criteria and requirements" of the ergonomic design criteria. If a requirement was to be compromised, the supplier was to so inform DaimlerChrysler and the matter would be resolved by a DaimlerChrysler ergonomics representative. DaimlerChrysler's Ergonomic Design Criteria are available at http://www.docstoc.com/docs/25321410/150-Ergonomic-Design-Criteria. The criteria can be downloaded for personal or business use. Taking into consideration what is now known about ergonomic design specifications, it is incomprehensible that employers continue to purchase equipment that is not ergonomically designed.

An adaptation of how another company developed extensive "General Design and Purchasing Guidelines" that serve both as design requirements to be met by its engineering staff and as the company's purchasing specifications appears as an addendum to the chapter on Procurement. My searching revealed that most companies consider their design and purchasing criteria to be proprietary and not distributable. But one such document that includes general design parameters is in the public domain and can be cited here.

MILITARY RESOURCES: DESIGN REQUIREMENTS

In a draft of MIL-STD-882E, an appendix contained suggested design guidelines that were not included in the final version. Nevertheless, they are quite good, and a brief version is given here. They promote thinking about and giving guidance on writing design specifications to fit the needs of a particular entity.

Example General Safety Design Requirements

1. Hazardous material use is to be eliminated or minimized. After considering material selection and substitution of lesser hazardous materials, the remaining risks are to be reduced in the design process.
2. Hazardous substances, components, and operations are to be isolated from other activities, areas, personnel, and incompatible materials.
3. Equipment is to be located so that access during operations, servicing, repair, or adjustment minimizes personnel exposure to hazards (e.g., hazardous substances, high voltage, electromagnetic radiation, and cutting and puncturing surfaces).
4. Power sources, controls, and critical components of redundant subsystems are to be protected by physical separation or shielding, or by other acceptable methods.

5. For hazards that cannot be eliminated, consideration is to be given to safety devices that will minimize mishap risk (e.g., interlocks, redundancy, fail safe design, system protection, fire suppression, and protective measures such as clothing, equipment, devices, and procedures).

6. Provisions for the disposal of systems are to be considered in the design process.

7. Warning signals are to be standardized within like types of systems: they are to be designed to minimize the probability of incorrect personnel reaction to them.

8. Warning and cautionary notes are to be provided in assembly, operation, and maintenance instructions; and distinctive markings are to be provided on hazardous components, equipment, and facilities to ensure personnel and equipment protection when no alternate design approach can eliminate a hazard. Use standard warning and cautionary notations where multiple applications occur. Standardize notations in accordance with commonly accepted commercial practice. Warnings, cautions, or other written advisories are not to be used as the only risk reduction method for hazards assigned Catastrophic or Critical mishap severity categories.

9. If safety critical tasks require personnel to have specific proficiency, a certification process for that proficiency should be used.

10. In the design process, specific consideration should be given to the minimization of injury or damage to equipment or the environment as a result of a mishap.

11. Inadequate or overly restrictive safety requirements are not to be included the system design specifications.

12. Acceptable risk is mishap risk that is as low as reasonably practicable (ALARP) within the constraints of operational effectiveness, time, and cost.

Although the foregoing outline doesn't cover much space, it represents an extensive body of knowledge. The outline is basic, and right on.

In the *Air Force System Safety Handbook* developed by the Air Force Safety Agency at Kirtland Air Force Base in New Mexico, a section opens with this statement:

As an accident avoidance measure, one develops strategies for energy control similar to those listed below.

The controls shown are an adaptation of Haddon's unwanted energy release theory, which is discussed later in this chapter. The controls listed in the *Air Force System Safety Handbook* are worthy of consideration.

a. Prevent the accumulation by setting limits on noise, temperature, pressure, speed, voltage, loads, quantities of chemicals, amount of light, storage of combustibles, heightof ladders, etc.

b. Prevent the release through engineering design, containment vessels, gas inerting, insulation, safety belts, lockouts, etc.

   c. Modify the release of energy by using shock absorbers, safety valves, rupture discs, blowout panels, less incline on the ramps, etc.

   d. Separate assets from energy (in either time or space) by moving people away from hot furnace, limiting the exposure time, picking up with tongs, etc.

   e. Provide blocking or attenuation barriers, such as eye protection, gloves, respiratory protection, sound absorption, ear protectors, welding shields, fire doors, sunglasses, machine guards, tiger cages, etc.

   f. Raise the damage or injury threshold by improving the design (strength, size), immunizing against disease, warming up by exercise, getting calluses on your hands, etc.

   g. And by establishing contingency response such as early detection of energy release, first aid, emergency showers, general disaster plans, recovery of system operation procedures, etc. (p. 65)

A General Design Safety Checklist appears later in this chapter. Also, several other design guidelines and checklists that can be adapted for design purposes appear in this book.

A SAFETY DESIGN REVIEW AND OPERATIONS REQUIREMENTS GUIDE

A composite is provided here of the procedures in place in three companies for design reviews and for a safety sign-off before new or modified equipment can be released into normal operations. Those procedures serve to avoid bringing hazards and risks into the workplace. In a way, these requirements represent culture statements: Managements have decided that hazards and risks are to be dealt with as equipment is designed and before it can be placed in operation. This composite presents a basis for thought as safety professional pursue having similar concepts and procedures adopted.

REQUIREMENTS: EQUIPMENT AND PROCESS DESIGN SAFETY REVIEWS

A. Purpose

To establish procedures to ensure that hazards are analyzed and that risks are at an acceptable level when considering new, redesigned, and relocated equipment or processes.

B. Scope

These procedures apply to all equipment and processes that may present risks of injury to people or damage to property or the environment. They pertain to the design or redesign of all new, transferred, and relocated equipment and processes. For all aspects of these procedures, documentation shall be appropriate to the activity.

   Safety, as the term is used here, encompasses risks of injury or damage to personnel (employees and the public), property, and the environment.

C. Responsibilities

Location Manager. The responsibility for safety rests with the location manager in the operations for which he or she has authority.

Project Manager. The project manager is responsible for assuring that:

- Corporate safety requirements are met
- Safety documentation to accompany capital expenditure requests is prepared
- Preliminary and subsequent design safety reviews are conducted
- Appropriate coordination and communication take place with outside design and engineering firms to assure that specifications are met
- Proper consideration is given to any safety problems identified by the staff during their visits to vendors and design and engineering firms prior to delivery of equipment

Design Engineers. Principal responsibility of the design engineers is to design inherently safer equipment and processes. Whether employees or contractors, design engineers shall assure that the considerations necessary for safety have been given in the design process. They will provide the Project Manager and the Safety Review Team documentation including:

- Detailed equipment design drawings
- Equipment installation, operation, preventive maintenance, and test instructions
- Details of and documentation for codes and design specifications
- Requirements and information needed to establish regulatory permitting and/or registrations

Safety Review Team. This team will conduct preliminary and subsequent design safety reviews for equipment and processes, or have design reviews made by outside consultants for particular needs. In addition to the Project Manager, members will include the project design engineers, production and maintenance personnel, the facilities engineer, selected disinterested engineers, the safety professional, and others (financial, purchasing) as needed. The Safety Review Team will also be responsible for:

- Arranging and conducting safety walk-downs of new projects
- Determining when visits are to be made at vendor design and engineering locations, and selecting the personnel with the necessary skills to make the visit

Safety Professional. The safety professional will:

- Serve as a member of the Safety Review Team and assist in identifying and evaluating hazards in the design process and provide counsel as to their avoidance, elimination, or control
- Visit design engineering firms and other vendors, when so requested by the Safety Review Team, to assure that safety problems are identified and corrected prior to shipment of equipment

---

Equipment Acceptance – Safety Review Form

Dept._____ Control No._____

Equipment description_____

This form must be completed prior to equipment being released to normal production. It is applicable to:

1. All newly installed equipment or processes

2. Changes made in the use of equipment or processes

3. Modifications of existing equipment or processes

Preliminary approval indicates that the equipment is ready for initial production trials, but needs additional work for safety as listed in a memo attachment titled "Safety items needing attention." Final approval indicates that the preliminary findings have been addressed satisfactorily and that the equipment can be released to normal production.

	Preliminary	Final
Signed:_____	_____	_____
Engineer-in-charge	Date	Date
Signed:_____	_____	_____
Dept. Mgr. or Supervisor.	Date	Date
Signed:_____	_____	_____
Safety Manager	Date	Date

The original copy of this form shall be retained by the Department Manager or supervisor. The Engineer-in-charge and the Safety Manager will retain copies. The company's file retention policy shall apply.

FIGURE 15.1   Safety Design Reviews

- Be a signatory on Equipment Acceptance-Safety Review Form [Figure 15.1] prior to newly installed or altered equipment or processes being released to normal production.

Department Managers and Supervisory Personnel. Department managers and supervisors will give support to the project manager for the activities that come under their jurisdiction. They will be signatories, along with engineering personnel and the safety professional, on sign-off forms before newly installed or altered equipment is released for normal operation.

D. Initial Capital Expenditure Safety Review

Capital expenditure requests at financial levels requiring divisional or corporate approval are to be accompanied by a Preliminary Safety Review Form, which is to be completed by the Safety Review Team. Comments would be included in the form, giving assurance that the hazards and risks identified can be dealt with properly.

For new or altered equipment or processes at a financial level not requiring divisional or corporate approval, formal Preliminary Safety Reviews are to be made at the discretion of the Safety Review Team.

E. Design Reviews

When designs have been completed and drawings and specifications are available, the Safety Review Team will hold one or more design review meetings to:

- Identify hazards not given appropriate attention, and recommend solutions to attain acceptable risk levels
- Assure that corporate safety requirements are being met
- Avoid the cost of risk-reduction retrofitting as the project moves forward
- Assure compliance with applicable regulations, codes, and standards

F. Safety Walk-Downs

When a project reaches completion at approximately a 70% level, the Safety Review Team will arrange and conduct a safety walk-down to provide an opportunity to:

- Assure that specifications have been met
- Determine that hazards identified in the preliminary safety review and the subsequent design review have been addressed properly
- Identify hazards that may have inadvertently been built into the project, and to arrange for the necessary action to be taken

Similarly, a final safety walk-down will be arranged as the project nears completion.

H. Equipment and Process Release Requirements

For newly purchased, redesigned, or relocated equipment or processes—before release for normal production:

- Task analyses shall be made to identify hazards and risks, and the hazards identified are to be dealt with properly.
- Any revisions necessary in the written job procedures shall be made.
- Retraining, as required, shall be given.
- An Equipment Acceptance-Safety Review Form shall be completed, the signatories to which shall be the engineer in charge, the department manager or supervisor, and the safety professional.

I. Design Reviews at Vendor and Design Engineering Locations

When considered advantageous by the Safety Review Team, arrangements will be made for personnel with the appropriate skills to visit vendors and design engineering firms to:

- Assure that vendors are building equipment to specifications
- Determine whether hazards exist that were not identified in the design review process, or by the vendor or design engineering firm that need attention
- Avoid the high cost of retrofitting for safety matters during installation, testing, and debugging

For these reviews, a specifically tailored Vendor/Design Engineering Review Form is to be created from sections of the General Design Safety Checklist that relate to the task. Reports are to be made available to the Safety Review Team and the Project Manager.

EQUIPMENT ACCEPTANCE: SAFETY REVIEW FORMS

Drafting equipment acceptance safety review forms specific to every piece of equipment or process would be a mammoth undertaking. Nevertheless, industry-specific safety review checklists do exist, and they should be used where applicable.

The example of an equipment acceptance review form shown in Figure 15.1 assumes the existence of a General Design Safety Checklist that can serve as a reference basis for the review process. Also, the example presumes that there will be two levels of review: a preliminary review to identify items needing attention, and a final sign-off.

PRELIMINARY SAFETY DESIGN REVIEW FORMS

A Preliminary Safety Design Review goes by several names. In one company its purpose is to meet "Fitness for Use Criteria." In another company it is referred to as a "Hazards Screening Analysis." A Preliminary Safety Review Form is a list of subjects pertaining to equipment, facilities, or processes that aids the reviewers in identifying hazards and risks that must be addressed. Completion of the form produces, in effect, an early design review. The review team indicates whether a subject needs further consideration.

No single list can be suitable for all needs. In drafting such a list, a safety professional should use a general design safety checklist as a reference and include some, all, or more than the subjects listed in Table 15.1.

GENERAL DESIGN SAFETY CHECKLIST

A detailed, specifically referenced design checklist covering all workplace safety needs would fill thousands of pages. The design safety checklist presented here is a brief composite taken from several sources. Some safety professionals will view it as

TABLE 15.1 Topics To Be Considered: Preliminary Design Safety Review

Ergonomics	Material handling
Machine guarding	Illumination needs
Fire protection	Means of egress
Walking and working surfaces	Confined spaces
Use of hoists, cranes, etc.	Work at heights
Electrical potentials	Lockout/tagout
Confined spaces	Temperature extremes
Hazardous or toxic materials	Noise or vibration
Personal protective equipment	Nonionization emitters
Environmental concerns	Sanitation

excessive; others will find that it does not address all their needs. Those who use it as a reference should be aware that there are many subject-specific and industry-specific checklists to which they should also refer. For example:

- The *Guidelines For Hazard Evaluation Procedures, 2nd Edition*, issued by the Center for Chemical Process Safety, includes a checklist–questionnaire for chemical operations that fills 45 pages.
- The checklist in Addendum B at the conclusion of Chapter 11, "A Primer on Hazard Analysis and Risk Assessment," is adapted from ISO 14121, *Safety of Machinery—Principles of risk assessment*, a standard issued by the International Organization for Standardization. It is to serve as a guide for those who design and manufacture equipment and machinery that goes into European workplaces.

This General Design Safety Checklist is presented intentionally as a list of questions without the boxes and lines for separation that are typically found in checklists. Also, the response that would be placed to the right of a design checklist in which users would enter check marks for "yes," "no," or "not applicable" has been eliminated.

GENERAL DESIGN SAFETY CHECKLIST

Preface

This checklist begins with a preface that brings attention to Haddon's unwanted energy release theory. That theory is: For all injuries or illnesses, property damage, and environmental damage, an unwanted and harmful transfer of energy or exposure to a harmful environment is a factor.

William Haddon espoused the theory that unwanted transfers of energy can be harmful (and wasteful) and that a systematic approach to limiting their possibility should be taken. Thus, it is proposed that a systematic approach be taken in the design process to limit harmful transfers of energy and exposures to harmful environments.

The questions in Section A, "Introduction: Basic Considerations," relate to Haddon's theory and are presented as general concepts to be considered when using the checklist, for which "yes" or "no" answers are to be obtained. They emphasize that the two distinct aspects of risk are to be considered in the design process:

- Avoiding, eliminating, or reducing the *probability* of a hazard-related incident or exposure occurring
- Reducing the severity of harm or damage if an incident or exposure occurs

A. Introduction: Basic Considerations

1. Can production of hazardous materials or energy be eliminated?
2. Will the amount of the hazardous materials or energy be limited?
3. Can less-hazardous materials be substituted?

4. Can hazardous material or energy buildup be prevented?
5. Can the release of hazardous materials or energy be slowed down?
6. Can unwanted energy release be separated in space or time from that which is susceptible to harm or damage?
7. Can barriers be interposed to separate the unwanted energy release from that which is susceptible to harm or damage?
8. Will surfaces with which people come in contact be modified to reduce the risk of injury?

B. Designing for Those with Disabilities

1. Do the designs take into consideration the requirements of the Americans with Disabilities Act (ADA)?
2. Are reasonable accommodations made for the disabled?

C. Confined Spaces

1. Have confined spaces been eliminated by design where practicable?
2. Are any confined spaces to be permit required? 1910.146(c)(1)
3. Have confined spaces been designed for easy of ingress, prompt egress, and where practicable, elimination of hazardous atmospheres?
4. Can confined spaces be designed with multiple, large accesses?
5. Are accesses provided with platforms that will support all required personnel and equipment?
6. Will access ports be large enough to permit entry when personnel are using personal protective equipment?
7. Will pipes or ducts limit entry to access ports?
8. Are locations of ladders and scaffolds in the space identified?
9. Are fall protection needs fulfilled (such as anchorage points)?
10. Can the necessary equipment be moved through accesses?
11. Does the design provide for isolation of the confined space from hazardous energy (i.e., electrical, chemical, etc.)?
12. Does the design provide for isolation by valve blocking, spools, double blocks and bleeds, flanges, and flushing connections?
13. Can spaces be designed so that maintenance and inspection can be performed from outside or by self-cleaning systems?

D. Electrical Safety

1. Overall, will the electrical system meet OSHA/NEC standards?
2. Will the system be sufficiently flexible to allow for future expansion?
3. Will emergency power be provided for critical systems?

4. Is grounding adequate?

5. Are ground fault interrupter circuits to be installed where needed? 1910.304(f)(7)

6. Are grounding connections to piping and conduits eliminated to prevent accumulation of static electricity?

7. Is grounding provided for lightning protection on all structures?

8. Are accommodations made for special-purpose or hazardous locations? 1910.307

9. Is the design adequate where there may be combustible gases or vapors?

10. Is high-voltage equipment isolated by enclosures such as vaults, security fences, and lockable doors and gates?

11. Are nonisolated conductors such as busbars on switchboards or high-voltage equipment connections that are located in accessible areas protected to minimize hazards for maintenance and inspection personnel?

12. Where injury to an operator may occur if motors were to restart after power failure, are provisions made to prevent automatic restarting upon restoration of power? 1910.262(c)(1)

13. Are electrical disconnect switches lockable, readily accessible, and labeled? 1910.303(f)

14. Are breakers/fuses sized properly? 1910.303(b)

15. Has the polarity of all circuits been checked? 1910.403(a)(2)

16. Do electrical cabinets and boxes have appropriate clearances? 1910.303(g) and (h)

17. Are exposed live electrical parts operating at 50 volts or more guarded against accidental contact by approved cabinets or enclosures, by location, or by limiting access to qualified persons? 1910.303(g)(2)(i)

18. Are rooms or enclosures containing live parts or conductors operating at over 600 volts, nominal, designed to be kept locked, or provisions made to be under the observation of a qualified person at all times? 1910.303(h)(2)

19. Are the electrical wiring and equipment located in hazardous (classified) locations intrinsically safe, approved for the hazardous location, or safe for the hazardous location? 1910.307(b)(1–3)

E. Emergency Safety Systems: Means of Egress

1. Are means of egress adequate in number, remote from each other, properly designated, marked, lighted, and easily recognized?

2. Has emergency lighting been provided for means of egress, and elsewhere where needed?

3. Does the design contemplate emergency lighting where workers may have to remain to shut down equipment?

4. Do means of egress exit directly to the street or open space?

5. Are doors, passageways, or stairways that do not lead to an exit marked by signs reading "not an exit" or by a sign indicating actual use?

6. Does the design provide internal refuge areas for workers who cannot escape?

7. Will reliable emergency power be provided for critical and life support systems?

8. Will emergency safety showers and eyewash stations be adequate and placed properly?

9. Will adequate first-aid stations, spill carts, and emergency stations be provided?

F. Environmental Considerations (Some are Operational, Beyond Design)

1. Have waste products been identified and a means of disposal established?

2. Will provisions be made for responding to chemical spills (containment, cleanup, disposal)?

3. Is there an existing spill control plan for chemicals?

4. Have all waste streams been identified?

5. Are adequate pre-treatment facilities provided for process waste streams?

6. Will an adequate storage area be available for wastes held prior to treatment or disposal?

7. Will waste storage areas have adequate isolation, or containment for spills?

8. Will hazardous wastes be disposed of at approved treatment, storage, and disposal facilities?

9. Is special equipment or specially trained personnel provided for treatment operations?

10. Has the acquisition of permits been addressed for the treatment or disposal of waste streams?

11. Have state or local requirements for permitting been evaluated and factored into the project?

12. Can the facility meet regulations for reporting spills or the storage of chemicals?

13. Have adequate provisions been made for cleaning the process equipment?

14. Have provisions been made for a catastrophic release of chemicals?

15. Have provisions been made for any necessary demolition and the resulting waste?

16. Have requirements for remediation at the site prior to construction been addressed?

17. Will all practicable measures for waste minimization be implemented?

18. Have the processes that generate air pollution been evaluated for minimization potential?

19. Will adequate air pollution controls be installed (scrubbers, fume hoods, dust collectors)?

20. Have handling and cleaning of air pollution control systems been addressed?

21. Have the processes that generate wastewater been evaluated for reduction potential?

22. Will indoor spills be protected from reaching drains?
23. Will outdoor spills be protected from reaching stormwater drains and sewer manholes?
24. Are adequate water disposal systems available?
25. Will pre-treatment methods be necessary and provided?
26. Will the discharges of domestic and industrial wastewater be in accord with regulations?

G. Ergonomics: Workstation and Work Methods Design

1. Generally, have material-handling designs considered worker capabilities and limitations, to accommodate the employee population at the 95% level?
2. Do material-handling designs promote the use of mechanical material-handling equipment such as conveyors, cranes, hoists, scissor jacks, and drum carts?
3. Do design layouts reduce in so far as practicable:
   a. constant lifting
   b. twisting and turning of the back when moving an object
   c. crouching, crawling, and kneeling
   d. lifting objects from floor level
   e. static muscle loading
   f. finger pinch grips
   g. work with elbows raised above waist level
   h. twisting motions of hands, wrists, or elbows
   i. hyperextension or hyperflexion of wrists
   j. repetitive motion
   k. awkward postures
4. Are workstations designed to provide:
   a. adequate support for the back and legs?
   b. adjustable work surfaces that are easily manipulated?
   c. delivery bins and tables to accommodate height and reach limitations?
   d. work platforms that elevate and descend, as needed?
   e. powered assists and suspension devices to reduce the use of force?
5. Has adequate attention been given to:
   a. lighting (to Illuminating Engineering Society requirements),
   b. heat,
   c. cold,
   d. noise, and
   e. vibration?
6. Does the design accommodate the hazards inherent in servicing, maintenance, and inspection?

7. Will there be adequate clearance and ready access to equipment for servicing?
8. Will controls be efficiently located in a logical and sequential order?
9. Will indicators be easy to read, either by themselves or in combination with others?

H. Fall Avoidance

1. Overall, has the design reduced the need for ladders and stairs insofar as practicable?
2. Where works at heights is to be done, has adequate consideration been given to providing work platforms or fixed ladders?
3. Are parapets or guardrails provided at roof edges?
4. Is equipment designed to avoid fall hazards during maintenance, inspection, and cleaning?
5. Does the design provide for fall arrest measures, such as anchorage points and fall restraining systems?

I. Fire Protection

1. Overall, in the design, will national and local fire codes and insurance requirements be met?
2. Will fire pumps, water tanks/ponds, and fire hydrants be adequate?
3. Will risers and post valves be accessible, and protected from damage?
4. Will small hose standpipes be adequate?
5. Will sufficient hose racks be provided?
6. Will special fire suppression systems be provided?
7. Has containment of fire suppression water been addressed?
8. Will there be adequate external fire zones?
9. Will emergency vehicle access be adequate?
10. Will flame arresters be installed where needed on equipment vents?
11. Will fire extinguishers be of appropriate types, adequate, and mounted for easy access?
12. Will the design for location of flammables be appropriate?
13. For flammables, will storage rooms and cabinets meet national fire codes and insurance requirements?
14. For flammable liquid dispensing, will grounding, bonding, and ventilation be adequate?
15. Will fire sensors, pull stations, and alarms be adequate?
16. Are flooding systems designed to provide a pre-discharge alarm which can be perceived above ambient light or noise levels before the system discharges, giving workers time to exit from the discharge area?
17. Has the project been reviewed by insurance personnel?

J. Hazardous and Toxic Materials

1. In the design process, have all materials in this category been identified?
2. Have the physical properties of the individual chemicals been identified?
3. Have the most conservative exposure limits been established as the design criteria?
4. Has a determination been made to use intrinsically safe equipment?
5. Have Material Safety Data Sheets been obtained for all materials?
6. Are the reactive properties known for chemicals that will be combined or mixed?
7. Have measures been taken to eliminate, substitute for, or reduce the quantities of hazardous chemicals?
8. Does the design emphasize closed process systems?
9. Will the design properly address all occupational illness potentials and reduce the need for monitoring, testing, and personal protective equipment insofar as is practicable?
10. Are storage facilities designed to separate hazardous from nonhazardous substances?
11. Does the design consider the chemical compatibility issues?
12. Have adequate provisions been made for chemical release, fire, explosion, or reaction?
13. Have provisions been made to contain water used in hazardous release control?
14. Are ventilation systems adequate to handle an emergency release?
15. Is the storage of hazardous chemicals belowground avoided?
16. Is storage tank location such as to reduce facility damage or damage to the public in a catastrophic event insofar as is practicable?
17. Are adequate storage tank dikes provided?
18. Will emergency ventilation be provided for accidental releases?
19. For extraordinary releases, will special ventilation, relief, and deluge systems be provided?
20. Will the normal use of chemicals allow operating without personal protective equipment?
21. Will the design of bulk loading/unloading facilities contain anticipated leaks and spills?

K. Lockout/Tagout: Energy Controls

1. In the design process, has adequate attention been given to lockout/tagout requirements to prevent hazardous releases from these energy sources:
   a. electrical,
   b. mechanical,
   c. hydraulic,

> d. pneumatic,
>
> e. chemical,
>
> f. thermal,
>
> g. nonionizing radiation, or
>
> h. ionizing radiation?

2. Are lockout/tagout devices adequate in design and number, readily accessible, and operable?
3. Are lockout/tagout devices standard throughout the facility?

L. Machine Guarding

1. Overall, do the designs prevent workers' hands, arms, and other body parts from making contact with dangerous moving parts? 1910.212(a)(3)
2. Have the requirements of all applicable machine guarding standards of the American National Standards Institute been identified and met?
3. Are safeguards firmly secured and not easily removed? 1910.212(a)(2)
4. Do safeguards ensure that no object will fall into moving parts? 1910(a)(1)
5. Do safeguards permit safe, comfortable, and relatively easy operation of the machine? 1910.212(a)(2)
6. Can the machine be lubricated without removing safeguards? 1910.212(a)(2)
7. Does the design include a system that requires shutting down machinery before safeguards are removed?
8. Are fixed machines anchored soundly?
9. Are in-running nip points guarded properly? 1910.212(a)(1)
10. Will the design address point-of-operation exposure properly? 1910.212(a)(3)
11. Are all reciprocating parts guarded properly? 1910.212(a)(3)(iv)
12. Are all rotating parts guarded properly? 1910.212(a)(3)(iv)
13. Are all shear points guarded properly? 1910.212(a)(3)(iv)
14. Are exposed set screws, keyways, collars, etc. guarded properly? 1910.212(a)(3)(iv)
15. Does the design eliminate the potential for flying chips? 1910.212(a)(i)
16. Has the potential for sparking been eliminated? 1910.212(a)(i)
17. If robots are to be used, are they designed to ANSI/RIA R15.06-1999; the American National Standard for Industrial Robots and Robot Systems safety requirements?

M. Noise Control

1. In the design process, have maximum noise levels been established that are to be stipulated in specifications for new equipment?
2. Is emphasis given to controlling noise levels through engineering measures?

3. Are the size or shape of rooms and proposed layout of equipment, workstations, and break areas to be evaluated for noise levels?
4. Will workers be separated from noise by the greatest practicable distance?
5. Will barriers be installed between noise sources and workers?
6. Are enclosed control rooms to be provided for operators in areas where the noise is above trigger levels?
7. Are lower-noise-level processes to be selected where feasible?
8. Have equipment and workstations been located so that the greatest sources of noise are not facing operators?

N. Pressure Vessels

1. Will all pressure vessels be designed to ASME and insurance company requirements?
2. Will pressure vessels containing flammables or combustibles meet OSHA 1910.106 and NFC standards?
3. Will pressure relief valves be:
   a. correctly sized and set,
   b. suitable for intended use, and
   c. directed to discharge safely?

O. Ventilation

1. Have all sources of emission been identified and their hazards characterized?
2. Have ways to reduce personnel interaction with the emission sources (location, work practices) been considered in the design process?
3. Have assessments been made with respect to incompatible emission streams (cyanides and acids, etc.)?
4. Has consideration been given to weather conditions and seasonal variations?
5. Will the design requirements of the ANSI Z9 series, the ACGIH ventilation manual, ASHRAE guidelines, and NFPA 45 and 90 met?
6. Will local ventilation effectively capture contaminates at the point of discharge?
7. Will room static pressures be progressively more negative as the operation becomes "dirtier"?
8. Will the ventilation system provide a margin of safety if a system fails?
9. Will emergency power and lighting be provided on critical units?
10. Will the ventilation equipment be remote and/or "quiet"?
11. Will spray booths and degreasers meet OSHA standards?
12. Will laboratory or contaminated air be totally exhausted?
13. If contaminated air is cleaned and reused, will it meet good safety requirements?
14. Will the makeup air to hoods be clean and adequate?

15. Have flow patterns been established to prevent exposure to personnel?
16. Does the design provide for proper gauging and alarm systems with respect to a sudden pressure drop?
17. Are ventilation controls easily accessible to operators?

P. Walking and Working Surfaces, Floor and Wall Openings, Fixed Stairs and Ladders

1. Will aisles, loading docks, and through doorways have enough clearance to allow safe turns where material-handling equipment is used? 1910.22(b)(1)
2. In the aisles, are people and vehicles adequately separated?
3. Are permanent aisles to be marked with lines on the floor? 1910.22(b)(2)
4. Does the design provide for floors, aisles, and passageways being free from obstruction? 1910.22(b)(1)
5. Has a logistics study been made to provide a safe and efficient flow of people and materials?
6. Will the construction texture of walking surfaces be nonslip?
7. Will the floors be designed to stay dry?
8. Will water and process flows be designed to keep off the walkway?
9. Will the floors be sloped and drained?
10. Will utilities and other obstructions be routed off the walking surfaces?
11. Will the design allow future utility expansion, with added facilities not having to be above and thereby cross floors, and be obstructive?
12. Will designs for floor and wall openings meet the requirements of OSHA 1910.23?
13. Do the designs for fixed stairs and ladders meet the requirements of OSHA 1910.23, .24, and .27?

RESOURCES

Appendix L in Z10 provides guidelines on audits. Design review and management of change are mentioned briefly as subjects to be considered when audits are made. In Appendix O, "Bibliography and References," several resources are listed under the caption "Design Review and Management of Change."

Entering "Safety Design Reviews" into a search engine will bring up several publications relating to systems design that individual safety professionals—depending on their fields of interest—may find interesting.

Bruce W. Main is the president of design safety engineering, inc. He is a practitioner in risk assessment and design reviews. His book *Risk Assessment: basics and Benchmarks* contains a chapter entitled "Design Reviews." He addresses:

- Purpose of a Design Review
- Types of Design Reviews

- Timing of a Design Review
- Design Review Mechanics
- The Decision Making Process
- Separating Analysis and Review
- Types of Safety Analyses for Design Reviews
- Practical Considerations.

Safety professionals who are interested in honing their knowledge and skills with respect to risk assessment and safety design reviews will find this book interesting.

Chapter 18 in this book, "Lean Concepts: Emphasizing the Design Process," includes a discussion on how a company merged lean concepts and environmental, safety, and health needs into its design process. In that process, design reviews are made at several stages as the design progresses.

CONCLUSION

I have been a strong proponent of addressing the hazards and risks that derive from them early in the design process. And I recommend that safety professionals move toward including the design review element in Z10 in the safety management systems they influence. I stand by the premise that:

- Hazards and the risks that derive from them are most effectively and economically avoided, eliminated, reduced, or controlled if they are considered early in the design process, and where necessary, as the design progresses.
- Risks of injury, illness, or damage are reduced significantly if processes are in place to avoid bringing hazards and risks into the workplace.

REFERENCES

Air Force System Safety Handbook, Air Force Safety Agency, Kirtland AFB, NM, 2000. http://www.system-safety.org/Documents/AF_System-Safety-HNDBK.pdf.

ANSI/AIHA Z10-2012. American National Standard, *Occupational Health and Safety Management Systems*. Fairfax, VA: American Industrial Hygiene Association, 2012. ASSE is now the secretariat. Available at https://www.asse.org/cartpage.php?link=z10_2005.

ANSI/ASSE Z590.3-2011. *Prevention through Design: Guidelines for Addressing Occupational Hazards and Risks in Design and Redesign Processes*. Des Plaines, IL: American Society of Safety Engineers, 2011.

Brander, Roy. "The Titanic and Risk Management." http://axion.physics.ubc.ca/titanic/ (Item 30).

Christensen, Wayne C. and Fred A. Manuele, Editors. *Safety Through Design*. Itasca, IL: National Safety Council, 1999.

DaimlerChrysler. *Ergonomic Design Criteria*. https://gsp.extra.daimlerchrysler.com/mfg/amedd/tooldesign/textsection15.htm.

Design for EHS: Equipment Design Philosophy. Chandler, AZ: Intel Corporation, 1999.

Guidelines for Hazard Evaluation Procedures, 2nd ed. New York: Center for Chemical Process Safety of the American Institute of Chemical Engineers, 1992.

Haddon, William, Jr. "On the Escape of Tigers: An Ecologic Note." *Technical Review*, May 1970.

ISO 14121. *Safety of Machinery—Principles for risk assessment.* Geneva, Switzerland, International Organization for Standardization, 2003.

Main, Bruce W. *Risk Assessment: basics and benchmarks.* Ann Arbor, MI: Design Safety Engineering Inc., 2004.

Main, Bruce W. *Risk Assessment: Challenges and Opportunities.* Ann Arbor, MI: design safety engineering inc., 2012.

MIL-STD-882E. *Standard Practice for System Safety.* Washington: DC: Department of Defense, 2000. Also at http://www.systemsafetyskeptic.com/yahoo_site_admin/assets/docs/MIL-STD-882E_final.135152939.pdf.

OSHA. Occupational Safety and Health Administration Standard 1910. Accessible at http://www.osha.gov/pls/oshaweb/owastand.display_standard_group?p_toc_level=1&p_part_number=1910.

CHAPTER 16

PREVENTION THROUGH DESIGN: SECTIONS 5.1.1 TO 5.1.4 OF Z10

This chapter, and ANSI/AIHA Z590.3-2011, the American National Standard titled *Prevention through Design: Guidelines for Addressing Occupational Hazards and Risks in Design and Redesign Processes*, relate to several provisions in Z10: risk assessment, Section 5.1.1; hierarchy of controls, Section 5.1.2; design requirements, Section 5.1.3; and procurement, Section 5.1.4. It also pertains to the prevention of serious injuries and fatalities.

We said earlier that applying prevention through design principles early in the design and redesign processes reduces the potential for incidents resulting in serious harm or damage. In Chapter 3 "Innovations in Serious Injury and Fatality Prevention", I presented "A Socio-Technical Model for An Operational Risk Management System" which gives prominence and significance to prevention through design. The following principles are offered in support of the premise that prevention through design should have such importance.

1. Hazards and risks are most effectively and economically avoided, eliminated, or controlled in the design and redesign processes.
2. Hazard analysis is the most important safety process in that, if it fails, all other processes are likely to be ineffective (Johnson, p. 245).
3. Risk assessment should be the cornerstone of an operational risk management system.

Advanced Safety Management: Focusing on Z10 and Serious Injury Prevention,
Second Edition. Fred A. Manuele.
© 2014 John Wiley & Sons, Inc. Published 2014 by John Wiley & Sons, Inc.

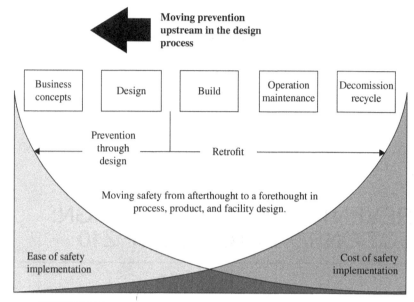

FIGURE 16.1   Theoretical ideal prevention through the design process.

4. If, through the hazard identification and analysis and risk assessment processes, specifications are developed that are applied in the procurement process so as to avoid bringing hazards and their accompanying risks into a workplace, the potential for serious injuries is reduced greatly.

5. The entirety of purpose of the personnel responsible for safety, regardless of their titles, is to manage their endeavors with respect to hazards so that the risks deriving from those hazards are acceptable.

6. To achieve acceptable risk levels, elements of a hierarchy of controls should be applied sequentially in the risk avoidance, elimination, reduction, and control endeavors.

The practice of safety is hazards-based. Thus, Johnson wrote appropriately that hazard analysis is the most important safety process. Since all risks in an operational setting derive from hazards and since the intent of an operational risk management system is to achieve acceptable risk levels, it follows that risk assessment should be the cornerstone of an operational risk management system. The core of prevention through design is hazard identification and analysis and risk assessment.

Figure 16.1 depicts the theoretical ideal. Prevention through Design is moved upstream in the design process. The intent is to have hazards and risks dealt with in the conceptual and design steps, but that requires unattainable perfection from the people involved. Hazards and risks will also be identified in the build, operation, and maintenance steps, for which redesign is necessary in a retrofitting process.

I was privileged to be the chair of the committee that wrote the Z590.3 standard. It will be obvious that what I have written here shows a bias in favor of its implementation. In this chapter; we:

- Provide a history of the safety through design/prevention through design movement.
- Present highlights of Z590.3, emphasizing the applicability of the provisions of the standard to all hazards-based initiatives: environmental controls, product safety, safety of the public, avoiding property damage and business interruption, and so on.
- Discuss the culture change necessary in most organizations as prevention through design concepts are implemented within an operational risk management system.
- Encourage safety professionals to become involved in prevention through design for job satisfaction and to be perceived as providing additional value in the organizations to which they give counsel.

HISTORY

In the early 1990s, several safety professionals recognized through their studies of investigation reports for occupational injuries and illnesses that design causal factors were not addressed adequately. For example, a study made by this author at that time indicated that although there were implications of workplace and work method design inadequacies in over 35% percent of the investigation reports analyzed, the corrective actions proposed did not relate to the design implications. That study is supported by a later analysis made in Australia.

In *Guidance On The Principles Of Safe Design For Work*, issued in 2006, comments are made on the "contribution that the design of machinery and equipment has on the incidence of fatalities and injuries in Australia."

Of the 210 identified workplace fatalities, 77 (37%) definitely or probably had design-related issues involved. Design contributes to at least 30% of work-related serious nonfatal injuries. (p. 6)

To provide the necessary education for designers, safety professionals are encouraged to develop their own supportive data on incidents in which design shortcomings are identified. That initiative should be followed by a major effort to have Z590.3-2011 be accepted as a design guide.

It was also noted in the early 1990s that designing for safety was addressed inadequately in safety-related literature, and that the safety and health management systems that organizations had in place infrequently included safety through design procedures.

Decision makers at the National Safety Council (NSC) gave authority to this author to form a committee to study the feasibility of the Council promoting the idea of having safety and environmental needs incorporated in the design process. In 1995, the outcome was that the NSC established The Institute For Safety Through Design.

An advisory committee for the institute was formed, whose members represented industry, academia, organized labor, and other interested persons. This is the mission established by the advisory committee:

> To reduce the risk of injury, illness and environmental damage by integrating decisions affecting safety, health, and the environment in all stages of the design process.

The message the advisory committee wanted to convey when the term *safety through design* was used is contained in its definition:

> The integration of hazard analysis and risk assessment methods early in the design and engineering stages and taking the actions necessary so that risks of injury or damage are at an acceptable level.

In the literature developed by the Institute it was said that the following benefits would be obtained by applying safety through design concepts:

- Improved productivity
- Decreased operating costs
- Significant risk reduction
- Avoidance of expensive retrofitting

Two groups were identified by the Institute to be given primary attention: Academia—professors and their students; and practicing engineers in industry and safety professionals.

Much was accomplished by the Institute. Seminars, workshops, and symposia were held; proceedings were issued; presentations were made at safety conferences; and a book entitled *Safety Through Design* was published. Operations of the Institute were discontinued in 2005, in accord with a previously established sunset provision.

In 2006, several of the participants in the activities of the Institute For Safety Through Design, and others, received an e-mail from an executive at the National Institute for Occupational Safety and Health (NIOSH) encouraging our participation in an initiative for Prevention through Design. In July 2007, NIOSH held a workshop to obtain the views of a variety of stakeholders on a major initiative to create a sustainable national strategy for Prevention through Design.

Some of the participants expressed the view that the long-term impact of the NIOSH initiative could be "transformative," meaning that a fundamental shift could occur in the practice of safety, resulting in greater emphasis being given to the higher and more effective decision levels in the hierarchy of controls.

In 2008, NIOSH announced that one of its major initiatives was to "Develop and approve a broad, generic voluntary consensus standard on Prevention through Design that is aligned with international design activities and practice." I volunteered to lead that endeavor.

Support was obtained from the Standards Development Committee at the American Society of Safety Engineers (ASSE). It was decided to develop a technical report first, for the learning experience that would provide. So TR-Z790.001, *A Technical Report on Prevention Through Design,* was issued in 2009.

On September 1, 2011, the American National Standards Institute (ANSI) approved the standard ANSI/ASSE Z590.3-2011, *Prevention Through Design: Guidelines for Addressing Occupational Hazards and Risks in the Design and Redesign Processes.*

It must be understood that activities at NIOSH are limited to occupational safety and health. Thus, the focus of Z590.3 is work-related. But, by intent, the terminology in Z590.3 was kept broad enough so that the guidelines could be applicable to all hazards-based fields: product safety, environmental control, property damage that could result in business interruption, and so on. In the standard, the definition of prevention through design is work-related and identical with that in the NIOSH literature.

Prevention through design: Addressing occupational safety and health needs in the design and redesign process to prevent or minimize the work-related hazards and risks, associated with the construction, manufacture, use, maintenance, retrofitting, and disposal of facilities, processes, materials, and equipment.

Promoting the acquisition of knowledge of prevention through design concepts is in concert with the ASSE Position Paper on Designing for Safety approved by its board of directors in 1994, the opening paragraph of which follows.

Designing For Safety (DFS) is a principle for design planning for new facilities, equipment, and operations (public and private) to conserve human and natural resources, and thereby protect people, property and the environment. DFS advocates systematic process to ensure state-of-the-art engineering and management principles are used and incorporated into the design of facilities and overall operations to assure safety and health of workers, as well as protection of the environment and compliance with current codes and standards.

HIGHLIGHTS OF ANSI/ASSE Z590.3

It is not my intention here to duplicate the Z590.3 standard; only highlights are given. In some instances, number or letter designations may not be exactly the same as in the standard. It is expected that safety professionals who become involved in design and redesign processes will develop a familiarity with Z590.3 in detail.

1. Scope, Purpose, and Application

This is the *Scope* of Z590.3: This standard provides guidance on including prevention through design concepts within an occupational safety and health management system. Through the application of these concepts, decisions pertaining to occupational hazards and risks can be incorporated into the process of design and redesign of the work premises, tools, equipment, machinery, substances, and work processes, including their construction, manufacture, use, maintenance, and ultimate disposal or reuse. This standard provides guidance for a life-cycle assessment and design model that balances environmental and occupational safety and health goals over the life span of a facility, process, or product.

Although the *Purpose* statement indicates that the standard pertains principally to the avoidance, elimination, reduction, or control of occupational safety and health hazards and risks in the design and redesign processes, this important extension follows the *purpose* statement.

> *Note*: Incidents that have the potential to result in occupational injuries and illnesses can also result in damage to property and business interruption, and damage to the environment. Reference is made in several places in this standard to those additional loss potentials which may require evaluation and resultant action.

Largely, this standard is applicable to all hazard-based operational risks. Particularly, it relates directly to all of the four major stages of occupational risk management.

a. Pre-operational stage: in the initial planning, design, specification, prototyping, and construction processes, where the opportunities are greatest and the costs are lowest for hazard and risk avoidance, elimination, reduction, or control.
b. Operational stage: where hazards and risks are identified and evaluated and mitigation actions are taken through redesign initiatives or changes in work methods before incidents or exposures occur.
c. Post-incident stage: where investigations are made of incidents and exposures to determine the causal factors that will lead to appropriate interventions and acceptable risk levels.
d. Post-operational stage: where demolition, decommissioning, or reusing/rebuilding operations are undertaken.

The *application* goals of utilizing prevention through design concepts in an occupational setting are to:

a. Achieve acceptable risk levels.
b. Prevent or reduce occupationally related injuries, illnesses, and fatalities.
c. Reduce the cost of retrofitting necessary to mitigate hazards and risks that were not addressed sufficiently in the design or redesign processes.

Addendum A outlines the risk assessment process, and Addendum B gives the progression of occupational hygiene issues flow.

2. Referenced and Related Standards

Relative American National Standards and other standards and guidelines that pertain to the purpose of the standard are listed.

3. Definitions

The definition list grew to 27 in response to suggestions made by commenters. Since one of the standard's principal goals is to achieve acceptable risk levels throughout the design and redesign processes, the definitions relative to that goal only are listed here.

a. *Acceptable risk.* That risk for which the probability of an incident or exposure occurring and the severity of harm or damage that may result are as low as reasonably practicable (ALARP) in the setting being considered.

b. *As low as reasonably practicable* (ALARP). That level of risk which can be lowered further only by an increase in resource expenditure that is disproportionate in relation to the resulting decrease in risk.

c. *Hazard.* The potential for harm.
   Note: Hazards include all aspects of technology and activity that produce risk. Hazards include the characteristics of things (e.g., equipment, technology, processes, dusts, fibers, gases, materials, chemicals) and the actions or inactions of people.

d. *Hierarchy of controls.* A systematic approach to avoiding, eliminating, reducing, and controlling risks, considering steps in a ranked and sequential order, beginning with avoidance, elimination, and substitution.

e. *Probability.* An estimate of the likelihood of an incident or exposure occurring that could result in harm or damage for a selected unit of time, events, population, items, or activity being considered.

f. *Residual risk.* The risk remaining after risk reduction measures have been taken.

g. *Risk.* An estimate of the probability of a hazard-related incident or exposure occurring and the severity of harm or damage that could result.

h. *Safety.* Freedom from unacceptable risk.

i. *Severity.* An estimate of the magnitude of harm or damage that could reasonably result from a hazard-related incident or exposure

4. Roles and Responsibilities

Top management shall provide the leadership to institute and maintain effective systems for the design and redesign processes. Key points: anticipate hazards and risks; assess risks; apply the hierarchy of controls to achieve acceptable risk levels. The following note is significant in the roles and responsibilities section.

Note: The processes of identifying and analyzing hazards and assessing risks improve if management establishes a culture where employee knowledge and experience is valued and respected and they collaborate in significant aspects of the design and redesign activities. Employees who do the work can make valuable contributions in identifying and evaluating hazards, in risk assessments, and in proposing risk reduction measures.

5. Relations with Suppliers

It became apparent as suggestions were received that commenters wanted help in creating procedures to avoid bringing hazards into the workplace. That help is provided in this section. The discussions and arrangements that organizations should have with suppliers are outlined and guidance on the specifics that should be expected of suppliers is provided. Procedures recommended and the related addendum pertains directly to the procurement requirements in Z10. Addendum C provides procurement guidelines that are to assist in making arrangements with suppliers.

One of the provisions in this section recommends that suppliers of equipment, technologies, processes, and materials provide documentation establishing that a risk assessment has been conducted and that an acceptable risk level, as outlined by the procuring organization, has been achieved.

Addendum D is an example of a basic risk assessment report that can be used as a guide for that purpose.

6. Design Safety Reviews

In the design process, risk assessments would be made as often as needed, on a continuum. In addition, a formal design safety review procedure should be put in place. This section is supported by Addendum E, a safety design review guide. Chapter 15 of this book is devoted entirely to design reviews.

7. The Hazard Analysis and Risk Assessment Process

This is the longest section in the standard. First, an outline of the hazard analysis and risk assessment process is given. That is followed by the "how" for each element in the outline. The outline follows.

- Select a risk assessment matrix.
- Establish the analysis parameters.
- Identify the hazards.
- Consider failure modes.
- Assess the severity of consequences.
- Determine occurrence probability.
- Define initial risk.
- Select and implement hazard avoidance, elimination, reduction, and control methods.
- Assess the residual risk.
- Risk acceptance decision making.
- Document the results.
- Follow-up on actions taken.

For many hazards the proper level of acceptable risk can be attained without bringing together complex teams of people. Safety and health professionals and design engineers with the proper experience and education can reach the proper conclusions of what constitutes acceptable risk. For more complex risk situations, management should have processes in place to seek the counsel of experienced personnel who are particularly skilled in risk assessment for the category of the situation being considered.

Reaching group consensus is a highly desirable goal. Sometimes, for what an individual considers obvious, achieving consensus is still desirable, so that buy-in is obtained for the actions taken. Addenda A and B serve as examples of suggested approaches to the risk assessment processes.

It is strongly urged that an appropriate risk assessment matrix be selected for the hazard analysis and risk assessment process. A risk assessment matrix provides a method to categorize combinations of probability of occurrence and severity of harm, thus establishing risk levels. A matrix helps in communicating with decision makers on risk reduction actions to be taken. Also, risk assessment matrices assist in comparing and prioritizing risks, and in effectively allocating mitigation resources. Addendum F provides several examples of risk assessment matrices and descriptions of terms to serve as a base for an organization to develop a matrix suitable for its operations.

8. Hazard Analysis and Risk Assessment Techniques

Top management shall adopt and apply the hazard analysis and risk assessment techniques suitable to the organization's needs and provide the training necessary to employees who will be involved in the process. Descriptions of eight selected techniques are presented in Addendum G. Addendum H is a failure mode and effects analysis form.

As a practical matter, having knowledge of three risk assessment concepts will be sufficient to address most, but not all, risk situations: preliminary hazard analysis and risk assessment, what-if/checklist analysis methods, and failure mode and effects analysis.

9. Hierarchy of Controls

Top management shall achieve acceptable risk levels by adopting, implementing, and maintaining a process to avoid, eliminate, reduce, and control hazards and risks. The process shall be based on the hierarchy of controls outlined in this standard, which is:

a. Risk avoidance
b. Elimination
c. Substitution
d. Engineering controls
e. Warning systems
f. Administrative controls
g. Personal protective equipment

This is a "prevention" standard. Research was done to develop a variation in the hierarchy of controls to adapt to the meaning of "prevention." Elimination, which is the first action to be taken in many hierarchies of controls, did not seem to fit with the meaning of prevention.

- *Elimination* means removal; purging; taking away; to get rid of something. To eliminate, there has to be something in place to remove.
- *Avoidance* means to prevent something from happening; keeping away from; averting.

Designers start with a blank sheet of paper or an empty screen in a computer-aided design system. Designers have opportunities to avoid hazards in all design stages: conceptual, preliminary, and final. In the early design phases, there are not yet any hazards to be eliminated, reduced, or controlled, so avoidance is a better match for a hierarchy of controls for a prevention standard.

Addendum I includes extensive comments on each of the elements in the hierarchy. Addendum J is the Bibliography.

GOALS TO BE ACHIEVED

Z590.3 says that insofar as is practicable, the goal shall be to assure that for the design selected:

- An acceptable risk level is achieved, as defined in this standard.
- The probability of personnel making human errors because of design inadequacies is as low as is reasonably practicable.
- The ability of personnel to defeat the work system and the work methods prescribed is as low as is reasonably practicable.
- The work processes prescribed take into consideration human factors (ergonomics)—the capabilities and limitations of the work population.
- Hazards and risks with respect to access and the means for maintenance are at as low as is reasonably practical.
- The need for personal protective equipment is as low as is reasonably practical, and aid is provided for its use where it is necessary (e.g., anchor points for fall protection).
- Applicable laws, codes, regulations, and standards have been met.
- Any recognized code of practice, internal or external, has been considered.

GETTING INVOLVED

It is recommended that safety professionals use actual cases in which they are involved to support the premise that addressing hazards and risks in the design and redesign processes will result not only in achieving acceptable risk levels but also higher productivity and operational efficiency. That may be done easier in the redesign process. I say to safety professionals:

- Try to convince designers to allow you to work with them on a project so that you can demonstrate your capability and value.
- Show the value you bring to the discussion.
- When teams are considering a given design or redesign situation, try to encourage input from all present who have knowledge of the process.
- Take a holistic, macro view; include all hazard-based subjects if you can.
- Try to involve operations personnel at all levels. Assume respectfully that those who do the work have knowledge and skill that can contribute to workplace and work method redesign and encourage their input.
- Do your homework; include alternative solutions and costs if that can be done.
- Understand that you are proposing a culture change—that designers may presume that you are intruding into their territory and will resist your involvement to maintain their territorial prerogative.
- Do not play down the fact that implementing a new program can be challenging and frustrating. Be patient. Utilize change management concepts. Be a good listener. Be open to comment and criticism.
- Study the incident history in the entity to which counsel is being given and that of its industry for support data showing that design shortcomings were among the contributing factors for incidents that have occurred.

A goal is to achieve buy-in by those who are involved, particularly at the senior executive level. Training programs should be considered for the value they provide. Also, ask if it will be advantageous in a particular situation for a prevention through design system to be written.

PATIENCE AND UNDERSTANDING

Implementing the concepts of risk assessment and risk avoidance or reduction in the design and redesign processes is a long-term effort. On that point, Bruce Main is eloquent, as in the Introduction to his book *Risk Assessment: Challenges and Opportunities*, reproduced here with his permission.

This is an exciting time in risk assessment. Whether machinery uptime, process throughput, cost savings, new ideas for features, patentable innovations, or simply documenting that a company makes really good products with acceptable risks, the opportunities abound to apply the process and drive improvements. The opportunities exist because risk assessment works.

The risk assessment process is a journey rather than an event. Companies that are just starting to complete risk assessments find their first efforts will require more time and will be less complete than later efforts. As personnel involved receive training and become more familiar with the risk assessment process, more hazards will be identified, more risk reduction methods deployed and the risk assessment process will improve and hasten in pace.

As lessons are learned and experience is gained, risk assessment becomes more refined. However, some time and experience is required for the risk assessment process to become fully integrated into a company. How much time depends on the company and its circumstances, but it typically takes months, not weeks. Eventually the risk assessment process will become a part of normal business procedures. Until then, industry needs time to fully and formally implement these concepts (Introduction).

For safety professionals who are interested in a well-written dissertation on "Implementing and Deployment" of a risk assessment system that affects all design decisions, Chapter 16 in *Risk Assessment: Challenges and Opportunities* is recommended. Bruce Main has given permission to duplicate the following key points, all of which should be considered in relation to procedures in place and the culture of an organization.

1. Leadership is a key and critical factor in successfully implementing and deploying the risk assessment process.
2. Integrating risk assessment in an organization is a process that generally follows a sequence of phases. (Three frameworks are discussed.)
3. Engineering design needs to change to include the risk assessment process to more effectively move safety into design. Only by changing the design process will risk assessment efforts succeed.
4. Introducing the risk assessment process will explicitly change the design process, allowing hazards to be identified and risk reduction methods to be incorporated early in the design process. As with any new process or substantial change, people may resist.
5. To be effective, the company culture must be willing to embrace the risk assessment process, and cultural acceptance stems from management leadership.
6. In industrial product or process applications, both equipment suppliers and users should perform risk assessments and be involved in the risk assessment process.
7. In consumer product and component product applications, the manufacturer is responsible for conducting the risk assessment, if applicable. Product users typically have no risk assessment responsibilities beyond using the product in conformance with the product information.
8. Practical guidance *is* shared to help companies get started and make progress in the risk assessment process. Topics addressed include: when to stop risk assessment, the time to complete an assessment, leaders in best practices, what to do in cross industry situations, when to revise an existing risk assessment, making changes to the method, results of risk assessment, and others.
9. To integrate risk assessment into the design process, engineers will likely need education and training on risk assessment in some form. (p. 230)

REVIEW OF ACTIVITIES AT NIOSH ON PREVENTION THROUGH DESIGN

Pamela Heckel, is the coordinator of prevention though design at NIOSH. Z590.3 would not exist if personnel at NIOSH had not concluded that one of its goals was to have a Prevention through Design (PtD) standard. Heckel was asked to provide a summary of the relative activities at NIOSH, which are extensive. This is what was written and approved at NIOSH for inclusion in this chapter.

Prevention through Design (PtD) was initiated by the National Institute for Occupational Safety and Health in 2007 to eliminate hazards through the design and redesign of facilities, processes, tools, equipment, products, and the organization of work. Strategic partners included the American Industrial Hygiene Association (AIHA), the American Society of Safety Engineers (ASSE), CPWR—The Center for Construction Research and Training, Kaiser Permanente, Liberty Mutual, the National Safety Council (NSC), the Occupational Safety and Health Administration (OSHA), ORC World-wide, and the Regenstrief Center for Healthcare Engineering.

The first PtD workshop was held in July 2007. Attendees included representatives from industry, labor, government, and academia. Proceedings were published in 2008 in a dedicated issue of the *Journal of Safety Research*. A PtD Council was formed to guide the new initiative. Council members were specialists in occupational safety and health and arranged strategic goals around the themes of Research, Education, Practice, and Policy. The Plan for the National Initiative (http://www.cdc.gov/niosh/docs/2011-121/) was published in 2009.

In 2010, the PtD role of the designer/engineer was investigated by benchmarking PtD regulations of designers in the construction industry in the United Kingdom (UK) to further understand the potential impacts and opportunities for implementation of the PtD concept in the US. The Education and Information Division of NIOSH coordinated a workshop, titled "Making Green Jobs Safe," which developed 48 compelling activities for including worker safety and health into green jobs and sustainable design. A summary of the workshop was published in 2011. http://www.cdc.gov/niosh/docs/2011-201/pdfs/2011-201.pdf

"Prevention through Design: A New Way of Doing Business," was held in August 2011, included supportive business leaders, noted safety experts, and academic researchers. At this stage, PtD concepts were included in 12 drafts or published consensus standards. Two booklets, three textbooks, and 50 peer-reviewed papers had been published. Four case studies had been created to demonstrate the business value of PtD. Presenters shared lessons learned from a Master's-level degree program at the University of Alabama–Birmingham and a PtD course at Virginia Tech. Half a dozen universities offered courses containing PtD content. These and other success stories were highlighted at the conference.

In 2011, the American Society of Safety Engineers (ASSE) obtained approved from the American National Standards Institute (ANSI) for ANSI/

ASSE Z590.3—the standard titled "Prevention through Design: Guidelines for Addressing Occupational Hazards and Risks in Design and Redesign Processes." This standard provides guidance on including Prevention through Design concepts within an occupational safety and health management system, and can be applied in any occupational setting. The standard focuses specifically on the avoidance, elimination, reduction, and control of occupational safety and health hazards and risks in the design process.

In 2011, NIOSH met with the US Green Building Council (USGBC), agreed to collaborate on the inclusion of worker health and safety into Leadership in Energy and Environmental Design (LEED) credits for certification. In 2012, NIOSH representatives met with the Construction User's Roundtable to identify opportunities for collaboration. Additional emphasis was placed on the development of business case studies. Three papers summarizing the 2011 conference presentations in the areas of practice, policy, and research were published in the January 2013 issue of *Professional Safety*. Three more were published in the March 2013 issue, including one focused on business value, one pertaining to OSH management systems, and the third, education. PtD has been the topic at more than 55 professional development courses, webinars, and roundtable presentations since 2008.

In August 2012, PtD and the NIOSH Nanotechnology Research Center (NTRC) collaborated with the State University New York at Albany, College of Nanoscale Science & Engineering, to hold a Safe Nano Design workshop. The purpose was to discuss the value of applying PtD to safely synthesize engineered nanoparticles and safely commercialize nano-enabled products. Applying PtD concepts in organizations handling engineered nanomaterials assures that worker health and safety is considered at each step in the supply chain, resulting in sustainable health and safety performance.

By the spring of 2013, two additional textbooks were published. A total of 93 first generation publications were cited by 721 second generation publications. Fifteen consensus standards containing PtD concepts have been published. More than two dozen universities have expressed interest in the PtD Education Modules for existing undergraduate classes. The first to be published was the Architectural Design and Construction Instructor's Manual (http://www.cdc.gov/niosh/docs/2013-133).

NIOSH is encouraging research in developing the business case for PtD. The focus is now on developing the process and related tools for determining business value. Areas of interest include the methods to measure the impact of PtD concepts on actually reducing injury and illness.

NIOSH is currently developing a systematic Health Hazard Banding approach to assist safety and health professionals in providing guidance for chemicals without authoritative occupational exposure limits (OELs). The NIOSH Hazard Banding process can be used with limited information and resources and can be performed quickly by in-house industrial hygienists and health and safety specialists. The outcome of the Health Hazard Banding process is an occupational exposure band (OEB).

Health Hazard Banding can be used to supplement and support OEL development by facilitating a more rapid evaluation of health risks, providing guidance for chemicals without OELs, identifying hazards to be evaluated for elimination or substitution, providing the user with recommendations when minimal data is available, and a tool for the development of NIOSH Recommended Exposure Limits.

Education activities focus on providing educational materials for students, engineers, and designers. The impact of these tools must be assessed. Most graduating engineering students have little practical experience with hazard identification, risk assessment, and risk avoidance or control methods. NIOSH is working with ABET, the organization that accredits engineering and technology curricula, to encourage faculty to consider the safety and health of students and staff during the discovery phase of cutting-edge research.

There are several areas where the incorporation of Prevention through Design concepts into the curriculum dovetails with ABET objectives to improve student outcomes, specifically under Criterion 3c. Incorporating PtD concepts into the curriculum shows a commitment to continuous improvement from faculty and the learning community.

Practice activities focus on stakeholder ability to access, share, and apply successful PtD practices. In addition to our continued dialog with USGBC and CURT, NIOSH is collaborating with industry to identify Best Practices so these can be front loaded into the Capital Design Process.

When a root cause analysis uncovers a design-related hazard, the design solution should be documented for future reference during the conceptual design phase of similar facilities or equipment.

Policy activities are focused on developing a PtD culture in industry. Including PtD into consensus standards is the first step in the development of a PtD culture.

CONCLUSION

Involvement of safety professionals in the design processes is hugely more extensive now than it was 20 years ago. There is opportunity here for professional satisfaction through participation in the design processes and for being perceived as a valued contributor in support of operational efficiency as well as risk management.

REFERENCES

ANSI/ASSE Z590.3-2011. American National Standard, *Prevention through Design: Guidelines for Addressing Occupational Hazards and Risks in Design and Redesign Processes.* Des Plaines, IL: American Society of Safety Engineers, 2011.

Christensen, Wayne C. and Fred A. Manuele, Editors. *Safety Through Design.* Itasca, IL: National Safety Council, 1999.

"Designing for Safety." A position paper approved by the board of directors of the American Society of Safety Engineers, Des Plaines, Il, 1994.

Guidance On The Principles Of Safe Design For Work. Canberra, Australia: Australian Safety and Compensation Council, an entity of the Australian government, 2006.

Johnson, William. *MORT Safety Assurance Systems*. Itasca, IL: National Safety Council, 1980. (Also published by Marcel Dekker, New York.)

Main, Bruce W. *Risk Assessment: Challenges and Opportunities*. Ann Arbor, MI: Design Safety Engineering, Inc., 2012.

TR-Z790.001. *A Technical Report on Prevention Through Design*. Des Plaines, IL: American Society of Safety Engineers, 2009.

CHAPTER 17

A PRIMER ON SYSTEM SAFETY: SECTIONS 4.0, 4.2, 5.1.1, 5.1.2, AND APPENDIX F

Identifying and analyzing hazards and making risk assessments as early as practicable in the design and redesign processes, and additionally as needed throughout the design processes, are the bases on which system safety is built. The goal of system safety initiatives is to attain acceptable risk levels.

Consider the following sections in Z10 as they relate to hazards, risks, risk assessments, the design process, and acceptable risks. These citations are abbreviated substantially.

- Section 4.0, "Planning": The goal is to identify and prioritize system issues (defined as hazards, risks, etc.).
- Section 4.2, "Assessment and Prioritization": The process shall assess the level of risk for identified hazards, establish priorities based on factors such as the level of risk, and identify factors related to system deficiencies that lead to hazards and risks.
- Section 5.1.1, "Risk Assessment": The organization shall establish and implement a risk assessment process(es) appropriate to the nature of hazards and the level of risk.
- Section 5.1.2, "Hierarchy of Control": The organization shall establish a process for achieving feasible risk reduction based on a preferred order of controls.

Advanced Safety Management: Focusing on Z10 and Serious Injury Prevention,
Second Edition. Fred A. Manuele.
© 2014 John Wiley & Sons, Inc. Published 2014 by John Wiley & Sons, Inc.

- Section 5.1.3, "Design Review": The organization shall establish a process to identify and take appropriate steps to prevent or otherwise control hazards at the design and redesign stages.
- Appendix F, "Risk Assessment": The goal of the risk assessment process, including the steps taken to reduce risk, is to achieve safe working conditions with an acceptable level of risk

There is a direct relationship between system safety concepts and processes necessary to implement Z10—and the practice of safety as a whole. I believe that generalists in the practice of safety will improve the quality of their performance by acquiring knowledge of applied system safety concepts and practices.

It is not suggested that safety generalists must become supra-specialists in system safety, although trends indicate that they will be expected to apply at least the fundamentals of system safety. To influence safety generalists to acquire knowledge of and apply system safety concepts, in this chapter we:

- Relate the generalist's practice of safety to applied system safety concepts.
- Give a history of the origin, development, and application of system safety methods.
- Review several definitions of system safety.
- Outline The System Safety Idea in terms applicable to the generalist's practice of safety.
- Encourage safety generalists to acquire knowledge and skills in system safety.

RELATING THE GENERALIST PRACTICE OF SAFETY TO SYSTEM SAFETY

In an American Society of Safety Engineers publication entitled "Scope and Functions of the Professional Safety Position," the following major activities are listed.

Functions of the Professional Safety Position

The major areas relating to the protection of people, property, and the environment are:

- A. Anticipate, identify, and evaluate hazardous conditions and practices.
- B. Develop hazard control designs, methods, procedures, and programs.
- C. Implement, administer, and advise others on hazard controls and hazard control programs.
- D. Measure, audit, and evaluate the effectiveness of hazard controls and hazard control programs.

A significant point to be made is that the professional safety function encompasses all initiatives that are hazard and risk based—the protection of people, property, and the environment. According to item A, the safety professional is to "anticipate hazards." Item B indicates that safety professionals are to "develop hazard control designs."

To be in a position to anticipate hazards, one must be involved in the design process. To participate effectively in the design process, a safety professional must be skilled with respect to hazard analysis and risk assessment techniques. Influencing the design process and using hazard analysis and risk assessment techniques to achieve acceptable risk levels are fundamental in system safety.

Enterprising generalists in safety will become proficient with respect to the hazard identification and control aspects and the design aspects of the scope and function of a safety professional. That will be to their advantage as they give counsel to clients to achieve acceptable risks with respect to the protection of people, property, and the environment.

AFFECTING THE DESIGN AND REDESIGN PROCESSES

System safety professionals make a great deal of designing things right the first time and being participants throughout the design and redesign processes. Richard A. Stephans, the author of *System Safety for the 21st Century*, expresses that view well.

Safety Is Productive

Safety is achieved by doing things right the first time, every time. If things are done right the first time, every time, we not only have a safe operation but also and extremely efficient, productive, cost-effective operation (p. 12).

Safety Requires Upstream Effort

The safety of an operation is determined long before the people, procedures, and plant and hardware come together at the work site to perform a given task (p. 13).

For products, facilities, equipment, and processes, and for their subsequent alteration, the time and place to avoid, eliminate, reduce, or control hazards economically and effectively is in the design or redesign processes. Participating in those processes presents opportunities for upstream involvement by safety professionals using system safety concepts.

Also, there has been an extended recognition that applying design and engineering solutions is the preferred course of action in operational risk management. That extended recognition derives from several sources, among which is the involvement of safety professionals in:

- Applied ergonomics.
- Giving counsel to meet European requirements whereby risk assessments are to be made on goods that are to go into workplaces in EU countries.

- Applying the requirements of guidelines and standards that propose or require risk assessments, and presenting an ordered sequence of measures to be taken in a hierarchy of controls to achieve acceptable risk levels. Examples are:
  - ANSI-AIHA Z10-2012. *Occupational Health and Safety Management Systems.*
  - MIL-STD-882E-2012. Department of Defense *Standard Practice for System Safety.*
  - ANSI-ASSE Z590.3-2011. *Prevention Through Design: Guidelines for Addressing Occupational Hazards and Risks in Design and Redesign Processes.*
  - ANSI/PMMI B155.1-2011. *Safety Requirements for Packaging Machinery and Packaging-Related Converting Machinery.*
  - ANSI B11.0-2010. *Safety of Machinery—General Safety Requirements and Risk Assessments.*
  - BS OHSAS 18001:2007. *Occupational health and safety management systems—requirements.*
  - *Guidance On The Principles Of Safe Design For Work.* Canberra, Australia: Australian Safety and Compensation Council, an entity of the Australian Government, 2006.
  - *Machine Safety: Prevention of Mechanical Hazards.* Quebec, Canada: The Institute for research for safety and security at work and The Commission for safety and security at work in Quebec, 2009.
  - *Risk Assessment.* The European Union, 2008.
  - CSA Z1002-12. *Occupational health and safety—Hazard identification and elimination and risk assessment and control.* Toronto, Canada: Canadian Standards Association, 2012.
  - EN ISO 12100–2010. *Safety of Machinery. General principles for Design. Risk assessment and Risk reduction.* Geneva, Switzerland: International Organization for Standardization, 2010.
- Meeting the requirements for hazards analysis in OSHA's standard Process Safety Management of Highly Hazardous Chemicals and in EPA risk management program requirements.

Of all of the foregoing references, the Z590.3 standard gets closest to applied system safety. It has been referred to as "system safety light."

DEFINING SYSTEM SAFETY

Unfortunately, the term *system safety* does not convey a clear meaning of the practice as it is applied. Published definitions of system safety are of some help in understanding the concept, but they do not communicate clearly. To give indications of the differences in the definitions of system safety, and to move this discussion forward, six sources are cited.

In MIL-STD-882E-2012, the Department of Defense *Standard Practice for System Safety*, system safety is defined as:

The application of engineering and management principles, criteria, and techniques to achieve acceptable risk within the constraints of operational effectiveness and suitability, time, and cost throughout all phases of the system life-cycle. (p. 8)

In *System Safety Primer*, Clifton A. Ericson II gave this definition of system safety in his 2011 book:

System safety is an engineering methodology employed to intentionally design-in safety into a product or system through the identification and elimination/mitigation of hazards. (p. 6)

In GEIA-STD-0010, the *Standard Best Practices for System Safety Program Development and Execution*, approved in 2008, this definition is given:

System safety is the application of engineering and management principles, criteria, and techniques to achieve mishap risk as low as reasonably practicable (to an acceptable level), within the constraints of operational effectiveness and suitability, time, and cost, throughout all phases of the system life cycle. (p. 8)

Richard A. Stephans' book *System Safety for the 21st Century* was published in 2004. He defines system safety as follows:

System safety: The discipline that uses systematic engineering and management techniques to aid in making systems safe throughout their life cycles. (p. 11)

System Safety and Risk Management, NIOSH Instruction Module, A Guide for Engineering Educators was developed for the National Institute for Occupational Safety and Health by Pat L. Clemens and Rodney J. Simmons in 1998. They write as follows:

What Is System Safety? System safety has two primary characteristics: (1) it is a doctrine of management practice that mandates that hazards be found and risks controlled; and (2) it is a collection of analytical approaches with which to practice the doctrine. (p. 3)

In *System Safety Engineering and Management*, 2nd ed., Harold E. Roland and Brian Moriarty asked in 1990: What is System Safety? In response to their own question, they give two meaningful comments and then establish the system safety objective.

The system safety concept is the application of special technical and managerial skills to the systematic, forward-looking identification and control of hazards throughout the life cycle of a project, program, or activity. The concept calls for safety analyses and hazard control actions, beginning with the conceptual phase of a system and continuing through the design, production, testing, use, and disposal phases, until the activity is retired. (p. 8)

The system safety concept involves a planned, disciplined, systematically organized and before-the-fact process characterized as the identify–analyze–control method of safety. The emphasis is placed upon an acceptable safety level designed into the system prior to actual production or operation of the system. (p. 9)

Using those definitions as a base, and with some extensions, the following outline of "The System Safety Idea" is presented for consideration by safety generalists to emphasize what system safety encompasses, relate system safety to the relative provisions in Z10, and connect system safety with the scope and function of a safety professional.

THE SYSTEM SAFETY IDEA

1. System safety is hazards, risks and design based.
2. Hazards are most effectively and economically anticipated, avoided, or controlled in the initial design processes or in the redesign of existing facilities, equipment, and processes.
3. Applied system safety requires a conscientious, planned, disciplined, and systematic use of special engineering and managerial tools.
4. Applying specifically developed hazard analysis and risk assessment techniques is a necessity in system safety applications.
5. Applied system safety begins in the conceptual design phase and continues into all subsequent design phases, production, and testing—through to the end of a system's life cycle. .
6. On an anticipatory and forward-looking basis, hazards are to be identified and analyzed, avoided, eliminated, reduced, or controlled so that, within operational constraints, safety can practicably be designed into systems and acceptable risk levels can be attained.
7. In applied system safety, the emphasis is on having acceptable risk levels designed into systems before actual production or operation of a system.
8. If action is needed to reduce risks to an acceptable level, the steps in the hierarchy of controls are to be taken sequentially: No lower-level step is to be taken until those above it are considered.
9. When trade-offs are made in the design process, and the needs of such as the utility and manufacturability of the end product, weight, operability, maintainability or cost have to be considered, the conclusion must, nevertheless, be at an acceptable risk level.
10. System safety applications apply to all aspects of an operation, including facilities, logistic support, storage, packaging, handling, and transportation entities.

11. System safety concepts promote the establishment of policies and procedures that are to achieve an effective, orderly, and continuous risk management process for the design, development, installation, and maintenance, of all facilities, materials, hardware, tooling, equipment, and products, and for their eventual disposal.

12. In the system design process, consideration is to be given to the interactions among humans, machines, and the environment, and the capabilities and limitations of people and their penchant for unpredictable behavior.

13. An overall requirement is that acceptable risk levels are to be attained, defined as follows: *Acceptable risk* is that risk for which the probability of a hazard-related incident or exposure occurring and the severity of harm or damage that may result are as low as reasonably practicable in the setting being considered.

This outline of "The System Safety Idea" encompasses most of the definitions given previously and goes beyond several. Safety generalists should ask: How closely does the system safety idea come to the results expected in applying the provisions in Z10? Do safety professionals serve themselves well by becoming knowledgeable and skilled in system safety?

System safety begins with hazard identification and analysis and risk assessment. So do all hazards and risk-based activities, whatever they are called. This author is confident that application of system safety concepts in the business and industrial setting will result in significant reductions in injuries and illnesses, damage to property, and environmental incidents.

HAZARD IDENTIFICATION AND ANALYSIS AND RISK ASSESSMENT TECHNIQUES

It is not surprising that many safety generalists are turned away from system safety when they encounter the number of analytical techniques that have been developed, the complexity of some of them, and the skill necessary to apply them. Earlier in this chapter it was made clear that safety generalists need not become supra-specialists in system safety. Nevertheless, the trends indicate that they will be expected to be skilled in applying basic system safety methods.

How many analytical systems are there? The second edition of the *System Safety Analysis Handbook* fills 626 pages and contains a compilation of 101 analysis techniques and methodologies. That handbook serves as a desk reference for the accomplished system safety professionals who may have to resolve highly complex or infrequently encountered or unusual situations.

Three national standards that constitute a set should be of interest to safety generalists who want to become familiar with system safety techniques. The American Society of Safety Engineers is the secretariat.

ANSI/ASSE Z690.1-2011. *Vocabulary for Risk Management* (National Adoption of ISO Guide 73:2009). This standard provides definitions of terms that, the originators hope, will be used in other standards.

ANSI/ASSE Z690.2-2011. *Risk Management Principles and Guidelines* (National Adoption of ISO 31,000:2009). The intent of this standard is to provide a broad-ranged primer on risk management systems that could be applied in any type of organization. The need to make risk assessments is introduced in Section 5.4, Risk Assessment.

ANSI/ASSE Z 690.3-2011. *Risk Assessment Techniques* (National Adoption of IEC/ISO 31,010:2009). For safety generalists who want an education in risk assessment concepts and methods and a ready reference, this standard is worth acquiring. It begins with a 15-page dissertation on risk assessment concepts and methods. In five pages, Appendix A provides brief comparisons of 31 risk assessment techniques. Reviews of the 31 techniques—Overview, Use, Inputs, Process, Strengths and Limitations—are provided in Annex B, which covers 79 pages.

ANSI/ASSE Z 690.3-2011 is a valuable resource. A list of the 31 risk assessment techniques follows. Some could be applied only by experienced system safety professionals. But knowledge of a few of them will serve a huge percentage of the needs of a safety generalist.

B01	Brainstorming	B02	Structured or Semi-Structured
B03	Delphi		Interviews
B05	Preliminary Hazard Analysis	B04	Checklists
B07	Hazard Analysis and Critical Control	B06	Hazard and Operability Studies
	Points	B08	Environmental Risk Assessment
B09	Structure—What if Analysis	B10	Scenario Analysis
B11	Business Impact Analysis	B12	Root Cause Analysis
B13	Failure Mode Effect Analysis	B14	Fault Tree Analysis
B15	Event Tree Analysis	B16	Cause and Consequence Analysis
B17	Cause-and-Effect Analysis	B18	Layer Protection Analysis
B19	Decision Tree	B20	Human Reliability Analysis
B21	Bow Tie Analysis	B22	Reliability Centered Maintenance
B23	Sneak Circuit Analysis	B24	Markov Analysis
B25	Monte Carlo Simulation	B26	Bayesian Statistics and Bayes Nets
B27	FN Curves	B28	Risk Indices
B29	Consequence/Probability Matrix	B30	Cost Benefit Analysis
B31	Multi-Criteria Decision Analysis		

In ANSI/ASSE Z590.3-2011, the *Prevention through Design* standard, Addendum G, comments on eight hazard analysis and risk assessment techniques:

- Preliminary Hazard Analysis
- What-If Analysis
- Checklist Analysis
- What-If Checklist Analysis
- Hazard and Operability Analysis
- Failure Mode and Effects Analysis

- Fault Tree Analysis
- Management Oversight and Risk Tree (MORT).

It was also said in Z590.3 that:

As a practical matter, having knowledge of three risk assessment concepts will be sufficient to address most, but not all, risk situations. They are Preliminary Hazard Analysis and Risk Assessment, the What-If/Checklist Analysis methods, and Failure Mode and Effects Analysis. (p. 23)

Having knowledge and capability with respect to the above-mentioned three standards will be sufficient to deal with a huge majority of the needs of Z10. Additional comments on risk assessment techniques appear later in this chapter in the section "Recommended Reading."

THE HAZARD ANALYSIS AND RISK ASSESSMENT PROCESS

Section 7 in ANSI/ASSE Z590.3 is devoted to the hazard analysis and risk assessment process. It is the core of the prevention through design standard. The following process outline is a recent work, having been approved by the American National Standards Institute on September 1, 2011. In the standard, the narrative for each subject is extensive and is recommended reading. Under management direction, the hazard analysis and risk assessment process follows.

1. Select a risk assessment matrix
2. Establish the analysis parameters
3. Identify the hazards
4. Consider failure modes
5. Assess the severity of consequences
6. Determine occurrence probability
7. Define initial risk
8. Select and implement hazard avoidance, elimination, reduction and control methods
9. Assess the residual risk
10. Risk acceptance decision making
11. Document the results
12. Follow-up on actions taken

Reaching group consensus in the risk assessment process is a highly desirable goal. Sometimes, for what an individual considers obvious, achieving consensus is still desirable, so that buy-in is obtained for the actions taken.

RISK ASSESSMENT MATRICES

It would be highly unusual for a text or standard on system safety not to include risk assessment matrices and provide examples. There are many, many variations of matrices, and the definitions of the terms used in them vary greatly. The matrix used should be the one that management and users in an organization decide is best for them. It is strongly recommended that a suitable matrix be chosen because of its value in risk decision making.

A risk assessment matrix provides a method of categorizing combinations of probability of occurrence and severity of harm, thus establishing risk levels. Z10 requires that priorities be established in the application of its requirements. A matrix helps in communicating with decision makers on risk reduction actions to be taken. Also, risk assessment matrices assist in comparing and prioritizing risks and in effectively allocating mitigation resources.

All personnel involved in the risk assessment processes must understand the definitions used for occurrence probability and severity and for risk levels in the risk assessment matrix chosen. Examples of risk assessment matrices are shown in several chapters in this book.

THE HIERARCHY OF CONTROLS

It was said in the "The System Safety Idea" that if risk reduction was necessary after a risk assessment, the steps to follow were those in the hierarchy of controls. That subject is covered in Chapter 14 in this book. Decision makers should understand that with respect to the six levels of control shown in Z10's hierarchy of controls, the ameliorating actions described in the first, second, and third control levels are more effective because they:

- Are preventive actions that eliminate or reduce risk by design, elimination, substitution, and engineering measures
- Rely the least on human behavior—the performance of personnel
- Are less defeatable by unit managers, supervisors, or workers

Actions described in the fourth, fifth, and sixth levels are contingent actions and rely greatly on the performance of personnel for their effectiveness. Inherently, they are less reliable.

In applying the hierarchy of controls, the expectation is that consideration will be given to each step in descending order, and that reasonable attempts will be made to avoid, eliminate, reduce, or control hazards and their associated risks through steps higher in the hierarchy before lower steps are considered. A lower step in the hierarchy of controls is not to be chosen until practical applications of the preceding level or levels are considered. It is understood that for many risk situations, a combination of the risk management methods shown in the hierarchy of controls is necessary to achieve acceptable risk levels.

In applying the hierarchy of controls, the outcome should be an acceptable risk level. In achieving that goal, the following should be taken into consideration.

- Avoiding, eliminating, or reducing the probability of a hazard-related incident or exposure occurring.
- Reducing the severity of harm or damage that may result if an incident or exposure occurs.
- The feasibility and effectiveness of the risk reduction measures to be taken, and their costs, in relation to the amount of risk reduction to be achieved.
- All of the requirements of Z10.

WHY SYSTEM SAFETY CONCEPTS HAVE NOT BEEN WIDELY ADOPTED

At least one other author expected a more widespread adoption of system safety concepts beyond the use by the military, aerospace personnel, and nuclear facility designers. He also had to recognize that it wasn't happening. In *The Loss Rate Concept In Safety Engineering*, R. L. Browning wrote this:

As every loss event results from the interactions of elements in a system, it follows that all safety is "systems safety." The safety community instinctively welcomed the systems concept when it appeared during the stagnating performance of the mid-1960s, as evidenced by the ensuing freshet of symposia and literature. For a time, it was thought that this seemingly novel approach could reestablish the continuing improvement that the public had become accustomed to; however, this anticipation has not been fulfilled.

Now, some three decades later, although systems techniques continue to find application and development in exotic programs (missiles, aerospace, nuclear power) and in the academic community, they are seldom met in the domain of traditional industrial and general safety. (p. 12)

Although there were countless seminars and a proliferation of papers on system safety, the generalist in the practice of safety seldom adopted system safety concepts. In response to his own question—Why this rejection?—Browning expressed the view that system safety literature and seminars on system safety may have turned off generalist safety professionals because of the "exotica" they usually presented. I believe that to be so.

Nevertheless, Browning went on to build *The Loss Rate Concept In Safety Engineering* on system safety concepts. He also gave this encouragement:

We have found through practical experience that industrial and general safety can be engineered at a level considerably below that required by the exotics, using the mathematical capabilities possessed by average technically minded persons, together with readily available input data. (p. 13)

There is a reality in Browning's observations: System safety literature at the time he wrote his book was loaded with governmental jargon, and it easily repelled the uninitiated. It made more of the highly complex hazard analysis and risk assessment techniques requiring extensive knowledge of mathematics and probability theory than it did of concepts and purposes.

Some system safety literature did, and still does, give the appearance of exotica. That is changing. Texts on system safety that are truly primers and slanted toward the neophyte have been written.

PROMOTING THE USE OF SYSTEM SAFETY CONCEPTS

With the hope of generating further interest by generalist safety professionals in system safety concepts, it is suggested that they concentrate on those basics through which gains can be made in an occupational, environmental, or product design setting and avoid being repulsed by the more exotic hazard and risk assessment methodologies. Ted Ferry said it well in the Preface he wrote for Richard Stephans' book *System Safety for the 21st Century*:

> Professional credentials or experience in "system safety" are not required to appreciate the potential value of the systems approach and system safety techniques to general safety and health practice. (p. xiii)

To paraphrase Browning: All hazards-related incidents result from interactions of elements in a system. Therefore, all safety is system safety. Therein lies an important idea. In *Safety Management*, John V. Grimaldi and Rollin H. Simonds wrote:

> A reference to system analysis may merely imply an orderly examination of an established system or subsystem. (p. 287)

Applying system safety as "an orderly examination of an established system or subsystem" to identify, analyze, avoid, eliminate, reduce, or control hazards can be successful in the less complex situations without using elaborate analytical methods.

Repeating for emphasis: Applying the fundamentals of system safety can meet a huge percentage of the provisions of Z10. For safety generalists who take an interest in system safety concepts, the following reading list is offered, from which selections can be made.

RECOMMENDED READING

Clifton Ericson's book is what the title says it is, a *System Safety Primer*. This paperback, published in 2011, is only a 140-page read, yet it covers the system safety subject very well as a primer. It is easy to read and is recommended.

System Safety for the 21st Century is an update by Richard Stephans of *System Safety 2000* by Joe Stephenson. Stephans followed the advice given to "keep it as a primer." The book begins with a history of system safety. Then the author moves into system safety program planning and management and system safety analysis techniques. About half of the book is devoted to those techniques. Stephans says:

This book is specifically written for:

- Safety professionals, including people in industrial and occupational safety, system safety, environmental safety, industrial hygiene, health, occupational medicine, fire protection, reliability, maintainability, and quality assurance
- Engineers, especially design engineers and architects
- Managers and planners
- Students and faculty in safety, engineering, and management (p. xv)

A safety generalist would find this book to be a valuable source and not too difficult to read.

Basic Guide to System Safety was written by Jeffrey W. Vincoli. These two sentences are taken from the Preface: "It should be noted from the beginning that it is not the intention of the *Basic Guide to System Safety* to provide any level of expertise beyond that of novice. Those practitioners who desire complete knowledge of the subject will not be satisfied with the information contained on these pages."

Vincoli also says: "The primary focus of this text shall be the advantages of utilizing system safety concepts and techniques as they apply to the general safety arena. In fact the industrial workplace can be viewed as a natural extension of the past growth experience of the system safety discipline." (p. 5) Vincoli fulfilled his purpose. He has written a basic book on system safety that will serve the novice well.

MIL-STD-882E, *Standard Practice for System Safety*, issued by the Department of Defense, serves well as a primer. It is available on the Internet and can be downloaded without charge at http://www.system-safety.org/. Click on 882E in the right-hand column.

The Federal Aviation Administration's *System Safety Handbook* is also on the Internet as a free download. This is really a training manual. There are 17 chapters and 10 appendices—all individually downloadable as a separate PDF file. Enter "Federal Aviation Administration System Safety Handbook" into a search Engine, or go to http://www.faa.gov/library/manuals/aviation/risk_management/ss_handbook/.

The Loss Rate Concept in Safety Engineering, by R. L. Browning, is a small but good book that I have referred to several times. Browning believes that one can apply system safety concepts in an industrial setting without necessarily delving into exotic mathematics. He builds on "The Energy Cause Concept", and works through qualitative and quantitative analytical systems.

System Safety Engineering and Management, 2nd ed., by Harold E. Roland and Brian Moriarty is a good but more involved book. It provides an extensive review of the concepts of system safety and their methods of application. An overview of a system safety program is given. The descriptions of several analytical techniques are valuable. For the application of some of them, quite a bit of knowledge about mathematics is necessary.

PROGRESS REVIEW

For an assessment of how system safety principles relate to the requirements for an occupational risk management system, readers are asked to relate the provisions in Z10 with what authors in system safety are writing.

Clifton Ericson in *System Safety Primer*:

A known and acceptable level of safety can be achieved when the system safety process is continuously and unconditionally applied. (p. 2)

Relative safeness is calculated by the metric of risk. Risk is the estimated value of a potential event, based on the event's gain or loss and the event's likelihood of occurrence. Thus, how safe something is becomes a function of the amount of risk involved. (p. 3)

Since 100% freedom from hazards and risk is not possible, safety is more effectively defined as freedom from unacceptable risk. (p. 4)

System safety is a form of preemptive forensic engineering, whereby potential mishaps are identified, evaluated and controlled before they occur. Potential mishaps and their causal factors are anticipated during the design stage, and then design safety features are incorporated into the design to control the occurrence of the potential mishaps—safety is intentionally designed-in and mishaps are designed-out. This proactive approach to safety involves hazard analysis, risk assessment, risk mitigation through design and testing to verify design results. (p. 7)

Jeffrey Vincoli in *Basic Guide to System Safety*:

The process of system safety revolves around a desire to ensure that jobs or tasks are performed in the safest manner possible, free from unacceptable risk or harm or damage. This forward-looking process occurs within a working environment where people, operating procedures, equipment/hardware, and facilities all are integral factors that may or may not affect the safe and successful completion of the job of task. (p. 12)

The Hazard Risk Matrix incorporates the elements of the Hazard Severity table and the Hazard Probability table to provide an effective tool for approximating acceptable and unacceptable levels or degrees of risk. Obviously, from a systems standpoint, use of such a matrix facilitates the risk assessment process. (p. 12)

Richard Stephans in *System Safety in the 21st Century*:

The first and most effective way to control identified hazards is to eliminate them through design or redesign changes. (p. 14)

Design and build safety into a system rather than modifying the system later in the acquisition process when any changes are increasingly more expensive. (p. 22)

CONCLUSION

The principal intent in this chapter is to establish that fundamental system safety concepts can be applied by generalists in the practice of safety to meet the provisions in Z10, outline "The System Safety Idea," and encourage generalists who have not adopted system safety concepts to begin the inquiry and education to do so.

I sincerely believe that generalists in the practice of safety can learn from system safety successes and be more effective in their work through adopting system safety concepts. Their application in the occupational, environmental, and product safety settings would result in significant reductions in incidents having adverse effects.

In summation: The entirety of purpose of those responsible for safety, regardless of their titles, is to manage their endeavors with respect to hazards so that the risks deriving from those hazards are acceptable.

Note: The substance of this chapter appears in the fourth edition of my book *On the Practice of Safety*. It has been modified significantly to relate particularly to the provisions in Z10.

REFERENCES

ANSI/AIHA Z10-2012. *Occupational Health and Safety Management Systems*. Des Plaines, IL: American Society of Safety Engineers, 2012. Also at https://www.asse.org/cartpage.php?link=Z10_2005&utm_source=ASSE+Members&utm_campaign=3677c44444-Z10_Standard_Announcement_9_17_129_13_2012&utm_medium=email.

ANSI/ASSE Z590.3-2011. *Prevention through Design: Guidelines for Addressing Occupational Hazards and Risks in Design and Redesign Processes*. Des Plaines, IL: American Society of Safety Engineers, 2011.

ANSI/ASSE Z690.1-2011. *Vocabulary for Risk Management*. Des Plaines, IL: American Society of Safety Engineers, 2011.

ANSI/ASSE Z690.2-2011. *Risk Management Principles and Guidelines*. Des Plaines, IL: American Society of Safety Engineers, 2011.

ANSI/ASSE Z690.3. *Risk Assessment Techniques*. Des Plaines, IL: American Society of Safety Engineers, 2011.

ANSI B11.0-2010. *Safety of Machinery—General Safety Requirements and Risk Assessments*. Leesburg, VA: B11 Standards, Inc., 2010.

ANSI/PMMI B155.1-2011. *American National Standard for Safety Requirements for Packaging Machinery and Packaging-Related Converting Machinery*. Arlington, VA: Packaging Machinery Manufacturers Institute, 2011.

Browning, R. L. *The Loss Rate Concept In Safety Engineering*. New York: Marcel Dekker, 1980.

BS OHSAS 18001:2007. *Occupational health and safety management systems—Requirements*. London: BSI Group, 2007.

Clements, P.L. and Rodney J. Simmons. *System Safety and Risk Management, NIOSH Instructional Module, A Guide for Engineering Educators*. Cincinnati, OH: National Institute for Occupational Safety and Health, 1998.

CSA Z1002-12. *Occupational health and safety—Hazard identification and elimination and risk assessment and control*. Toronto, Canada: Canadian Standards Association, 2012.

EN ISO 12100–2010. *Safety of Machinery—General principles for Design. Risk assessment and Risk reduction*. Geneva, Switzerland: International Organization for Standardization, 2010.

Environmental Management Systems: An Implementation Guide for Small and Medium-Sized Organizations, 2nd ed. Access at http://www.fedcenter.gov/_kd/Items/actions. cfm?action=Show&item_id=598&destination=ShowItem. Copyright is held by NSF International Strategic Registrations, Ltd., Ann Arbor, MI.

Ericson, Clifton A. II. *System Safety Primer*. Self-published, 2011. Available through Internet booksellers.

Federal Aviation Administration System Safety Handbook. Enter the title into a search engine, or access at http://www.faa.gov/library/manuals/aviation/risk_management/ss_handbook/.

GEIA-STD-0010-2008. *Standard Best Practices for System Safety Program Development and Execution*. Arlington, VA: Information Technology Association of America, 2008.

Grimaldi, John V. and Rollin H. Simonds. *Safety Management*. Homewood, IL: Irwin, 1989.

Guidance On The Principles Of Safe Design For Work. Canberra, Australia: Australian Safety and Compensation Council, an entity of the Australian government, 2006.

Machine Safety: Prevention of mechanical hazards. (2009). Quebec, Canada: The Institute for research for safety and security at work and The Commission for safety and security at work in Quebec, 2009. Also at www.irsst.qe.ca.en/home.html.

MIL-STD-882E. Department of Defenses *Standard Practice System Safety*, 2012. It is available on the Internet and can be downloaded at http://www.system-safety.org/. Click on 882E in the right-hand column.

OSHA's Rule for Process Safety Management of Highly Hazardous Chemicals, 29 CFR 1910.119. Washington, DC, OSHA, 1992.

Risk Assessment. The European Union, 2008. http://osha.europa.eu/en/topics/riskassessment.

Roland, Harold E. and Brian Moriarty. *System Safety Engineering and Management*, 2nd ed. Hoboken, NJ: Wiley, 1990.

"Scope and Functions of the Professional Safety Position" brochure. Des Plaines, IL. American Society of Safety Engineers, 1998.

Stephans, Richard A. *System Safety in the 21st Century*. Hoboken, NJ: Wiley, 2004.

System Safety Analysis Handbook (for which Warner Talso and Richard A. Stephans provided stewardship). Unionville, VA: International System Safety Society, 1999.

Vincoli, Jeffrey W. *Basic Guide to System Safety*. Hoboken, NJ: Wiley, 1993.

CHAPTER 18

LEAN CONCEPTS—EMPHASIZING THE DESIGN PROCESS: SECTION 5.1.3 OF Z10

Colleagues who are much involved in lean concepts were asked if they had experience with their being applied in the original design process. The answer was no. Surely, there is such activity, but I have located only one initiative in which lean—and safety, environmental, and health—considerations were integrated into the design process.

Even though applying lean concepts to eliminate waste, improve efficiency, and lower production costs has become popular at the senior management level, it seems that most of the initiatives relate to redesigning existing systems and work methods. Nevertheless, that activity is a match for the provision in Z10 in Section 5.1.3, whereby management is to have processes in place "to prevent or otherwise control hazards in the design and redesign stages."

Safety professionals have opportunities to make contributions to operational results as they tactfully but forcefully bring to management's attention that:

- An element of waste that should be addressed in the lean process is the waste arising from the direct and ancillary costs of accidents.
- As is the case with hazards and their accompanying risks, operational waste can be most economically and effectively avoided in the design process.

Reducing waste is the base on which the lean concept is built. Simply stated, *lean* means creating more value for customers with fever resources. In a lean venture, activities or processes that consume resources, add cost, or require unproductive time

Advanced Safety Management: Focusing on Z10 and Serious Injury Prevention,
Second Edition. Fred A. Manuele.
© 2014 John Wiley & Sons, Inc. Published 2014 by John Wiley & Sons, Inc.

without creating value are eliminated. Direct accident costs are substantial, and those costs are a form of waste. Ancillary accident costs, such as those deriving from interruption of work, facility and equipment repair, idle time of workers, training of replacements, and investigation and report preparation time, may represent an amount of waste equal to or greater than the direct costs. For incidents resulting in serious injury, particularly when property damage and business interruption are extensive, the ancillary cost and accompanying waste can be substantial.

To encourage safety professionals to seek meaningful involvement in lean initiatives, in this chapter we:

- Comment on the absence of recognition in the literature that accidents produce waste and that their outcomes are a form of waste to be eliminated
- Discuss the origin of lean concepts and how broadly they are being applied
- List definitions pertinent to lean and relate them to injury and illness prevention
- Discuss a successful merging of lean, safety, health, and environmental concepts into a design process
- Illustrate how the 5S concept is foundational in a lean application and how hazards and risks are reduced through 5S applications
- Comment on lean implementations in which hazards and risks were not addressed, the result being greater risks of injury and illness
- Discuss a major educational work
- Provide a simplified model of a value stream map that recognizes hazards and waste potential

ON THE LEAN LITERATURE

There is plenty to read regarding lean, but there is a dearth of information in the literature on how the waste deriving from accidents should be addressed. It is a rarity when consultants who advertise their lean capabilities include the outcome of accidents as a form of waste. An example of that scarcity is demonstrated in one of most popular books on lean, *Lean Thinking*, written by James P. Womack and Daniel T. Jones. This excerpt is taken from the book's jacket: "This book has been sold in the hundreds of thousands of copies in a dozen countries." How good is the book? This is what an executive who managed a process that adopted lean concepts says about it:

> While there are many good books about lean techniques, *Lean Thinking* remains one of the best sources to understand "what is lean" because it describes the thought process, the overarching key principles that must be used as guides in the application of lean techniques and tools. [Comments made to me confidentially.]

Nevertheless, the many readers of *Lean Thinking* would not be informed that the outcomes of accidents are to be included in a list of the forms of waste to be reduced.

But a fairly recent publication that I recommend highly can be of great value to safety professionals and other management personnel. It is unique. Robert B. Hafey is the author of *Lean Safety: Transforming Your Safety Culture with Lean Management*, published in 2010. Hafey builds his case in support of the interconnection of lean and safety from 40-plus years in various management roles in manufacturing, *none of which was as a safety director*.

This is a different and interesting book in that it gives guidance, from an operations management viewpoint, on how to make progress toward achieving a world-class safety management system by applying lean concepts. Hafey emphasizes that safety must be considered as a value within an organization's culture to achieve world-class performance. He makes it clear that to eliminate waste and to improve safety, the focus at all operations levels has to be on improving the work process. He writes:

> The Key to understanding this lean leadership style is the acceptance of the fact that the process is the problem, not the person. (p. 19)

The book contains a large number of practical examples about process improvement for lean and safety. It also contains many lean and safety forms. Thus, this is as much a how-to book as it is a concept book. Throughout the book, Hafey promotes the five-why system for solving problems of every type, including accident investigation. In applying the five-why system, users ask why sequentially until the problem's real causal factors are identified.

This is a valuable book for safety professionals who want to become familiar with lean basics so as to be able to give guidance as a contributor on a lean process team. It is a book of fundamentals and is not as complex as other texts on lean with which I am familiar. It is a good investment for continuing education. It is unique.

Progressive safety professionals will recognize this shortcoming—the non-recognition of accidents as a source of waste by the appliers of lean concepts—as an opportunity to educate all levels of management on the advantage of including safety considerations as the lean process is applied.

ORIGIN OF THE LEAN CONCEPT

In much of the literature on lean, Taiichi Ohno is recognized as the originator of the lean concept about 50 years ago while at Toyota; and Toyota has been a highly successful applier of the concept. But, occasionally reference is made in the literature to concepts utilized early in the twentieth century by Henry Ford, who created the first "lean" auto production line; to Frank Bunker Gilbreth, who was a proponent of scientific management and motion study; to Walter Shewhart, a pioneer in statistical control; and to W. Edwards Deming, who achieved world renown for his work on quality management.

Whatever the origins of lean, the leaders at Toyota—as they strove for operational excellence—combined, refined, and converted their waste elimination concepts into what is called *lean* in the United States.

LEAN CONCEPTS ARE BROADLY APPLICABLE

While the original literature on lean concepts describes applications in manufacturing, the concepts have been adopted to reduce waste in a variety of situations, such as for accounting systems, a large spectrum of service businesses, transportation companies, warehousing, construction, health care facilities (including entities as small as group physician practices), product quality improvement, and environmental management, and to eliminate non-value-added e-mails. How broadly have lean concepts been employed? Entering "lean concepts" into a search engine brings up over 13,000,000 references.

Safety professionals, particularly those who have environmental management responsibilities, may want to look into the EPA entry on the Internet entitled *Lean and Environment Toolkit*. A major section in the *Toolkit* is captioned "How to Incorporate Environmental Considerations into Value Stream Mapping." The methodology shown is the same as would be utilized for all aspects of safety.

DEFINITIONS

Understandably, several terms associated with the lean concept are Japanese. Abbreviated definitions of those terms follow as they are applied in the design process discussed later in the chapter, as well as some other definitions that are used. This list is somewhat lengthy, intentionally. In reality, safety professionals need to be familiar with the terms used in the organizations to which they give counsel, and their meanings. They would probably not need to be thoroughly familiar with all of the terms in this list.

> *Flow*, as a goal in the lean process, is achieved after waste is removed from the system and the improved process (value stream) runs smoothly and efficiently with very little waste in the work of personnel or in equipment downtime.
>
> *Jidoka* refers specifically to machines or the production line itself being able to stop automatically in abnormal conditions (e.g., when one machine breaks down, when heat rises beyond a set limit). Jidoka applications do not allow defective parts or products to go from one workstation to another.
>
> *Kaizen* in Japanese means "change for the better". In American English, the term has come to mean continual improvement. For the purpose of this chapter, the emphasis in applying the continual improvement process is to eliminate waste, meaning those activities that add to costs but do not provide value.
>
> *Muda* encompasses all activities that consume resources wastefully, therefore not adding value. Seven types of waste were identified at Toyota for which continual reduction was to be obtained.

> *The Seven Wastes*
> > *Defects* in products or services are obviously wasteful in that they consume material and require additional production and correction time.

Overproduction is the excess production or acquisition of items beyond what is actually needed. Where overproduction occurs, additional capital investment is necessary, and costs are increased without adding value since more storage space and material handling are necessary. Overproduction that results in excessive material handling adds to risks.

Transportation wastes are those that require additional and unproductive moving of a product in process. Each time a product is moved there is added risk of damage to the product, equipment, and facilities, and harm to personnel. In the moving process, the product fills valuable space and requires time expenditures without adding value.

Waiting refers to both the unproductive time spent by workers waiting for material or components in a process to arrive and the time required for excess production to flow through the system. An additional example is material or information waiting to be worked on to complete a customer order. Similar wastes occur when incidents happen that could result in injury or damage to property.

Inventory buildup in excess of what is needed requires an additional capital outlay and produces waste because of the need for additional storage space and handling time. Frequent handling of the inventory adds to the risk of injury.

Motion refers to worker unproductive time and movement where the process is cumbersome, inefficient, and wasteful. This implies that the process may also be hazardous.

Overprocessing means using a more expensive or otherwise valuable resource than is needed for the task. Overprocessing also includes costly rework.

Poka yoke means "mistake-proofing" or "fool-proofing", the purpose being to design work and processes so that it is nearly impossible for people to make mistakes. An example is designing hose connections or electrical connections so that they can be put together in one way only, thereby reducing risk. This is an important but often neglected concept with respect to employee and product safety.

Mura pertains to unevenness in the work flow: The goal is steady work flow.

Muri relates to avoiding overburdening equipment or employees: The goal is to reduce the workload to acceptable levels. For equipment, that might mean operating at 80% of the maximum specified limit. For employees, designing work methods that are overly stressful and working excessive hours are to be avoided.

Pull defines the operational situation after which much has been accomplished in applying the lean process and inventories can be maintained in relation to the "pull" as represented by customer orders. Waste from having excessive product in inventory, and all that implies, is to be as low as reasonably practicable (e.g., the cost of excess space, the financing of the excess inventory, the cost and risk of additional handling of inventory).

Total Productive Maintenance is to assure that all equipment used in a process is always able to perform its tasks so that production or work processes will not be interrupted.

MERGING LEAN AND DESIGN CONCEPTS

The one exception with which I have become familiar, in which lean, safety, and environmental concerns were merged into the original design process, involves a pharmaceutical company. For a major project, new equipment was to be acquired and installed in an existing facility. In lean language, that would be a "brownfield" application. In a "greenfield" application, design engineers incorporate lean concepts into the design of an entirely new facility.

What this pharmaceutical company has done is an excellent example of how lean and safety can be addressed in the design process concurrently. This is a unique, imaginative, and creative methodology. In a discussion with the senior executive who managed this project, he made the following comments.

> Often within an organization there are separate Quality, EHS and Operational Excellence/Lean functions. Each of these functions has its own distinct vocabulary, metrics, and evaluation processes and procedures. Viewed from above, the purpose, motivation, and objectives of these functions have considerable overlap: Do it once; do it safely; do it with quality; do it cost-effectively.
>
> Responsibility for the overall management and delivery of a Design/Capital project is usually in the Operations function. Past practice often has been for Operations to interface with Quality, EHS and Lean functions separately. This creates a significant redundancy of effort and can raise the risk of an issue "falling in the cracks" between these functions.
>
> In a lean application, it was decided to gather all of the functions and construct a Value Stream Map covering the entire project. This creates a "visual" map of the project from beginning to end. It allows a clear identification of functions and who is responsible and accountable for each step of the project. It identifies opportunities to perform tasks in such a way that all of the oversight and approval criteria of each function are satisfied simultaneously.
>
> A cross-functional integrated risk assessment tool was developed. Our theme was—Do it once: Do it right. If this cross-functional team is created and managed properly, the expected communication and responsibility requirements are clearly established. If the team is well managed, the result is significant cross-functional cooperation and excellent results.
>
> Lean requires front-line operator participation. Significant valuable input and buy-in can be achieved with their active participation in the project.
>
> Lean terminology was the base language for the management and measurement of the project. [From my notes].

All operations personnel at this pharmaceutical location have had lean training. An abbreviated version of this company's process follows. It is close to the theoretical ideal. Safety professionals can learn from it. But first, the relationship to Z10 of the lean-design process is shown.

TABLE 18.1

Provision in Z10	Section Designation
Risk assessments	5.1.1
Hierarchy of controls	5.1.2
Design reviews	5.1.3
Management of change	5.1.3
Procurement	5.1.4

RELATION TO Z10

The writers of Z10 were prudent when they said in E.1.3 that:

> This standard is designed so it can be integrated with quality, environmental, and other management systems within an organization.

Such integration would enhance the probability of a lean process being highly successful. That was done in the case being discussed. This company's initiative is particularly noteworthy in that it incorporates several of the provisions in Z10. Those provisions are noted in Table 18.1.

CRITERIA FOR APPLYING THE LEAN-DESIGN PROCESS

Use of this company's Lean-Design process begins when an assumption is made that a project is of such magnitude that it will require following the steps outlined in its Request for Capital Expenditure Procedure. For purchases below the capital expenditure request level, the basics in the process are applied, but not as extensively.

For example, if the machine shop supervisor put through a request to purchase a metal-cutting saw:

- Safety considerations would be established and they would be reviewed by environmental, health, and safety professionals as well as by more than one level of management.
- Those safety requirements would be included in the purchase order.
- After receipt and installation of the equipment, a safety validation would be made.

THE COMPANY'S LEAN-DESIGN PROCESS

The Concept Stage

From any source—research and development, engineering, any operations department, a cross-functional group, maintenance, individual workers—an idea may be proposed for process improvement. A broad range of brainstorming by a team takes place. If it is concluded that the idea should be moved forward and the expenditure level requires following the organization's Request for Capital Expenditure Procedure, a review and tentative approval is sought at a senior management level. A project manager is assigned.

Capital Expenditure Request and Element Champion Review

The Capital Expenditure Request would describe the design objectives of the project generally, make the business case for it, and request the necessary funding. In this company, each of the 26 elements in its safety management system is assigned to a Champion, most often someone at an upper management level.

For example, the chief executive assumes direction and accomplishment responsibility for two of those elements; four are assigned to another senior manufacturing executive. At this stage, all of the safety management system element champions have become aware of the project, and a sign-off is required by each of them.

Identify the Customers/Users

In this context customer or user is every employee who may be affected by the revision in a process being proposed. It really means everyone. The purpose is to assure that all persons who could be affected are aware of the process change proposed and can provide input as the activity proceeds. Identifying the customers and users is considered a very important step in the lean process.

With respect to external customers, the characteristics of the products manufactured have been agreed upon, close estimates are made of the product amounts that will be purchased and over what time spans, inventories are kept under tight control, and the delivery methods and times of delivery are arranged.

Project Customers/Users Requirement Specification

At this point a senior-level manager prepares a document expanding on the original idea. The document contains enough detail to specify the outcomes expected, and some criteria are established. Customers and users (employees) may submit their specifications and their suggestions on how waste can be eliminated.

Value Stream Map

A value stream map is created at this point. This is a preliminary flowchart that includes every step of the production process as conceived at this time, from raw material receipt to product going out the door. It is an important step in that it documents the processes to be considered in the waste elimination initiative. The value stream includes every step in a process to produce a product or provide a service.

Value stream mapping is a vital step in the lean concept, in that it provides an opportunity for team brainstorming to identify activities that do not add value. Addendum A in this chapter describes "A Simplified Initial Value Stream Map". Lean practitioners use value stream mapping to:

- Identify major sources of non-value-added time in a value stream.
- Envision a less wasteful future state.
- Develop an implementation plan for future lean activities.

Project Conceptual Design

All that preceded this step in the process influences the drafting of the project conceptual design. It shows the layout proposed, and building and utility impacts, and contains specifics on the major equipment needed. Environmental, health, and safety considerations are addressed in this concept stage. All of the following personnel review and sign off on the concept design: operation executives; subject matter experts; environmental, health, and safety professionals; engineers; maintenance personnel; and the building manager.

Since I knew that management did not rely entirely on the drawings that came out of the CAD (computer-aided design) system as the sources in the development of operating procedures, the executive who managed the process was asked how he came to that conclusion. This is what he said.

One of the shortcomings of CAD/Computer design and printed blueprints is that they do not allow for full real-life visualization of the process in actual physical/human terms. The creation of physical mock-ups of key components of the plant, particularly where critical operator interface is involved, can greatly improve the design.

A physical mock-up will allow line-of-sight, ergonomic, safety, simplicity of maintenance, and lean productivity issues to be identified that otherwise would be missed by relying on computer-generated designs. It is important to understand that people may have difficulty visualizing a computer-based design in real life.

Mock-ups can be constructed out of plywood or cardboard for relatively little cost. They allow the actual plant operators to see and feel the critical steps of the process and apply lean concepts to the development of work instructions for each step of the process. Mock-ups allow for front-line input and participation. Our experience was that many lean and safety improvements came from suggestions made by operators as they worked in the simulated mock-ups. That participation creates a significant and valuable "buy-in."

Experience has shown that without exception, a mock-up will reveal a number of critical issues that otherwise would have been missed in the design process when using CAD as the only means of illustration. Further, mock-ups allow for the development of lean work practices and training procedures parallel to the actual building and installation of the plant and speeds the startup and debugging process significantly. [From my notes].

Change Control Provisions

This company operates under the regulations of several governmental entities. Therefore, a rigid change control system is in place to assure that all quality, safety, and environmental requirements are met. At a senior management level, a change control document is produced requiring approval by all department heads. At this point, the head of the compliance group is particularly interested in seeing that all regulations are met. In Z10, the comparable requirement is to have a management of change process in place: Section 5.1.3.

Project Safety Clearance and Lean Review

This is a summation step with respect to all of the foregoing. The design document is reviewed by the environmental, health, and safety group and by the compliance group. Determinations are made with respect to the need for further safety analysis in individual pieces of the process or because of their interrelationships. Safety specifications are expanded and become more specific.

Although lean considerations have been a part of this process from the beginning, applying lean concepts is stressed more rigorously here by the project manager. The purpose is error-proofing, waste elimination, to have the process stop when the equipment recognizes a fault, and to avoid rejects. All or some of the lean systems mentioned previously—Poka Yoke, Jidoka, Kaizen, or Muda—may be brought into play, but the Muda concepts prevail throughout. Waste is to be as low as reasonably practicable.

Also, since this company has been a meticulous applier of the 5S system (defined in their usage as – Sorting, Simplification, Systematic Cleaning, Standardization, and Sustaining), the 5S system concepts are overriding in the lean process. It was said by a senior executive at this location that "If the staff has not been educated in 5S concepts and believe that their substance is a core value, you can forget about lean. You must have established a stable environment in which waste elimination is a fundamental to move into the next step to accomplish lean."

Drafting Vendor Specifications

Engineering personnel draft vendor specifications. Manufacturing, environmental, health, and safety, and operating personnel may also be involved. At this stage, communication begins with a selected vendor. Subject matter experts employed by the vendor may assist in drafting specifications for the project.

Conceptual Design Risk Assessment

This review takes place at the concept and drawing level. Formal risk assessment methods, qualitative or quantitative, are used as required. The risk assessments are documented and approved by a multifunctional team, of which the environmental, health, and safety personnel are a part.

An independent reviewer, not a part of the project team, must also sign off on the risk assessments. Several people at this location have been trained to do Failure Mode and Effects Analyses.

The Preliminary Design

Project team members work with the vendor to assure that the users' requirements are met. The receivables from the vendor include schematics, flow diagrams, drawings, further component specifications, and operating procedures and training manuals.

Value Stream Map: Waste Scavenger Hunt (Muda Check)

A value stream map was created when the project was in the concept stage. At this developmental phase, an additional flowchart is made to depict the design proposed. As stated previously, Muda encompasses all activity that wastefully consumes resources but does not add value.

A Muda check takes place as a waste scavenger hunt to further reduce product defect possibilities, overproduction, excessive product handling, idle and waiting time by operating personnel, excessive inventory, and to achieve efficiency in processing and the best probable use of employee skills. All personnel levels are involved.

Proposed Design Safety/Risk Assessment: Create System Drawings

Now that a proposed design is available, additional risk assessments as needed are made, prior to building the system. The environmental, health, and safety staff is prominently active in the risk assessments, along with other involved personnel. Use of formal risk assessment methods is more frequent at this stage. A final sign-off by the independent reviewer is necessary.

At this point, the design is frozen, the vendor creates system drawings, and the vendor builds to drawings.

Safety, Operational and Lean Review

At the vendor's location, before the equipment can be shipped, the purchaser's environmental, health, and safety personnel assure that all safety-related specifications have been met.

Factory acceptance testing takes place at the vender's location and members of the review team (engineering, operations, maintenance, validation, et al.) determine that the equipment operates as expected and that waste is as low as reasonably practicable.

This is a large part of the approval process prior to shipment of the equipment. The staff has found that testing at the vendor's location has avoided many issues that would have to be resolved later on their shop floor.

Review by maintenance is especially important here, as their sign-off affects the company's ability to apply a Total Preventive Maintenance initiative. With approval, the equipment may be shipped to the purchaser.

Standard Operating Procedures

In reality, this function is done in parallel with the previous steps. It involves writing standard operating procedures, developing training modules, defining record keeping needs, drafting production records, and so on.

Facility Review and Approval

After installation, with which the vendor is involved extensively, site acceptance tests are performed. Approval in operation is needed by the project team, including environmental, health, and safety personnel, before acceptance. The purpose is to validate that the equipment performs as intended, that the quality level expected is achieved, and that environmental, health, and safety specifications have been met.

In Production

At this stage, Kaizen—continual improvement—is a governing concept. Superior quality is maintained. Adherence to standard operating procedures, including safe practices, is the norm. Waste is constantly sought after and reduced.

5S Review

Since this organization has made applying the 5S system a core value, a final review is made to assure that all 5S system elements have been maintained: Sorting, Simplifying, Systematic Cleaning, Standardization, and Sustaining.

The 5S Concept

Originators of this Lean/Design process were asked to critique this chapter for technical accuracy. This is one of the comments made: "We have found that 5S is one of the foundations of lean. As far as safety is concerned, nothing makes hazardous conditions and practices stick out more than a well-organized facility. You should expand on 5S and how it can help improve safety performance."

His premise required further inquiry into how the 5S program operates in his facility. They say that their 5S program is an underlying reason for their receiving a bundle of awards on employee safety, environmental management, and product quality. Personnel who critiqued this chapter say that the 5S concept can have a solid impact on worker safety and that it is folly to expect good work practices and outstanding performance from workers if the work environment is dismal, messy, and disorderly, and operational discipline is lacking. Comment on the 5S system follows.

Sorting, the first step in a 5S application, is to get rid of everything not needed, all the cluttering, and to achieve an atmosphere of orderliness. When that orderliness is achieved in operational and storage areas—both for work in process and equipment needed to do the work—efficiency and housekeeping are improved, hazards and risks are reduced, and time wasted searching for work items is eliminated.

Simplifying is the next step in the 5S process. If there is a place for everything retained, and those places are well marked and labeled and known to the staff, it is easier to find tools, parts, and the equipment needed to do a job and to keep things orderly. Simplifying in a disciplined manner promotes identification of hazardous situations and makes it easier to get things done with less risk.

Systematic cleaning is the third step in 5S. Everyone is to be involved in the systematic cleaning endeavor. Workers in a unit are assigned ownership of and responsibility for the cleaning tasks. The purpose is to produce orderliness: Dirt, disorder, or things stored in aisles and getting in the way or stored in a manner that makes their recovery hazardous are not tolerated. The cleaning processes are to add to operational efficiency, to eliminate waste, and to reduce risk.

Standardization, the fourth step in 5S, is to adopt the best practices for equipment and machinery layout, and the design of equipment and work practices for productivity, mistake-proofing, and continual improvement. Workers at all levels have opportunities for input into the standardization procedure. Comments are sought on the design of the work methods for efficiency as well as to avoid risky situations.

Since at this location, accidents are recognized as a form of waste, safety is an integral part of the standardization process. Performance standards and expectations for predictable results are set. Operational breakdowns are to be few and far between. Causal factors for operating problems are studied and largely eliminated on an anticipatory basis. Up-front prevention is the thinking. Methods to identify possible breakdowns and how to respond with as little waste as practicable when they occur are a part of the standardization procedure.

For maintenance personnel, that makes their work easier: Thus, they are exposed to fewer hazardous situations; jerry-rigging for unusual work is not condoned. It is emphasized that maintaining tight control over the management of change procedures is an integral part of the standardization element in 5S.

Sustaining what has been accomplished in the four previous steps is the fifth step in the 5S concept. This, they say, is the most difficult step after superiority is attained in the first four steps. It is expected that some workers might revert to previous practices, particularly with respect to cluttering the workplace and avoiding cleanliness. Sustaining the concept can be achieved only by continuous management leadership.

The CEO in this company says that he knows he must, continuously and personally, embrace the 5S concept and both talk the talk and walk the talk, repeatedly. He is visible and involved as he holds his staff accountable for sustaining what they have achieved—an orderly and stable work environment in which efficiency is at a high level, waste is as low as reasonably practicable, and hazards and risks are at an acceptable level.

Some organizations have added a sixth S to their system to stress safety. But in reality, if a 5S system is installed and managed properly, safety is integrated and intertwined into all of the first five steps. It can be argued that adding a separate S to the system for safety creates the impression that safety is separate from the management system and could produce adverse consequences.

REAL-WORLD OBSERVATIONS

Unfortunately, some attempts by organizations to improve operations by applying lean concepts have not included safety considerations. Worse yet, existing systems to control risks have been overridden in some lean applications, the result being that hazardous situations that were addressed previously reappear. Retrofitting for correction of those hazards may be difficult but is certainly wasteful and expensive.

I have observed situations similar to those for which Kevin Newman and Theodore Braun offer caution in "Advice on incorporating ergonomic safety initiatives into your continuous improvement process." They say:

> Unfortunately, "Lean" doesn't necessarily mean safer though the two should go hand in hand. After all, a poorly designed task that requires a worker to reach excessively is not only inefficient, requiring more time and motion than needed, but is also likely to cause injury. Similarly, a worker lifting materials beyond his or her own capabilities takes more time and energy to perform the task and runs the risk of overexertion.
>
> In the worst-case scenario, an overzealous company may implement extreme Lean strategies where safety is not merely overlooked, but compromised. In the end, increasing efficiency without incorporating safety will cost far more than it saves.

Newman and Braun imply that ergonomics risks occur if the work requires "a worker to reach excessively" and if there is "the risk of overexertion." If ergonomics-related injuries occur, one of the central themes in lean—avoiding interruptions in the work flow—is negated. While all hazards and risks should be addressed in a lean process, applying ergonomics principles fits particularly well within a lean initiative.

A MAJOR WORK ON SAFETY AND LEAN

Other safety professionals have recognized the dearth of information in the lean literature about how safety and lean can be integrated. They have also encountered situations where safety concepts and lean applications were in conflict, with the results being far from satisfactory.

The Association For Manufacturing Technology (AMT) has published ANSI B11. TR7 2007, *Designing for Safety and Lean Manufacturing: A guide on integrating safety and lean manufacturing principles in the use of machinery.*

Although the purpose is to address lean and safety concepts in the use of machinery, this technical report can be valuable to all safety professionals who become involved in lean. Its content is largely generic and the principles apply to all enterprises.

TR7 provides guidance on how a low waste level at a low-risk level can be achieved and helps fill the gap in the technical literature on lean and safety. This is how the abstract reads.

Lean manufacturing includes a variety of initiatives, technologies and methods used to improve productivity (better and faster throughput) by reducing waste, costs and complexity from manufacturing processes. However, the effort to get lean has too frequently led to the misapplication of lean manufacturing principles in ways that result in significant risks to worker safety and to the goal of lean manufacturing.

Safety is a critical element in the lean manufacturing effort to yield processes that are better, faster, less wasteful and safer. This document provides guidance for persons responsible for integrating safety into lean manufacturing efforts. This integration is only possible if lean manufacturing concepts and safety concerns of machinery are addressed concurrently.

A brief overview of lean manufacturing concepts is presented. The challenge of concurrently addressing safety and lean is described and examples demonstrate situations where this has not occurred.

A process model for safe and lean is presented. A risk assessment framework is outlined that demonstrates how lean manufacturing concepts and safety can be implemented concurrently. Examples where safety and lean have been successfully applied are shared. This document also provides design guidelines on how to meet lean objectives without compromising safety.

Safety professionals were prominent in the development of this technical report. It is a good resource for concurrently addressing safety and lean.

CONCLUSION

Because of the base on which the lean concept is built—removing waste from a system—lean will probably have staying power. Since accidents and their consequences are so fundamentally wasteful, preventing them should be an integral part of lean applications. From the very beginning, when an organization begins to discuss adopting lean concepts, safety professionals should step forward to become members of the lean team.

There is opportunity here to address hazards and the risks that derive from them as processes are designed and redesigned. To be meaningful participants, safety professionals must become familiar with lean concepts. Several helpful resources on lean are listed following the References for this chapter.

Also, an Internet search indicates that several courses on the lean concept are available. For example, the Society of Manufacturing Engineers (SME) has developed courses that award lean certificates on three levels.

Each progressive level of lean certification requires continuing education—either academic coursework or structural classroom training. Information about the SME courses and its Lean Certification Body of Knowledge may be found at http://www.sme.org/. As the SME literature says, "Lean thinking requires Lean learning."

REFERENCES

ANSI/AIHA Z10-2012, American National Standard, *Occupational Health and Safety Management Systems*. Fairfax, VA: American Industrial Hygiene Association, 2012. ASSE is now the secretariat. Available at https://www.asse.org/cartpage.php?link=z10_2005.

ANSI B11.TR7 2007. *Designing for Safety and Lean Manufacturing: A guide on integrating safety and lean manufacturing principles in the use of machinery.* Secretariat and Accredited Standards Developer, B11 Standards, Inc., POB 690905, Houston, TX 77269. Also at American National Standards Institute – www.ansi.org.

EPA Lean and Environment Toolkit. "How to Incorporate Environmental Considerations into Value Stream Mapping." http://www.epa.gov/epainnov/lean/toolkit/ch3.htm#definition.

Hafey, Robert B. *Lean Safety: Transforming Your Culture with Lean Management*. New York: Production Press, 2010.

Newman, Kevin and Theodore Braun. "Advice on Incorporating Ergonomic Safety Initiatives into Your Continuous Improvement Process." *Occupational Hazards*, Aug. 2, 2005.

Womack, James P., and Daniel T. Jones. *Lean Thinking: Banish Waste and Create Wealth in Your Corporations*, 2nd ed. Northampton, MA: Free Press, 2003.

ADDITIONAL READING

Dennis, Pascal. *Lean Production Simplified: A Plain-Language Guide to the World's Most Powerful Production System*. Shelton, CT: Productivity Press, 2002.

Hallowell, Matthew R., Anthony Veltri, and Stephen Johnson. "Safety & Lean." *Professional Safety*, Nov. 2009, pp. 22–27.

Imai, Masaaki. *Gemba Kaizen: A Commonsense Low-Cost Approach to Management*. New York: McGraw-Hill, 1997.

Main, Bruce, Michael Taubitz, and Willard Wood. "You Cannot Get Lean Without Safety." *Professional Safety*, Jan. 2008, pp. 38–42.

Rother, Mike and John Shook. *Learning to See: Value Stream Mapping to Create Value and Eliminate Muda*. Cambridge, MA: Lean Enterprise Institute, 2003.

Taubitz, Michael. "Lean, Green & Safe." *Professional Safety*, May 2010, pp. 39–46.

ADDENDUM A

A SIMPLIFIED INITIAL VALUE STREAM MAP USED TO IDENTIFY WASTE (MUDA) AND OPPORTUNITIES FOR CONTINUOUS IMPROVEMENT (KAIZEN)

A   B   C   D   E   F

Defects: The machinery at station A is old and worn. Regardless of the amount of tinkering, it cannot achieve a quality defect level lower that 3 parts per 10,000. Producing defects at that level, below some customer specifications, is wasteful.

Motion: Adjustments of the machinery at Station A and die changes must be made frequently. That is wasteful motion and adds risk. Also, the lockout/tagout device is over 100 feet from the machine. An arrangement of that sort is error-provocative and promotes risk taking. Getting to and from the device wastes time.

Because of customer specifications, all parts processed at Station A are inspected at station B. Parts are moved to Station B in carts. Since the casters on the carts are too small, moving them is cumbersome and time consuming, and they are tippy. They have tipped over, injuring workers and damaging parts. This inspection motion is expensive, wasteful, boring, and adds elements of risk.

Overproduction: At Station C the machinery processes parts faster than can be handled by the remainder of the production line. Thus, materials in progress get stacked in aisles until they are transferred to a storage area. Having excess materials in process is wasteful. An additional result is overly stressful manual material handling and the ergonomic risks that implies.

Transportation: Station D represents the wastes deriving from the additional storage space and material handling needed because of overproduction at

Advanced Safety Management: Focusing on Z10 and Serious Injury Prevention,
Second Edition. Fred A. Manuele.
© 2014 John Wiley & Sons, Inc. Published 2014 by John Wiley & Sons, Inc.

Station C. The storage configuration is not conducive to efficiency. Aisles are narrow. Powered vehicles have collided, have struck workers, and goods have been damaged.

Waiting: Although overproduction occurs at Station C, personnel at Station E often are not fully occupied, and waste occurs while they are waiting for other components to be delivered. Inventory controls are inadequate, and the motorized delivery system is inefficient and risky.

Inventory: The inventory at Station D is greater than needed, and thereby wasteful. Excessive material handling is necessary.

Overprocessing: Because the quality level achieved at Station A is inadequate for some customers, considerable parts rework is necessary at Station F. That wastes resources, and use of the machinery in the process adds risk.

CHAPTER 19

MANAGEMENT OF CHANGE: SECTION 5.1.3 OF Z10

Section 5.1.3 in ANSI/AIHA Z10-2012 is entitled "Design Review and Management of Change." As was said in Chapter 15, "Safety Design Reviews," the processes for design reviews and for management of change are major elements in a safety and health management system. Although they have some common characteristics, they are implemented through distinctively separate management processes.

One of the reasons that the management of change process in addressed separately is to promote an understanding and application of the change analysis concept on which it is based. In this chapter we:

- Make the case that having an effective Management of Change System (MOC) in place as a distinct element in a safety and health management system will reduce the potential for serious injuries and fatalities, one of the focus points in this book. (Effective MOC systems also reduce the potential for injuries, environmental damage, and other forms of damage at all levels of severity.)
- Introduce the change analysis concept and relate it to the management of change provisions in Z10.
- Cite statistics in support of having effective MOC systems in place.
- Define the purpose and methodology of a management of change system.
- Establish the significance of management of change as a method to prevent serious injuries and fatalities.

Advanced Safety Management: Focusing on Z10 and Serious Injury Prevention,
Second Edition. Fred A. Manuele.
© 2014 John Wiley & Sons, Inc. Published 2014 by John Wiley & Sons, Inc.

- Outline management of change procedures, keeping in mind the staffing limitations at moderate-sized locations and their need to avoid burdensome paperwork.
- Provide guidelines on how to initiate and utilize a MOC system.
- Emphasize the significance of communication and training.
- Include examples of four management of change systems in place and provide access to six other real-world MOC systems.

ON CHANGE ANALYSIS

Change analysis is a commonly used process. Inquiry through an Internet search engine will show that the literature on change analysis is abundant. A few examples related to safety follow. OSHA says this about change analysis in *Safety & Health Management System eTool—Worksite Analysis*:

> Anytime something new is brought into the workplace, whether it be a piece of equipment, different materials, a new process, or an entirely new building, new hazards may unintentionally be introduced. Before considering a change for a worksite, it should be analyzed thoroughly beforehand. Change analysis helps in heading off a problem before it develops.

In the *Aviation Ground Operations Safety Handbook*, 6th edition, change analysis is listed within "The Risk Management Process" as a method "to detect the hazard implications of both planned and unplanned change." (p. 10)

In *MORT Safety Assurance Systems*, William Johnson makes references to change analysis throughout the book as he discusses applying the "Management Oversight and Risk Tree (MORT)". In *System Safety for the 21st Century*, Richard Stephans has a chapter entitled "Change Analysis."

Provisions for design reviews and management of change are also contained in other standards and guidelines, perhaps by other names. For example, in *Quality management systems—Requirements*, ANSI/ASQ Q9001-2000, Section 7.3.7 is titled "Control of design and development changes." It reads as follows:

> Design and development changes shall be identified and records maintained. The changes shall be reviewed, verified and validated, as appropriate, and approved before implementation. The review of design and development changes shall include evaluation of the effect of the changes on constituent parts and product already delivered. Records of the results of the review of changes and any necessary actions shall be maintained.

Change analyses are made to assess the consequences of a change on a broad scale and to provide a base for planning the change to assure that success is achieved and that negative consequences are as few as reasonably practicable. Change analysis is the fundamental process on which the MOC provisions in Z10 are built. Literature on change analysis is abundant. Throughout the literature, emphasis is given to:

- Planning for the change process, as in Z10's Section 4.0
- Attempting to foresee the obstacles that could arise
- Repeated communication
- Involving all personnel who are affected by the change
- Being aware of the normal resistance to change that many of the persons involved may demonstrate and its probable intensity
- Going gradually, step by step, if that is practicable
- The additional training that may be required
- Appreciating the extent of the culture change that will be necessary if the process change is to be successful

STUDIES THAT SUPPORT HAVING A MOC SYSTEM IN PLACE

Three significant studies establish that having a management of change system as an element within an operational risk management system would serve well to reduce serious injury potential.

Reviews made by this author of over 1800 incident investigation reports, mostly for serious injuries, support the need for and the benefit of having MOC systems. They showed that a significantly large share of incidents resulting in serious injury occurs:

- When unusual and nonroutine work is being performed
- In nonproduction activities
- In at-plant modification or construction operations (replacing a motor weighing 800 pounds to be installed on a platform 15 feet above the floor)
- During shutdowns for repair and maintenance, and during startups
- Where sources of high energy are present (electrical, steam, pneumatic, chemical)
- Where upsets occur: situations going from normal to abnormal

Having an effective MOC system in place would have served to reduce the probability of serious injuries and fatalities occurring in the operational categories shown above.

A study led by Thomas Krause, chairman of the board at BST in 2011 produced results in support of having MOC systems in place. (Data were provided in personal communication with Krause. BST is to publish a paper including these data.) Seven companies participated in the study. Shortcomings in pre-job planning, another name for management of change, were found in 29% of incidents that had serious injury or fatality potential. Focusing on reducing that 29%, a noteworthy number, would be an appropriate goal.

Correspondence with John Rupp at United Auto Workers in Detroit confirmed the continuing history with respect to fatalities occurring in UAW-represented workplaces. A previously issued UAW bulletin (no longer available) indicated that from 1973 through 2007, 42% of fatalities occurred to skilled-trades workers, who represented about 20% of the membership.

Additional data provided by Rupp for the years 2008 through 2011 indicated that 47% of fatalities occurred to skilled-trades workers. The point is that skilled-trades workers are not in routine production jobs. They are often involved in unusual and nonroutine work, at-plant modification or construction operations, shutdowns for repair and maintenance, startups, and where sources of high energy are present. That is the sort of activity for which a MOC system would be beneficial.

PURPOSE OF A MANAGEMENT OF CHANGE SYSTEM

Although the term *management of change* is not defined in Z10, its purpose is clearly established. Appendix H—Management of Change—provides guidance on what the writers of Z10 intended. With respect to operational risks, the MOC process is to assure that:

- Hazards are identified and analyzed and risks are assessed.
- Appropriate avoidance, elimination, or control decisions are made so that acceptable risk levels are achieved and maintained throughout the change process.
- New hazards are not knowingly brought into the workplace by the change.
- The change does not have a negative impact on previously resolved hazards.
- The change does not make the potential for harm of an existing hazard more severe.
- The occupational safety and health management system is not affected negatively.

APPLICATION CONSIDERATIONS

In the following list of categories to which the management of change process could apply, subjects other than for the safety of employees are included to demonstrate the breadth of benefits that could be obtained from an effective MOC system. Consideration would be given, as applicable, to:

- The safety of employees making the changes
- The safety of employees in adjacent areas
- The safety of employees who will be engaged in operations after changes are made
- Environmental aspects
- Safety of the public
- Product safety and product quality
- Fire protection so as to avoid property damage and business interruption

As stated in E5.1.3 in Z10, the management of change provisions may apply when changes are made in technology, equipment, facilities, work practices and procedures, design specifications, raw materials, organizational or staffing changes affecting skill capabilities, and standards or regulations.

OSHA's Rule for Process Safety Management of Highly Hazardous Chemicals, 29 CFR 1910.119, issued in 1992, requires that an operation affected by the standard have a management of change process in place. Similar requirements do not appear in other OSHA standards.

However, as shown previously, change analysis (management of change) was an element in OSHA's *Safety & Health Management System eTool—Worksite Analysis*. Also, having a MOC system in place is a requirement to achieve OSHA's VPP designation (OSHA's Voluntary Protection Program).

ASSESSING THE NEED FOR A FORMALIZED MOC SYSTEM

Useful data can be developed on the need for a formalized MOC system through a risk assessment initiative and a study of an organization's incident experience and that of the industry of which the organization is a part. Workers' compensation claims experience can be a valuable resource for such a study.

To develop meaningful and manageable data, it is proposed that a computer run be made of an organization's claims experience covering at least three years, to select out all claims valued at $25,000 or more—paid and reserved. If experience in other organizations is a guide, the computer run will probably encompass between 6 and 8% of the total number of claims and from 65 to 80% of the total costs.

An analysis of the data acquired should be made to identify job titles and incidents that have occurred when changes were taking place, and to determine whether the resulting data support proposing a formalized MOC system. If available, industry experience, such as through a trade association, should also be part of the study.

If it is found that very few incidents resulting in serious injury occurred when changes were being made, that should not deter proposing that the substance of a MOC system be applied for particular changes that present serious injury potential.

EXPERIENCE OF OTHERS IMPLIES OPPORTUNITY

To test whether personnel in operations other than chemicals had recognized the need for and developed management of change systems, assistance of the staff at the American Society of Safety Engineers responsible for practice specialties was sought. A request was sent to members of the management specialty group asking that copies of MOC procedures for operations other than chemicals be forwarded. The response was overwhelmingly favorable. The number of documents received was more than could be practicably used.

Examples received demonstrate that managements in a variety of operations recognized the need to have MOC systems in place. Four real-world MOC systems are addenda to this chapter. Because of space considerations, all of the examples are not included here. But, as will be shown later, six additional MOC systems can be accessed by going on the Internet. These 10 MOC systems were selected specifically to show:

- The broad range of harm and damage categories covered
- Similarities with respect to subjects covered
- The enormity of the variations in how those subjects are addressed

These examples are real-world applications that display the substance of MOC systems in place.

HISTORY DEFINES MOC SYSTEM NEEDS AND DIFFICULTIES IN THEIR APPLICATION

At least 25 years ago, personnel in the chemical and process industries recognized the importance of having a management of change process in place as an element within an operational risk management system. That awareness developed because of the number of major accidents that occurred when changes were taking place.

In 1989, the Center for Chemical Process Safety issued *Guidelines for Technical Management of Process Safety*, which included a management of change element. In 1993, the Chemical Manufacturers Association published *A Manager's Guide to Implementing and Improving Management of Change Systems.*

In 2008, the Center for Chemical Process Safety issued *Guidelines for Management of Change for Process Safety*. It builds on and considerably extends the previous publications. These comments appear in the Preface:

> The concept and need to properly manage change are not new; many companies have implemented management of change (MOC) systems. Yet incidents and near misses attributable to inadequate MOC systems, or to subtle, previously unrecognized sources of change (e.g., organizational changes), continue to occur. To improve the performance of MOC systems throughout industry, managers need advice on how to better institutionalize MOC systems within their companies and facilities and to adapt such systems to managing non-traditional sources of change. (p. xiii)

Note that incidents and near misses (near hits) that are attributable to inadequate MOC systems continue to occur. Also, recognition is given to organizational changes as a previously unrecognized source from which management of change difficulties could arise. Authors of the *Guidelines* recognized that:

> Management of Change is one of the most important elements of a process safety management system. (p. 1)

THE MANAGEMENT OF CHANGE PROCESS

As is the case with all management systems, an administrative procedure must be written to communicate what the management of change system is to encompass and how it is to operate. Care should be taken to assure that the MOC system is designed to be compatible with:

- The organization's and industry's inherent risks
- Other relevant management systems in place
- The organizational structure
- The dominant culture
- The expected participation of the workforce

Although brevity is the goal, consideration should be given to the following as subjects are selected for inclusion in a MOC procedure:

1. Defining the need for and the purpose of the MOC system
2. Establishing accountability levels
3. Specifying the criteria that are to trigger the initiation of formal change requests
4. Making clear how personnel are to submit change requests, and specifying the change request form to be used
5. Outlining the criteria for request reviews and the responsibilities for reviews
6. Indicating that the MOC system is to encompass:
   a. The risks to the workers who are to do the work and other employees who are affected
   b. Possible damage to the property, and business interruption
   c. Possible environmental damage
   d. Product safety and quality
   e. The procedures to accomplish the change
   f. How the results are to be evaluated
7. Establishing that minute-by-minute control is to be maintained to achieve acceptable risk levels, and that risk assessments will be made as often as necessary as the work progresses
8. Giving instruction on the action to be taken if, as the work proceeds, unanticipated risks of concern are encountered
9. Assigning responsibility for acceptance or declination of the change request, including MOC approval form
10. Outlining a method to determine the actions necessary because of the effect of the changes (e.g., additional training of operators and maintenance workers; revision of standard operating procedures and drawings; updating emergency plans)
11. Indicating that a final review of the work will take place before startup of operations, and identifying the titles of the persons who are to make the review

RESPONSIBILITY LEVELS

In drafting an MOC system, responsibility levels must be defined and be in accord with an entity's organizational structure. This is a highly important step in developing a MOC system. In an entity where even minor changes in a process are considered critical with respect to employee injury and illness potential, possible environmental contamination, and the quality and safety of the product, the levels of responsibility can be many.

Responsibility levels are set forth clearly in some examples of MOC systems, but not in others. Examples of levels of responsibility, as outlined in an organization where the inherent hazards require close control, are shown here as reference points in drafting MOC systems.

Initiator: The initiator owns the change and is responsible for initiating the change request form. If the complexity of the proposed change requires, the initiator's responsibilities may be reassigned at any time during the change process. The initiator will fully describe and justify changes, ensure that all appropriate departments have assessed the changes, manage the execution of the change request, and ensure that the changes are implemented properly.

Department Supervisor: The department supervisor has the responsibility to assign qualified personnel to initiate change requests. The change control process is critical to the safety of our employees, an avoidance of environmental contamination, and the quality of our products. The departmental supervisor is responsible to ensure that the change request is feasible, and presented adequately for review.

Document Reviewers: Document reviewers will review and approve change request forms. The review/approval activities include a review of the document for accuracy and adequacy with respect to the changes proposed.

Approvers: Department managers will select pre-approvers with expertise related to the nature of the proposed change. Each reviewer will be responsible to evaluate and assess the impact of the proposed change on existing processes in his or her area of expertise. The reviewers must also review and approve the change request form and the implementation plan to evaluate the change and assure that the steps for implementation are appropriate. This is the final review before the proposed change is implemented.

Post-Implementation Approvers: Department managers will select post-implementation approvers who are to assure that the change has been implemented appropriately as indicated when approval for the requested change was given. This process is also to assure that only the changes shown on the change request form have been implemented.

ACTIVITIES FOR WHICH THE MOC PROCESS SHOULD BE CONSIDERED

Hazard and risk complexities of an organization in which change is to take place and the desirability of establishing a MOC system that is adequate but not overly complex should be kept in mind when selecting the activities that are to trigger activating the MOC system (see Appendix H in Z10). A list follows from which activity categories can be selected.

- Non-routine and unusual work is to be performed
- The work exposes employees to sources of high energy
- Types of maintenance operations for which pre-job planning and safety reviews would be beneficial because of inherent hazards
- Substantial equipment replacement work
- Introduction of new or modified technology
- Modifications are made in equipment, facilities, or processes
- New or revised work practices or procedures are introduced
- Design specifications or standards are changed
- Different raw materials are to be used
- Modifications are made in health and safety devices and equipment
- Significant changes occur in the site's organizational structure
- Staffing changes in numbers or skill levels are made that may affect operational risks
- A change is made in the use of contractors

MANAGEMENT OF CHANGE REQUEST FORM

Developing a management of change request form is a necessity, and the content of the form should be in concert with an organization's structure and other management systems in place (e.g., capital expenditure request, work orders, purchasing procedures). Saving the form as a computer file allows flexibility when descriptive data and comments are to be added. A list follows of subjects to be considered in drafting the request form.

- Name of person initiating the request.
- Date of request.
- Department, section or area.
- The equipment, facility, or processes affected
- Brief description of the proposed change and what is to be accomplished by it.
- Potential performance, safety, health, and environmental considerations.
- Titles and space for entering the names of personnel who are to review the change.
- The effect the change may have on standard operating procedures, maintenance, and training, etc.
- Space for reviewers to enter special conditions or requirements.
- Approvals and authorizations.
- Routing indicators or provisions for copies to be sent to personnel responsible for training and updating operating procedures, drawings, etc.

Management of Change Request Form

General Information

Date_____ Originator_____ Department_____

Sent to:_____

Equipment, facility or process effected_____

Urgency of change: ____Emergency   ____Priority   ____Routine

Basis for the Change (Check those applicable)

___ Improved safety—risk reduction

___ Improved performance—efficiency

___ Pollution prevention—waste minimization

___ Essential to operation

___ Other

Description of Proposed Change and Potential Hazards

Summarize the technical basis for the proposed change and any potential safety, health, or environmental impacts from the proposed change. Describe how the change will affect SOPs, maintenance, training, etc. State the change start and end dates.

_____

_____

_____

_____

_____ Approved or disapproved by

Name and date _____   Organization/Position _____

_____   _____

_____   _____

Comments

_____

_____

_____

FIGURE 19.1

Figure 19.1 is an example of a management of change request form. Others may be found in some of the examples in the addenda to this chapter. As would be expected, each form was developed to reflect the views and needs of a particular organization.

IMPLEMENTING THE MANAGEMENT OF CHANGE PROCESS

Senior management and safety professionals must appreciate the magnitude of the task they face when initiating activity to implement a MOC system. Expect pushback: egos get in the way; territorial prerogatives may be maintained; the power structure may present obstacles; and the expected and normal resistance to change can be huge since the people affected may have had little experience with the administrative systems being proposed.

Although MOC systems have been required in the chemical industries for many years, the literature indicates that difficulties have been encountered in their application. These comments occur in *Guidelines for Management of Change for Process Safety*:

> Even though the concept and benefits of managing change are not new, the maturation of MOC programs within industries has been slow, and many companies still struggle with implementing effective MOC systems. This is partly due to the significant levels of resources and management commitment that are required to implement and improve such systems. MOC may represent the biggest challenge to culture change that a company faces. (p. 10)

> Developing an effective MOC system may require evolution in a company's culture; it also demands significant commitment from line management, departmental support organizations, and employees. (p. 11)

Management commitment, evidenced by providing adequate resources and the leadership required to achieve the necessary culture change, must be emphasized. Stated or written management commitment that is not followed by providing the necessary resources is not management commitment. Because of the magnitude of the procedural revisions necessary when a MOC system is initiated, it must be recognized that methods to achieve a culture change are to be applied. A list follows of subjects to be considered in an attempt to successfully implement a MOC/Pre-Job Planning system.

- Management commitment and leadership must be obtained and demonstrated. That means providing personal direction and involvement in initiating the procedures, providing adequate resources, and making the appropriate decisions to achieve acceptable risk levels when there is disagreement in the change review process.
- Keep procedures as simple as practicable. A less complicated system that is actually utilized achieves better results that an unused complex system.
- A goal is to obtain widespread acceptance and commitment.
- Communicate extensively. Inform all affected employees in advance of a decision to initiate an MOC system, solicit their input, and respect their perspectives and concerns.
- Recognize the need for and provide the necessary training.
- If practicable, go slow—step by step.
- Field-test a system prior to implementation. Debugging pays off in the long run.
- After the system is refined through a field test, select a job or an activity for which a MOC system would be beneficial for both productivity/efficiency and safety.
- The purposes of testing the MOC system in a selected activity are to:
  - Demonstrate the value of the system
  - Achieve credibility for it
  - Create a demand for additional application of the system

- Emphasizing the potential efficiency benefit of a MOC system is encouraged. Doing so should result in more favorable interest in the system being proposed.
- Monitor the progress and performance of the system through periodic audits and through informal inquiry of employees on their perspectives.

MANAGING ORGANIZATIONAL CHANGE

In some of the examples given in this chapter, procedures require that risk assessments be made of the significance of organizational changes. Those procedures exist because it has been recognized that organizational and personnel changes can have a negative impact on the effectiveness of an operational risk management system.

Although there is considerable literature on the subject, I have chosen *Managing the Health and Safety Impacts of Organizational Change* as a reference because it fits closely with the intent of some of the examples of MOC system included in this chapter. The publication was issued by the Canadian Society for Chemical Engineering in 2004. It can be accessed, and downloaded at no cost, at http://psm.chemeng.ca/Products/OCM_Guidelines.pdf.

Types of organizational and personnel changes that can have a negative effect on operational risk management, as listed in this publication, follow.

- Reorganizing or reengineering
- Downsizing the workforce
- Attrition and aging of the workforce
- Outsourcing of critical services
- Changes that affect the competence or performance of other organizations providing critical services under contract (e.g., equipment design, process control software, hazard and risk assessment)
- Loss of skills, knowledge, or attitudes as a result of the above

Such changes, the authors say, are not as well addressed in applicable guidelines as are changes in equipment, tools, work methods, and processes. The purpose of the publication is to promote what is considered to be the appropriate consideration.

I now give more emphasis to the impact of organizational changes on operational risk management because incident reports on some serious injuries and fatalities indicate that a significant contributing factor was a reduction in staffing, as to both number of employees and talent level. As a result, unacceptable risk situations developed for such as:

- Inadequate maintenance
- Inadequate competency
- Workers being stressed beyond their mental and physical capabilities (two persons doing the work for which three had previously been assigned)
- A person working alone in a high-hazard situation for which the standard operating procedure calls for a work buddy

RISK ASSESSMENTS

Some of the MOC examples require that risk assessments be made at several stages of the change activity. The intent is to achieve and maintain acceptable risk levels throughout the work process. Thus, risk assessments are to be made as often as needed as changes occur and particularly when unexpected situations arise. In this regard, safety professionals who become skilled in making risk assessments can provide a consultancy that demonstrates significant value added.

Appendix F, Risk Assessment, in Z10 is a valuable resource. In addition, a recently developed American National Standard is recommended as a reference on risk assessment. ANSI/ASSE Z590.3 is titled *Prevention through Design: Guidelines for Addressing Occupational Hazards and Risks in Design and Redesign Processes.* This standard was approved by ANSI on September 1, 2011. Risk assessment is its core. Its content is applicable to management of change whether the contemplated change involves new designs or redesign of existing operations. Of particular interest should be the sections on:

- Relationships with Suppliers
- Safety Design Reviews
- The Hazard Analysis and Risk Assessment Process
- Hazard Analysis and Risk Assessment Techniques
- Hierarchy of Controls

RISK ASSESSMENT MATRICES

In Z590.3, the use of a risk assessment matrix in the risk assessment process is strongly recommended. It is also emphasized that all involved in risk assessments arrive at a common understanding of the meanings of the terms used in the matrix. Several examples of matrices are given in a Z590.3 addendum.

An example of a matrix is available in Chapters 2 and 11. It was selected because it was preferred by operating employees who were brought into the risk assessment process. They said that establishing a mental relationship between numbers such as 6 and 12, first, made it easier to understand the relation between terms such as Moderate Risk and Serious Risk.

THE SIGNIFICANCE OF TRAINING

To emphasize the significance of training in achieving a successful MOC system, reference is made again to the *Guidelines for Management of Change for Process Safety.*

> Training for all personnel is critical. Many systems failed or encountered severe problems because personnel did not understand why the system was necessary, how it worked, and what their role was in the implementation. (p. 58)

The culture change necessary to put a successful MOC system in place cannot be achieved without a training program that helps supervisors and workers understand the concepts to be applied. Where the MOC system relates to many risk categories (occupational, the public, environmental, fire protection and business interruption, product quality and safety), the training provided must be more extensive.

I would like to avoid the impression that training is not of great importance because of the brevity of the comments made here on the subject. It is a near absolute certainty that installation of MOC/pre-job planning systems will fail if the training given is not appropriate to the significance of the changes to be made.

DOCUMENTATION

The importance of maintaining a history of operational changes needs emphasis. It is vital that all modifications be recorded in drawings, prints, and appropriate files. They become the historical records that would be reviewed when changes are to be made at a later date. Comments on *changes made that were not recorded* in drawings, prints, and records are found too often in reports on incidents resulting in serious consequences. Examples found in incident investigation reports of unrecorded changes include the following:

- The system was rewired.
- A blank was put in the line.
- Control instruments were disconnected.
- Relief valves for higher pressures had been installed.
- Sewer line sensors to detect hazardous waste were removed.

ON THE MOC EXAMPLES

To display the substance and variety of the MOC systems that organizations have in place, very little change was made in the examples that I received. They vary greatly in content and purpose. Some are contained in one page. Others require several pages to cover the complexity of exposures and procedures. Some MOC procedures have introductory statements on policy and procedure. Some don't. Nevertheless, it is apparent that a MOC system need not necessarily have to meet a theoretical ideal to provide value.

Examples are to serve as references. It is highly recommended that none of the examples be adopted as presented. A MOC system should be drafted in accord with an organization's particular needs and its culture.

As each of the MOC examples was reviewed, it became apparent that some of the terms used in them were not readily understandable. But more than likely, the terms are understood in the organization that developed the MOC system. It was decided to leave those terms in the examples to emphasize that: whatever terminology is included in a MOC procedure, the terms must be in concert with the language commonly used within an organization; and the terms must be understood by all who become involved in a MOC initiative.

Four examples are included in this chapter as addenda. They are designated MOC Examples 1, 3, 4, and 10, the designations given to them in the total of 10 MOC systems available.

Access to six other MOC systems, some of which cover several pages, can be obtained at www.asse.org/psextra. This is an American Society of Safety Engineers site for which open access is provided. An earlier and briefer version of this chapter was published in the July 2012 issue of *Professional Safety*. Click on "Professional Safety Extras," scroll down to Archives, and go to July 2012. You can click on supplemental material for my article and bring up 43 pages of MOC system examples.

Example 1: For An Operation Producing Mechanical Components

The Pre-Job Planning and Safety Analysis system shown in Example 1, a one-page outline, was developed because of adverse occupational injury experience in work that was often unusual or one-of-a-kind or required extensive and complicated maintenance activity. (Example 1 is an addendum to this chapter.)

Its relative simplicity in relation to other examples will be obvious. But it was applied successfully for its purposes. It would be well to note that safety professionals:

- Prepared the data necessary to convince management and shop floor personnel to try the pre-job planning system they proposed
- Said that the training sessions held were highly significant in achieving success
- Emphasized that work situations discussed in the training sessions were real to that organization
- Addressed the benefits for both productivity/efficiency and risk control in their proposal and in the training sessions

For whoever initiates a MOC system, the procedure described in the following will be of interest.

At a location where the serious injury experience was considered excessive for non-routine work, safety professionals decided that something had to be done about it. As they prepared a course of action and talked it up at all personnel levels, from top management down to the worker level, they encountered the usual negatives and push-back: e.g., it would be time consuming, the workers would never buy into the program, and the supervisors would resist the change. The safety professionals considered the negatives as normal expressions of resistance to change.

Their program consisted, in effect, of indoctrinating management and the workforce in the benefits to be obtained by doing pre-reviews of jobs so that the work could be done effectively and efficiently while, at the same time, controlling the risks.

Eventually, management and the line workers agreed that classroom training sessions could be held. Later, the safety professionals said that the classroom training sessions and follow-up training were vital to their success.

At the beginning of each of those sessions, a management representative introduced the subject of Pre-Job Planning and Safety Analysis and discussed reasons why the new procedure was being adopted. Statistics on accident experience prepared by safety professionals were a part of that introduction. Then, safety professionals led a discussion of the outline shown in MOC Example 1. It set forth the fundamentals of the pre-job review system being proposed. After discussion of those procedures, attendees were divided into groups to plan real-world scheduled maintenance jobs that were described in scenarios that had previously been prepared.

At this location, supervisors took to the pre-job planning and safety analysis system when they recognized that the system made their jobs easier, improved productivity/efficiency, and reduced the risks. And they took ownership of the system. As one of the safety professionals said: "Our supervisors and workers have become real believers in the system." A culture change had been achieved.

Note the requirements under the caption "Upon Job Completion" in the Pre-Job Planning and Safety Analysis Form. The detail of the requirements reflects particular incidents with adverse results that occurred over several years. It is recommended that every MOC system include similar procedures to be followed before the work can be considered completed.

Example 2: Specialty Construction Contractor

This Field Work Review and Hazard Analysis system is encompassed within two pages. It was provided by a safety professional employed by a specialty construction contractor that has several crews active in various places at the same time. Note that the names of employees on a job are to be recorded as having been briefed on the work to be done. The checklist included in the form pertains to occupational, public, and environmental risks.

When asked what gave impetus to the development of the change procedure, the safety professional responded "We learned from costly experience." It was said that the procedures required by the change system are now embedded in the company's operations and that it is believed that the procedure results in greater efficiency. It is known that fewer costly incidents have occurred.

This example has a direct relation to the purposes of ANSI/ASSE 10.1, a standard for construction and demolition operations entitled *Pre-Project & Pre-Task Safety and Health Planning*. This standard is an excellent resource for contractors and for organizations that establish requirements contractors on their premises. Note the distinctions: Pre-Project Planning and Pre-Task Planning.

Example 3: Form for An Operation That Has Had Serious Injury Experience

This Pre-Task Analysis form makes much of obtaining required permits and of assuring that supervisors brief employees on the order of activities and of the risks to be encountered. Employees are required to place their signatures on the form, indicating that they have been so briefed. This is the only example requiring employees to acknowledge, by their signatures, that they have been informed about hazards and risks prior to the start of work. (Example 3 is an addendum to this chapter.)

Example 4: Paper Setting Forth a Management Policy and Procedure

This is a basic guidance paper on MOC concepts and procedures, condensed to three pages. It is a composite of several MOC policies and procedures in place issued by organizations in which operations were not highly complex. Reference is made in this example to a MOC "Champion." The point made is that someone has to be assigned the responsibility for the change to be accomplished and manage it through to an appropriate conclusion. (Example 4 is an addendum to this chapter.)

Example 5: MOC System With a Specifically Defined Pre-Screening Questionnaire

This MOC system, in three pages, begins with an interesting pre-screening questionnaire. If the answer is "No" to all the questions asked, the formal management of change checklist and approval form need not be completed for the work being proposed. The question often arises in discussions of management of change systems: To what work does the system apply? This organization developed a way to answer that question for its own operations.

Example 6: High-Risk Multiproduct Manufacturing Operation

This Management of Change Policy and Procedure reflects the high hazard levels in the organization. Captions in this four-page system are safety; ergonomics; occupational health; radiation control; security/property loss prevention; clean air regulations; spill prevention and community planning; clean water regulations; solid and hazard waste regulations; environmental, safety, and health management systems; and an action item tracking instrument. The following appears in the procedure paper announcing the system:

> If a significant change occurs with respect to key safety and health or environmental personnel, the matter will be reviewed by the S&H Manager and the Environmental Manager and a joint report including a risk assessment and their recommendations will be submitted to location management.

Example 7: Food Company

This four-page Management of Change policy includes product safety and product quality among the subjects to be considered. Detail on the subjects listed is highly technical and extensive. The safety director at this location informed this author that discipline in the application of this MOC system is rigid and that it reflects management's determination to avoid damaging incidents affecting personnel and the environment and variations in product quality and taste. Provisions for pre-startup and post-modification are extensive. A risk assessment is to be made after changes are made and prior to startup.

Example 8: A Conglomerate

Iota Corporation has a five-page management of change procedure outlined in four sections. It is worthy of review.

Section I requires completion of a Change Request and Tracking System Requirements form in which the change is described, a tracking number is assigned, and approval levels are established. Approval levels are numerable, including headquarters in some instances.

Section II outlines a Change Review and Approval Procedure that is extensive with respect to occupational safety and health and environmental concerns.

Section III is a Pre-Implementation Action Summary Form which lists subjects for which actions are necessary before work on the change can be begun, and the persons responsible for those actions.

Section IV lists 11 points in a Post Completion Form.

Example 9: An Extensive MOC System for a Particular Operation

Reading this Management of Change standard is recommended for educational purposes because of its structure and its content. It is somewhat different in relation to the other examples. In its seven pages:

- Requirements for technical changes and organizational changes are dealt with separately and extensively.
- Much is made about organizational changes for which risk assessments are required.
- Technical changes to which the standard applies are outlined in detail.
- Risk assessment is a separately listed item pertaining to all operations.
- After a lengthy discussion of general considerations, requirements for a six-point management of change standard are outlined. They are: Management Process; Capability; Change Identification; Risk Management; The Change Plan; and Documentation. Each of these subjects is discussed thoroughly.

This is a concept and procedural paper. It does not include the forms used to implement the procedures.

Example 10: An International Multioperational Entity

Application of this MOC system, a Management of Change Policy for Safety and Environmental Risks, extends the activities of safety and environmental professionals beyond that of any other example. It is an informative read. The example shown here covers 10 pages. (Example 10 is an addendum to this chapter.) The bulletin, issued by the Safety and Industrial Hygiene entity within the organization, is much longer than is shown here. All exhibits could not be made available for proprietary reasons. Brief comments follow on the unique aspects of this MOC system.

- Due Diligence is included in a list of Definitions. Safety and environmental professionals are to assess acquisitions et al.

- Global Franchise Management Board Members are listed under "Responsibilities." They are to ensure compliance with the standard.
- A Preliminary Environmental, Safety & Health Assessment Questionnaire shall be initiated during the project planning stage.
- Under a Section titled Evaluating Change (Risk Assessment Guidelines) these subjects are included, which may not be included in other examples, at least not as extensively:
  - New Process Product and Development
  - Capital Non-Capital Project
  - External Manufacturing
  - Business acquisitions
  - Significant Downsizing/Hiring
- Conducting Risk Analyses is a major section.

This is a noteworthy MOC system because of its breadth. It is interesting that the system was issued by the Safety & Industrial Hygiene unit. That implies management support for personnel in that unit and superior operational risk management.

CONCLUSION

It is the intent of this chapter to provide a primer that can serve as a base from which a MOC system suitable for a particular entity can be crafted. I recommend that safety professionals consider whether the organizations to which they give counsel could benefit from having MOC systems within their overall Operational Risk Management Systems. Having such a system in place for changes that should be pre-studied because of their inherent hazards and risks and which may affect the safety, productivity, and environmental controls is good risk management.

REFERENCES

A Managers Guide to Implementing and Improving Management of Change Systems. Washington, DC: Chemical Manufacturers Association (now known as the American Chemistry Council), 1983.

ANSI/AIHA Z10-2012. American National Standard, *Occupational Health and Safety Management Systems*. Fairfax, VA: American Industrial Hygiene Association, 2012. ASSE is now the secretariat. Available at https://www.asse.org/cartpage.php?link=z10_2005.

ANSI/ASQ Q9001-2000. *Quality Management Systems—Requirements* Milwaukee, WI: American Society for Quality, 2000.

ANSI/ASSE Z590.3-2011. *Prevention Through Design: Guidelines for Addressing Occupational Hazards and Risks in Design and Redesign Processes*. Des Plaines, IL: American Society of Safety Engineers, 2011.

Aviation Ground Operations Safety Handbook, 6th ed. Itasca, IL, National Safety Council. 2007.

Guidelines for Management of Change for Process Safety. Hoboken, NJ: A joint publication of the Center for Chemical Process Safety of the American Institute of Chemical Engineers (New York) and John Wiley & Sons (Hoboken, NJ), 2008.

Guidelines for Technical Management of Process Safety. New York: Center for Chemical Process Safety, 1989.

Johnson, William. *MORT Safety Assurance Systems.* Itasca, IL: National Safety Council, 1980. (Also published by Marcel Dekker, New York.)

Managing the Health and Safety Impacts of Organizational Changes. Ottawa, Canada: Canadian Society of Chemical Engineering, 2004. Accessible at http://psm.chemeng.ca/Products/OCM_Guidelines.pdf.

OSHA's Rule for Process Safety Management of Highly Hazardous Chemicals. 29 CFR 1910.119. Washington, DC: U.S. Department of Labor, 1992.

OSHA's Voluntary Protection Program (VPP). Section C. in CSP 03-01-002–TED 8.4, Voluntary Protection Programs (VPP): Policies and Procedures, 2003. At http://www.osha.gov/pls/oshaweb/owadisp.show_document?p_table=DIRECTIVES&p_id=2976.

ANSI/ASSE A10.1-2011. *Construction and Demolition Operations: Pre-Project & Pre-Task Safety and Health Planning.* Des Plaines, IL: American Society of Safety Engineers, 2011.

Safety & Health Management System eTool—Worksite Analysis. U.S. Department of Labor, OSHA, at http://www.osha.gov/SLTC/etools/safetyhealth/comp2.html.

Stephens, Richard A. *System Safety for the 21st Century.* Hoboken, NJ: Wiley, 2004. (This is an updated version of *System Safety 2000,* by Joe Stephenson.)

MOC EXAMPLE 1

ALPHA CORPORATION

Pre-Job Planning and Safety Analysis Outline

1. Review the work to be done. Consider both productivity and safety:
   a. Break the job down into manageable tasks.
   b. How is each task to be done?
   c. In what order are tasks to be done?
   d. What equipment or materials are needed?
   e. Are any particular skills required?
2. Clearly assign responsibilities.
3. Who is to perform the pre-use of equipment tests?
4. Will the work require: a hot work permit; a confined entry permit, lockout/tagout (of what equipment or machinery), other?
5. Will it be necessary to barricade for clear work zones?
6. Will aerial lifts be required?
7. What personal protective equipment will be needed?
8. Will fall protection be required?

Advanced Safety Management: Focusing on Z10 and Serious Injury Prevention,
Second Edition. Fred A. Manuele.
© 2014 John Wiley & Sons, Inc. Published 2014 by John Wiley & Sons, Inc.

9. What are the hazards in each task? Consider –

Access	Work at heights	Work at depths	Fall Hazards
Worker position	Worker posture	Twisting, bending	Weight of objects
Elevated loads	Welding	Fire	Explosion
Electricity	Chemicals	Dusts	Noise
Weather	Sharp objects	Steam	Vibration
Stored energy	Dropping tools	Pressure	Hot objects
Forklift trucks	Conveyors	Moving equipment	Machine guarding

10. Of the hazards identified, do any present severe risk of injury?
11. Develop hazard control measures, applying the Safety Decision Hierarchy.
    - Eliminate hazards and risks through system and work methods design and redesign
    - Reduce risks by substituting less hazardous methods or materials
    - Incorporate safety devices (fixed guards, interlocks)
    - Provide warning systems
    - Apply administrative controls (work methods, training, etc.)
    - Provide personal protective equipment
12. Is any special contingency planning necessary (people, procedures)?
13. What communication devices will be needed (two-way, hand signals)?
14. Review and test the communication system to notify the emergency team (phone number, responsibilities).
15. What are the workers to do if the work doesn't go as planned?
16. Considering all of the foregoing, are the risks acceptable? If not, what action should be taken?

Upon Job Completion

17. Account for all personnel
18. Replace guards
19. Remove safety locks
20. Restore energy as appropriate
21. Remove barriers/devices to secure area
22. Account for tools
23. Turn in permits
24. Clean the area
25. Communicate to others affected that the job is done
26. Document all modifications to prints and appropriate files

ADDENDUM B

MOC EXAMPLE 3

GAMMA CORPORATION

Pre-Task Analysis

Completed form must be submitted and work must be authorized before activity is commenced

Submitted by _____ Date _____ Location _____
Description of work _____
_____
_____
_____
_____

Commencement date _____ Expected completion date _____

Permits, special skills and licenses: Work must not be commenced if required permits have not been received or if arrangements have not been made for the special skills and licenses required.

Advanced Safety Management: Focusing on Z10 and Serious Injury Prevention,
Second Edition. Fred A. Manuele.
© 2014 John Wiley & Sons, Inc. Published 2014 by John Wiley & Sons, Inc.

Permits	Required		Received	
Confined space	____ Yes	____ No	____ Yes	____ No
Electrical	____ Yes	____ No	____ Yes	____ No
Combustion equipment	____ Yes	____ No	____ Yes	____ No
Excavation	____ Yes	____ No	____ Yes	____ No
Hot work	____ Yes	____ No	____ Yes	____ No
Lifting-rigging	____ Yes	____ No	____ Yes	____ No
Fire systems	____ Yes	____ No	____ Yes	____ No

			Obtained	
Will special skills be required	____ Yes	____ No	____ Yes	____ No
Will special licenses be needed	____ Yes	____ No	____ Yes	____ No

Check off each of the following that apply. Each checked subject is to be addressed in the pre-task planning.

___ Lockout/Tagout	___ Pinch points	___ Inadequate access
___ Hot/cold/burns	___ Chemical exposure/spill	___ Electrical shock
___ Sharp objects	___ Excavations	___ Equipment loading/unloading
___ Asbestos	___ Falling objects	___ Particles in eye
___ Elevated work	___ Fall from height	___ Manual lifting
___ Lighting	___ Lifting or rigging	___ Mobile equipment
___ Isolated area	___ Radiation	___ Fire/explosion
___ Ladder usage	___ Confined space	___ Heat stress
___ High noise levels	___ Scaffolding	___ Inhalation hazard
___ Other	___ Other	___ Other
___ Other	___ Other	

Specify personal protective equipment needed _____

_____

_____

Specify special equipment needed if any _____

_____

_____

If unexpected and unacceptable risks are encountered, it must be understood that work is to be stopped until the situation is resolved. This provision must be made clear in pre-job discussions and in employee briefings.

Employee briefings. Supervisors are to assure that all employees involved in the work are briefed on the order of and the procedures to be followed and of risks considered in the pre-job discussions.

Name	Signature	Date	Name	Signature	Date

Additional space is provided for the briefings of personnel added after the work is commenced.

Name	Signature	Date	Name	Signature	Date

Notifications: Names of persons to be notified before the work is commenced and after it is completed.

Pre-work commencement
       Names   **Department**

_____   _____

_____   _____

_____   _____

Upon completion
       Names   **Department**

_____   _____

_____   _____

_____   _____

For emergencies encountered, list names and phone numbers of the people who should be notified.
       Names   **Department**   **Phone Number**

_____   _____   _____

_____   _____   _____

_____   _____   _____

Approval signatures

Safety and environmental personnel, who are to assure that foreseeable hazards have been identified and risk control methods are appropriate
_____ **Date** _____
_____   _____
_____   _____

Manager with authority to approve the project

_____   _____

Upon completion

All personnel, tools and equipment accounted for   _____ Yes   _____ No
Operation successfully restored   _____ Yes   _____ No

Signature: Project initiator _____ Date _____

ADDENDUM C

MOC EXAMPLE 4

A COMPOSITE MANAGEMENT SYSTEM GUIDE FOR MOC SYSTEMS

Using several resources on management of change and reflecting on my experience, the following general statement on a management of change policy and procedures is offered as a guide for reflection by those who initiate MOC systems. Note that it encompasses both occupational safety and environmental considerations. It is a brief management guide only and is presented as such. It does not contain any forms. This composite guide is worthy of thought. It will serve well as a resource in initiatives to adopt the management of change provisions in Z10.

Management of Change Policy and Procedures: Occupational Safety and Health and Environmental Considerations

1. Overview

This policy defines the requirements for a management of change process with respect to occupational safety and health and environmental considerations.

2. Purpose

This policy establishes a process for evaluating occupational safety and health and environmental exposures when operational changes are made so as to control the

Advanced Safety Management: Focusing on Z10 and Serious Injury Prevention,
Second Edition. Fred A. Manuele.
© 2014 John Wiley & Sons, Inc. Published 2014 by John Wiley & Sons, Inc.

internal risks during the change process and to avoid bringing new hazards and risks into the workplace.

3. Scope
This policy applies to all operations at this location: there are no exceptions.

4. Responsibility
a. The Management Executive Committee, with counsel from the senior safety professional, is responsible for establishing processes to determine when these management of change procedures shall apply, how they are to be implemented and by whom, and to follow the processes through to an effective conclusion.
b. Facility management is responsible for ensuring that these processes and procedures to address the safety, health, and environmental implications of operational changes are implemented within their areas of responsibility.
c. Employees at all levels—division managers, supervisors, line operators, and ancillary personnel—after receiving training on this policy and process, are responsible within their domains of influence for initiating communications on operational changes that may have an impact on safety, health, and environmental considerations and for the implementation of this policy and process.
d. All safety professionals have responsibility to identify operational changes that require study as to hazard and risk potential and to bring their observations to management's attention through their organizational structures.

5. Application
a. This policy applies to all operational changes that may potentially affect on the safety and health of employees and on our environmental controls.
b. This policy will be implemented as an adjunct to all issued Safety, Health, and Environmental policies and procedures, but particularly to our Design and Procurement of Equipment and Facilities Procedure.
c. The senior safety professional, and other safety professionals with particular skills shall participate routinely in management staff meetings when operational changes are discussed.
d. A safety professional will sign off on the change plans considering the provisions of our Design and Procurement of Equipment and Facilities Procedure and on the New Product Development process.
e. Examples of operational changes to which this policy and process may apply include:
   - Unusual, nonroutine, nonproduction work, work where high-energy exposures are contemplated, and maintenance projects for which the scope of the work requires a determination that pre-job planning and safety analysis would be beneficial
   - Revisions in operating methods and procedures
   - Revised production goals
   - Plans to lower operating costs
   - Revisions in staffing levels, upward or downward

- Organizational restructuring
- Revisions in the environmental management system
- New product development
- Adoption of new information technology that has an impact on operations
- Changes in safety, health, or environmental regulations
- Acquisitions, mergers, expansions, relocations, or divestitures

6. Management of Change System

When an operational change is identified that requires study as respects its impact on occupational safety, health, or environmental controls, a person at an appropriate management level shall be appointed the management of change champion to chaperon the review process to a conclusion.

    a. That person, having obtained counsel from safety professionals, will determine how extensive the review procedure will be and decide on whether:

- Completion of a Management of Change Request Form is necessary [Figure 19.1] within which the accountability and sign-off levels are set forth
- Multidisciplinary group discussions are to be held to encompass the content of the Pre-Job Planning and Safety Analysis Outline [Addendum B]
- The hazards and risks that may result as the operational change moves forward are of greater significance and require appointment of an ad hoc Management of Change Committee to oversee the project. Safety professionals having the necessary skills are to be members of such committees.
- Our Capital Expenditure Request Procedure is to be initiated

7. The Management of Change Champion shall:

    a. Assure that input on the operational change has been obtained from all who might be affected.

    b. Arrange for resources, staffing, and scheduling to accomplish the change.

    c. Schedule the necessary risk assessment.

    d. Obtain comments from line operating personnel on their views on how the hazards and risks can be ameliorated, and their concerns.

    e. Get safety personnel to sign off.

    f. Follow the review process to a logical conclusion.

    g. Arrange a final review of the changes made to assure that hazards and risks have been addressed properly.

    h. Determine that residual risks, after risk reduction and control measures have been taken, are acceptable.

    i. See that documentation is appropriate.

- The Management of Change Champion is to give emphasis to documenting all changes made that should be recorded in prints and appropriate files so that persons who make further changes at a later date will know precisely what was done.
- In making decisions on what documentation is to be made, this principle is to apply: Be supercautious and consider later needs.
- Risk assessments are to be retained in accord with our document retention policy.

j. Have Standard Operating Procedures modified as necessary.

k. Determine whether additional training is necessary.

8. Training

a. Personnel responsible for safety, health, and environmental training at the management, supervisory, line worker, and ancillary personnel levels shall incorporate this policy and process into the training curricula.

b. Employees shall receive training on the revised Standard Operating Procedures before changes are finalized.

c. In the training process, employees will be properly informed with respect to their assuming responsibility for the aspects of safety, health, and environmental matters over which they have control.

ADDENDUM D

MOC EXAMPLE 10

This is S&IH Bulletin 423221, effective date June 1, 2008.

LAMBDA CORPORATION

Management of Change Policy for Safety and Environmental Risks

1. Overview
This document outlines the requirements for Management of Change with respect to safety and environmental risks.

2. Purpose
2.1 To establish a process requiring early reviews of various business changes to diminish the probability of or prevent the occurrence of adverse effects that may impact the safety of associates, the public, equipment or facilities, and associated environmental aspects.

3. Scope
3.1 This policy applies to all domestic and international operations.
3.2 To address potential impact on safety or environmental issues when various changes outlined in this process are proposed, and the related management of those changes.

Advanced Safety Management: Focusing on Z10 and Serious Injury Prevention, Second Edition. Fred A. Manuele.
© 2014 John Wiley & Sons, Inc. Published 2014 by John Wiley & Sons, Inc.

3.3 For reference only, various forms may be found in Appendix Section 7.0. Others may be hyperlinked to other respective documents or Web pages as indicated in this document.

4. Definitions

4.1 Change Control—Mechanism or tool used by the organization to control identified changes in the organization.

4.2 Design for Environment (DfE)—An environmental review process for all products, packaging, and processes in the research and development phase—includes all new and changed products.

4.3 Due Diligence—A pre-purchase/pre-lease or exit audit for acquisitions, divestitures, leases or joint ventures of all business related aspects and real estate property to ensure that potential liabilities are identified.

4.4 Environmental Impact Assessment (EIA)—a reference tool for Environmental Professionals conducting Environmental Impact Assessments of facility-related projects, new or modified equipment, new materials and new or modified processes being implemented at the location.

4.5 Ergonomic Job Analysis/Analyzer (EJA)—An analysis tool used to evaluate jobs and tasks to rate them by level of ergonomic risk.

4.6 MAP (Management Action Plan)—A tracking tool that incorporates items identified during an annual operating location self-assessment, or of issues identified during any of the risk assessments as listed in this policy.

4.7 Production Standards—The value of the production work that has been determined through methodical evaluation to be reasonably expected during an allotted time period.

4.8 Project Leader—The term project leader is used in the generic sense. If a person is assigned the responsibility to manage a project, that person in essence is a project leader.

4.9 S&IH (Safety & Industrial Hygiene)—A Lambda organization within the technical Resource Group (TRG).

5. Responsibilities

The responsibilities of the personnel required to administer the management of change process are listed as follows:

5.1 Environmental Professional Resources
EPR shall be responsible for conducting environmental assessments as needed. This includes, but may not be limited to, an Environmental Impact Assessment.

5.2 S&IH Professional Resources
Maintains this policy and ensures its existence at the affected locations. S&IH shall be responsible for providing the technical resources to comply with the requirements of this policy. S&IH Professionals shall work with Environmental Professionals at the respective operating location to ensure compliance to applicable standards.

5.2.1 A review of this policy shall be conducted during the annual self-assessment or whenever this policy fails to serve the purpose of its intent.

5.2.2 Deficiencies noted in this policy shall be incorporated into the operating location MAP to ensure that corrective actions are tracked and completed.

5.3 Global Franchise Management Board Members

Shall ensure compliance with this standard for their respective organizations described in the scope of this document.

5.4 Line Management

Shall be responsible for ensuring that the requirements and procedures contained in this policy are implemented and adhered to within their respective departments.

5.5 Project Leader

Shall be responsible for ensuring that the S&IH Professionals and other appropriate members of management are informed of upcoming projects, and that any safety related issues are resolved prior to the project completion per the requirements of this policy.

6. Policy Statements

6.1 Due-Diligence Process

A Due-Diligence study of the prospective acquisition and divestitures shall be conducted to determine strengths and weaknesses of the business, and the impact on safety and environmental requirements.

6.2 Preliminary Environmental, Safety & Health Assessment Questionnaire

A Preliminary Environmental, Safety & Health Assessment Questionnaire shall be initiated during the project design/planning phase by the Project Leader and submitted to the primary S&IH Professional at the operating location.

The questionnaire will help ensure that the S&IH and Environmental professionals complete the appropriate risk assessment(s) and associated forms. Assessment information will be forwarded to the project leader for inclusion in the project plan. (See Form—Section 7.1, Appendix I, Preliminary Environmental, Health & Safety Assessment Questionnaire.)

6.2.1 Program Elements shall include the following:

6.2.1.1 What is the basis for the proposed change?

6.2.1.2 What is the time period for the change?

6.2.1.3 What assessments of the proposed change will be required?

6.2.1.4 From an Environmental, Health & Safety perspective, what are the "'things to consider" when making the change?

6.2.1.5 What are the related Environmental, Health & Safety authorization requirements for the proposed change?

6.3 Design for Environment (DfE)

Each operating location shall develop and implement a Management system to ensure that the Design for Environment (DfE) Tool Assessment is performed and documented for all products, packaging, and processes in the research and development phase (all new and changed products).

6.4   Evaluating Change (Risk Assessment Guidelines)—Requirements

The respective S&IH and Environmental professionals shall work with those responsible for implementing the change to ensure that the appropriate assessments are completed and to determine what actions need to be taken during the design phase.

6.4.1   Changes in any one of the following areas can have significant impact on the safety of associates, the public, the environment, equipment and facilities, and shall require one or more risk analyses to be conducted to determine needs early in the planning process. Identified needs shall be included in the operating location MAP to ensure that corrective actions are brought to closure.

6.4.1.1   New Process Product Development

Development of a new process or a product shall be reviewed for their potential impact on associates and the public relative to exposure, and safe handling of process components and equipment and products.

6.4.1.2   Facility (Construction/Renovation) Project

New facility construction and/or renovation to existing facilities shall be reviewed for their potential impact on life safety, environmental aspects, and other related needs prior to occupancy.

6.4.1.3   Capital/Non-Capital Project

New capital projects shall be reviewed to ensure that safety or environmental issues are identified. While "replacement in kind" (purchase of like equipment) typically does not impact change, capital projects should be evaluated so that advances in safety or environmental technology, or changes in related regulations can be incorporated as part of the project. This includes the purchase of computers, other similar equipment, and/or workstations to ensure that the workstations that will be used with the computers and other similar equipment is appropriate for the job/office.

6.4.1.4   Existing Product Process Modification

Upgrades or modifications to current operating processes or products such as improvements based on technological advancements or safer chemical components shall be reviewed for their inherent risk and how the change may affect associates or existing systems, processes, and/or equipment.

6.4.1.5   External Manufacturing

A manufacturer of Lambda products or components shall be evaluated for appropriate compliance to safety and environmental standards. (See Forms—Section 7.3 Appendix III, Safety & Industrial Hygiene External Manufacturing Checklist, and Section 7.4, Appendix IV Environmental External Manufacturing Checklist.)

6.4.1.6   Methods or Procedures & Production Standards Changes to production methods, procedures, or production standards, shall be reviewed for safety to ensure that the associate performing the work is not placed at risk. Documents relative to those methods or procedures, such as Job Safety Analysis (JSA), shall be revised and communicated to affected associates. The evaluation shall include an Industrial Hygiene review if chemicals are used and an Ergonomic Job Analysis (EJA) if production standards have been modified.

6.4.1.7   Product or Equipment Transfer
Product and/or equipment transfer from one operating location to another can adversely impact the receiving location in a number of ways. It is imperative that the location transferring the product/equipment obtains the services of the safety and environmental professional(s) at the operating location receiving the product/equipment to communicate anticipated needs or modifications at the receiving location. Operating locations transferring equipment shall either ensure compliance of the equipment with applicable standards or communicate and document those deficiencies to the receiving location.

6.4.1.8   Business Acquisitions
New businesses introduced to the Lambda family of companies typically are not familiar with Lambda's stringent safety or environmental requirements. It is therefore imperative to introduce the new business to Lambda requirements as early as possible after the acquisition.

6.4.1.9   Significant Downsizing/Hiring
In periods of significant downsizing or hiring, S&IH shall be included in management discussions in order to anticipate potential impact the change may have on the business.

6.4.1.10   Changes In Management
The attitudes, beliefs and values of management, particularly of top management, relative to safety and environmental issues, will set the standard for the operating location. New members of management who have been recruited from outside of Lambda must become familiar with their role and responsibilities to obtain their support in achieving desired results. Likewise, those who have been promoted from within should reaffirm their support to the resolution of safety and environmental issues.

6.5   Conducting Risk Analysis—Assessment Formats

6.5.1   Safety Strategy Checklist
The S&IH professional shall complete the Safety Strategy Checklist for all projects associated with new process/product development, facility construction or renovations, and capital projects. Upon completion, this document shall be reviewed with the project leader to

identify safety-related needs early in the process for inclusion in the Development Phase. (See Form—Section 7.2, Appendix II, Safety Strategy Checklist.)

6.5.2 Post Production Safety Qualification Operation Review The S&IH Professional, who completed the EIQ safety review, shall complete this document, which is then kept on file in the Safety Department at the operating location. Observations noted during the original safety EIQ process shall be documented, and action scheduled for a re-review within 180 days from the date of the original safety EIQ. The S&IH Professional is responsible for ensuring that the post review occurs within the prescribed time frame. Action items that have not been corrected are to be flagged as "repeat" observations, and communicated to the Qualification Coordinator for follow up.

6.5.3 Ergo Job Analysis/Analyzer (EJA)

This tool shall be used to ensure that modifications made to methods, procedures, or RE's do not pose an increased risk to the associate performing the job and/or task.

6.5.4 Environmental Impact Assessment

The Environmental Professional at the operating location where the modifications will be made, will include all environmental aspects of the modifications in the facility's assessment, and will use the results of the assessment to implement actions in accordance to the facility's Environmental Management System.

6.5.5 Due-Diligence

In the event of an acquisition, lease, joint venture or divestiture, an audit of the prospective business and real estate property shall be conducted by appropriate company individuals or a contracted third party. The purpose is to ensure that potential liabilities are identified prior to closure of contracts. The due-diligence report shall be generated by those individuals in order to develop a plan to bring the business under compliance of all Lambda guidelines within prescribed time frames from the date of the acquisition according to the level of requirement as listed in the Lambda Acquisition Guide.

6.5.6 Design for Environment (DfE)

The concept portion of this tool is to be completed during the Concept Stage of new product development. The design portion of this tool is to be completed during the Development Stage of new product development. It is the Project Team's responsibility to complete this tool for all new and changed products with the assistance and support of Environmental Professionals.

6.6 Procedure

6.6.1 At the beginning of the projects design/planning phase, the Project Leader will complete the Preliminary Environmental, Health & Safety Assessment Questionnaire (see Form—Section 7.1 Appendix D), and forward the completed form to the Environmental, Health & Safety Professionals located at the operating location that will be impacted by the change. If an Environmental, Health & Safety Professional is not physically located at the operating location, the completed Questionnaire shall be sent to the Environmental, Health & Safety Professional responsible for that location.

6.6.2 The Environmental, Health & Safety Professionals, as a minimum, shall contact the Project Leader to discuss the proposed project in order to ascertain the level of evaluation required for the project. Evaluations shall include, but are not limited to, the types of checklists found in the Appendices of this policy.

6.6.3 Assessment information will be documented and forwarded to the Project Leader for inclusion in the project plan.

6.7 Training

Management shall be trained in the operating requirements of this policy to ensure its consistent application to business change. Associates involved in operating a new process or technology, or whose safety would otherwise be affected by the change shall be informed of, and trained in, the change prior to its implementation. This training must be documented.

6.7.1 The associate training, at a minimum, shall include:

6.7.1.1 Overview of the changes that will affect them.

6.7.1.2 Any new or modified requirements governing those changes [i.e., personal protective equipment (PPE), chemicals, facility/ area layout, evacuation routes, ergonomic related issues, machine guarding, etc.].

6.7.1.3 A review of any procedures, i.e., operational and/or safety, i.e., JSA's, process specifications, etc., that have been changed.

6.7.1.4 Introduction of any new procedures that are under development in order to gain input from those whom the changes shall affect most.

7. Appendices

7.1 Appendix I—Preliminary Environmental, Health & Safety Assessment Questionnaire

7.2 Appendix II—Safety Strategy Checklist (Not available)

7.3 Appendix III—Safety & Industrial Hygiene External Manufacturing Checklist (Not available)

7.4 Appendix IV—Environmental External Manufacturing Checklist (Not available)

7.5 Appendix V—Environmental Impact Assessment Checklist

LAMBDA CORPORATION

PRELIMINARY ENVIRONMENTAL, HEALTH AND SAFETY ASSESSMENT QUESTIONNAIRE

Possibly Triggering a Safety Review

For Intra-Computer Access Enter—Lambda S&IH Appendix I

Project leader's name and phone number _____ Date _____

Enter an "X" in each box that pertains to this project and send the form to S&IH. If an "X" is entered in any box, a Safety Review by S&IH and Environmental Professionals.

1. _____ Project includes purchase of new equipment, including computers and associated workstations.
2. _____ Modification or addition to the existing facility.

Change

3. _____ To the current energy sources (electric, hydraulic, pneumatic, etc.)
4. _____ To machine guarding or electrical safety systems.
5. _____ Involves the use of a new chemical.
6. _____ Involves the use of a biological agent.
7. _____ Affects person protective equipment requirements
8. _____ Results in increased noise, vibration, fumes, vapors, radiation, temperature, etc.
9. _____ Results in increased chemical storage requirements or increased chemical utilization.
10. _____ Affects manual material handling requirements (weight, force, frequency, container capacity).
11. _____ Affects exhaust system requirements.
12. _____ Impacts on Operators work stations, tools, equipment, etc.
13. _____ Impacts on production procedures, production output expectations, or work methods.
14. _____ Results in an increase or decrease of Operators.
15. _____ Results in greater risk to employees who are not operators.

Project Leader signature _____

For Safety & IH Use Only

16. _____ Safety Assessment Required
17. _____ Ergonomics Job Analyzer Required
18. _____ No Further Required

Signature S&IH _____ Date _____

LAMBDA CORPORATION

ENVIRONMENTAL IMPACT ASSESSMENT CHECKLIST

For Intra Computer Access, Enter—Lambda S&IF Appendix V
<u>Check All That Are Impacted</u>

Air
____ Impacts on facility potential to emit
____ Impacts on existing permits
____ Requires permit applications
____ Implications – pollution control/MACT
____ Other

Wastewater
____ Impacts on existing permits
____ Requires permit applications
____ Affects treatment systems
____ Other

Asbestos
____ Fume hoods impacted
____ Survey verification required
____ Continuous sampling required
____ Other

Drinking Water
____ Impacts on cross connections
____ Treatment systems impacted
____ Sampling plans needed
____ Other

Community Impacts
____ Legal/Communications
____ Noise
____ Other

Facilities
____ PM/Calibrations
____ Water connections/Drains
____ Water conservation
____ Energy connections/Conservation
____ Ventilation connections
____ Drawings/Surveys
____ Other

Reporting/Reg. Triggers
____ FIFRA
____ TSCA
____ TRI
____ CRTK
____ p Listed waste
____ Risk Management/Planning
____ Other

Non-Hazardous or Hazardous Waste
____ Segregation
____ Collection
____ Shipments/Generation rate
____ Storage
____ Other

Chemical Management
____ Labeling
____ Storage/Grounding
____ Segregation/Spill prevention
____ Inventory
____ MSDS needed
____ Pharma/Controlled substances
____ Other

Emergency Response
____ Emergency preparedness
____ Plans/SPCC plan
____ Other

EMS
____ Management Review
____ Commitment and Policy
____ Environmental Planning
____ Implementation and Operations
____ Training (RCRA, DOT)
____ Other

Environmental Action Plan

Issue Description	Follow-up Action	Responsible Party	Schedule

Comments _____

_____

_____

_____

Date of Action Plan Completion _____

Project Leader Sign-Off _____ Date _____

Environmental Sign-Off _____ Date _____

CHAPTER 20

THE PROCUREMENT PROCESS: SECTION 5.1.4 OF Z10

1) intro

Although the requirements in Z10 for the procurement processes are plainly stated and easily understood, they are brief in relation to the enormity of what will be required to implement them. Little help is given as to utilization of the procurement provisions. Advisory notes are brief. Documenting the procurement process is the subject of E5.1.4. Appendix I is a one-page informative description of the procurement process.

As is the case for the provisions in Z10 on safety design reviews, the purpose of the procurement processes is to avoid bringing hazards and their accompanying risks into the workplace. The standard requires that processes be instituted so that reviews are made of products, materials, and other goods purchased and related services to identify and evaluate health and safety risks before their introduction into the work environment. To fulfill the procurement provisions, safety specifications must be included in purchase orders and contracts.

To assist safety professionals as they give advice on implementing those provisions, in this chapter we:

- Comment briefly on prevalent purchasing practices
- Establish the significance of the procurement processes
- Discuss the pre-work necessary to include safety specifications in the procurement process
- Provide some resources

Advanced Safety Management: Focusing on Z10 and Serious Injury Prevention,
Second Edition. Fred A. Manuele.
© 2014 John Wiley & Sons, Inc. Published 2014 by John Wiley & Sons, Inc.

- Comment on the paucity of publicly available occupational health and safety purchasing specifications
- Remark on the need, sometimes, to set specifications above published standards
- Give examples of design specifications that become purchasing specifications

THE REAL WORLD OF PURCHASING PRACTICES

Unfortunately, the practice in many companies in the bid process for acquiring machinery, equipment, and materials is that purchasing departments are to choose the lowest-qualified bidder. For many years, safety professionals have told stories about how purchasing personnel have accepted the lowest bid on safety-related products or materials only to find, after their receipt, that they did not fulfill operational expectations and that safety needs were not met. Expensive retrofitting for safety purposes was necessary.

Retrofitting to accommodate safety needs starts with evaluating the deficiencies in the equipment as it is in operation: that is, identifying what was overlooked in the design process. Unfortunately, the resulting level of risk when safety requirements are addressed through retrofitting may be higher than would be the case if safety specifications were included in the bid or purchasing papers. As retrofitting proceeds, it's easy for decision makers to rationalize acceptance of higher risk levels. Nevertheless, the goal should be to achieve acceptable risk levels.

Getting managements and purchasing personnel to adopt the procurement provisions in Z10 will not be easy. A safety professional who proposes adding procurement provisions as an element in safety and health management systems and to have safety specifications included in a company's purchasing practices should expect the typical resistance to change. In most places, a culture change will be necessary for success.

An oblique interpretation of the procurement requirements could be: Safety and health professionals, you are assigned the responsibility to convince managements and purchasing agents that, in the long term, it can be very expensive to buy cheap.

SIGNIFICANCE OF THE PROCUREMENT PROVISIONS

I place great emphasis on having the procurement provisions in Z10 becoming an element in safety and health management systems because doing so prevents introducing hazards and risks into a workplace. This is the thinking that supports that position.

Risks of injury derive from hazards. If hazards are properly addressed and avoided, eliminate, reduced or brought under control in the design process so that the risks deriving from them are at an acceptable level, the potential for harm or damage and operational waste are minimized.

The logical extension of addressing hazards and risks in the design process is to have the design specifications the organization decides upon included in purchase orders and contracts so that suppliers and vendors know what safety specifications are to be met. That reduces the probability of bringing hazards into the workplace.

Although having safety specifications included in purchase orders or contracts is not a practice applied broadly, safety professionals are encouraged to consider the benefit to be achieved if they are. If the ideal is attained in the purchasing process and hazards and risks brought into the workplace are as low as reasonably practicable, the result is significant risk reduction, which means fewer injuries.

Unfortunately, the only safety-related terminology in purchase orders and contracts may be to meet all OSHA and other governmental requirements.

PRE-WORK NECESSARY FOR PROCUREMENT APPLICATIONS

As we say in Chapter 15, "Safety Design Reviews," there is a close relationship between establishing safety design specifications and including safety specifications in procurement documents. The latter cannot be achieved successfully until the former has been accomplished.

Once safety design specifications are established, the next step is to have them applied internally. Then, an appropriate extension is to have them incorporated into purchase orders and in contracts.

It is common practice for vendors and suppliers of equipment to use their own, and possibly inadequate, safety specifications if a purchaser has not established its requirements. That could end up being costly for the purchaser, especially if production schedules are delayed and the retrofitting expense to get systems operating as designed and for safety purposes is substantial.

RESOURCES

In Annex E, "Objectives/Implementation Plans," the following appears under Objectives for the Procurement provision: Distribute approved policy; Train on policy and procurement procedure; and Distribute safety requirements to be included in standard contracts. (p. 44)

That presumes that a procurement policy and safety requirements have been established and distributed and that training on their implementation is to be given. Procedurally, that's a good and recommended practice. Procurement is listed in Appendix L, "Audit," as one of the subjects to be reviewed when a safety audit is made. This audit guide says that in the audit process, the following are to be considered as aspects of the procurement section in the standard.

Documents to Be Reviewed
• Procedures for selection, evaluation, and management

Records to Be Reviewed
• Selected supplier self-assessments
• Selected supplier audits and ratings

- Selected supplier contracts
- Incoming product inspection records
- Product risk analyses

Interviewee
- Selected supplier management personnel
- Hourly employees

Observations
- Selected purchased products to check associated procurement records

Although the foregoing provides some guidance, what is proposed lacks the most important element in a procurement process, which is to establish the specifications that suppliers are expected to meet. Section 5, "Relationships with Suppliers," in ANSI/ASSE Z590.3, *Prevention Through Design: Guidelines for Addressing Occupational Hazards and Risks in Design and Redesign Processes*, is informative with respect to the procurement provisions in Z10. For one example: the standard says that in relationships with suppliers, "top management shall establish and document its occupational hazard and risk specifications."

THE PAUCITY OF AVAILABLE HEALTH AND SAFETY PURCHASING SPECIFICATIONS

Several safety-related texts were reviewed to determine whether they give guidance on including safety specifications in the procurement process. They do not. Examples of some safety-related purchasing specifications posted on the Internet follow. There are others.

- University of California, Environmental, Health and Safety (EH&S) Laboratory Safety Design Guide, 2nd edition. At http://us.yhs4.search.yahoo.com/yhs/search?p=University+of+California%2C+Environmental%2C+Health+and+Safety+%28EH%26S%29+Laboratory+Safety+Design+Guide&hspart=att&hsimp=yhs-att_001&type=att_lego_portal_home
- Queen's University Environmental Health and Safety: Laboratory Flammable and Combustible Liquid Handling Procedures. At http://us.yhs4.search.yahoo.com/yhs/search;-ylt=A0oG7oNbrlhR.ysAu14PxQt.?p=%E2%80%A2%09University+of+California%2C+Environmental%2C+Health+and+Safety+%28EH%26S%29+Laboratory+Safety+Design+Guide&type=att_lego_portal_home&hsimp=yhs-att_001&hspart=att&pstart=1&b=11
- Duke University, Summary of Class 4 Laser Laboratory Design Guidance, at www.safety.duke.edu/RadSafety/laser-lab-design.pdf
- Yale University's Procedure 3220, "Purchases of Restricted Items," among which are:

- *Hazardous materials*: Materials that present special safety risks during transport, storage, use, or disposal. These include, but are not limited to, certain highly toxic, reactive, or otherwise hazardous chemicals, gases, and biological agents.
- Safety-Critical Equipment: Equipment that can present safety hazards to users (e.g., x-ray and laser equipment) as well as equipment used to control exposures to recognized hazards, and whose improper use could subject users to harm (e.g., fume hoods, biological safety cabinets, respirators, automated film processors). At http://us.yhs4.search.yahoo.com/yhs/search?p=%E2%80%A2%09 Yale+University%E2%80%99s+Procedure+3220+&hspart=att&hsimp= yhs-att_001&type=att_lego_portal_home
- University of Wollongong, New South Wales, Australia. WHS Purchasing Guidelines (WHS, Workplace Health and Safety). Available at http://us.yhs4. search.yahoo.com/yhs/search?p=HRD-WHS-GUI-070.9+WHS+Purchasing+ Guidelines+2013+&hspart=att&hsimp=yhs-att_001&type=att_lego_portal_ home. (The Guidelines are an addendum to this chapter.)
- The Bedfordshire Community Safety Guide. At http://www.bedford.gov.uk/ environment_and_planning/planning_town_and_country/what_is_planning-policy/ documents_of_the_bdf/bedfordshire_community_safety.aspx

It is interesting that the first five entities in the list above are universities, and the last is a municipality. Since they are governmental entities, the materials they produce are almost always in the public domain.

Examples of applications of safety-related design standards that become purchasing specifications are not easily acquired. Most companies consider their specifications proprietary and don't make them available to others freely. For a moderate-sized company with a limited engineering staff, writing design and purchasing specifications will not be easy to do.

It seems appropriate to suggest that organizations prevail on the business associations of which they are members to undertake writing generic design specifications and purchasing specifications that relate to the hazards and risks inherent in their operations. Safety professionals should consider this inadequacy as an opportunity to be of service and make their presence felt.

However, arrangements were made for an additional example of internal design specifications that are also purchasing specifications to be included as an addendum to this chapter.

OPPORTUNITIES IN ERGONOMICS

As is the case with Z10's safety design review provisions, safety professionals who are not involved in the design or purchasing processes should consider ergonomics as fertile ground in which to get started. Some of the comments made in Chapter 15, "Safety Design Reviews," are repeated here because they apply equally to Z10's design and procurement provisions.

Musculoskeletal injuries, ergonomically related, are a large segment of the spectrum of injuries and illnesses in all industries and businesses. Since they are costly, reducing their frequency and severity will show notable results. Furthermore, it is well established that successful ergonomics applications result not only in risk reduction, but also in improved productivity, lower costs, and waste reduction.

Ergonomists know how to write design specifications for the workplace and the work methods that take into consideration the capabilities and limitations of workers. A company that established detailed ergonomics design criteria to be followed by its own engineers and by its vendors and suppliers was DaimlerChrysler. The following introduction to the DaimlerChrysler ergonomic design criteria demonstrates the relationship between writing design specifications and including them in purchasing requirements.

> This document attempts to integrate new technology around the human infrastructure by providing uniform ergonomic design criteria for DaimlerChrysler's manufacturing, assembly, power train and components operations, as well as part distribution centers. These criteria supply distinct specifications for the Corporation, to be used by all DaimlerChrysler engineers, designers, builders, vendors, suppliers, contractors etc. providing new or refurbished/rebuilt materials, services, tools, processes, facilities, task designs, packaging and product components to DaimlerChrysler.

In effect, the ergonomic design criteria to be used internally at DaimlerChrysler also became the ergonomic specifications that vendors and suppliers had to meet. In a section on supplier roles and responsibilities, it is made clear that all suppliers were to "make all reasonable efforts to implement all of the criteria and requirements" of the ergonomic design criteria. If a design requirement was to be compromised, the supplier had to so inform DaimlerChrysler and the matter would be reviewed to a conclusion by a DaimlerChrysler ergonomics representative.

DaimlerChrysler's ergonomic design criteria are available at http://www.docstoc.com/docs/25321410/150-ergonomic-design-criteria. The criteria may be downloaded for personal use or business use.

GENERAL DESIGN AND PURCHASING GUIDELINES

Addendum A to this chapter is a combination of design and purchasing guidelines currently in use. Note again that design specifications were developed that became purchasing specifications. The document is presented here as a reference from which engineering personnel and safety professionals can make selections and add, subtract, or alter items to suit location needs. It would be inappropriate to implement these guidelines without study and adjustment to reflect the hazards and risks inherent in a particular entity.

Adoption of a modification of the guidelines will usually require persuasive discussion with the purchasing staff. Getting a procurement process as outlined in the guidelines applied will require a culture change in all but a few organizations. Safety professionals must understand the enormity of what is being undertaken when they try to have safety specifications included in purchasing documents when that is not the current practice. Nevertheless, the productivity, risk reduction, and waste-saving benefits of a process that avoids bringing hazards and risks into the workplace cannot be refuted.

The Guidelines begin with sections on general safety requirements, machine guarding, industrial hygiene, ergonomics, machine and process controls, and environmental impact/hazard evaluation. Section 8 sets forth a procedure for which the instruction is: "Use this document as a guide whenever purchasing new (or modifying existing) equipment."

Major parts of this section deal with codes and Standards, equipment/fixture design; mechanical design and construction, electrical design and construction, pneumatics design and construction, software, and machine guarding.

These guidelines are quite broad. They relate to occupational safety and health, environmental concerns, productivity and avoiding events that result in business interruption.

DESIGNING AND WRITING SAFETY SPECIFICATIONS BEYOND THE LEVEL OF STANDARDS

In the safety standards writing process, it is common for contributions to be made by many participants, and compromises are made in the deliberations to accommodate the variety of views expressed on the subject being considered. The result often is a standard that includes minimum requirements. The following appears. In Chapter 12: Provisions for Risk Assessments in Standards and Guidelines.

> A supplementary and advisory document to SEMI S2 (*Environmental, Health, and Safety Guideline for Semiconductor Manufacturing Equipment*) is titled *Related Information 1 – Equipment/Product Safety Program*. It makes an interesting statement, cited below, about the need, sometimes, to go beyond issued safety standards in the design [and purchasing] process.
>
> Compliance with design-based safety standards does not necessarily ensure adequate safety in complex or state-of-the-art systems. It often is necessary to perform hazard analyses to identify hazards that are specific with the system, and develop hazard control measures that adequately control the associated risk beyond those that are covered in existing design-based standards.

Two subjects come to mind that encourage designing beyond compliance standards. Systems designed in accord with OSHA's lockout/tagout and confined space standards may be error-provocative.

- Assume that an electrical system is designed to OSHA lockout/tagout requirements and to the requirements of the National Electrical Code but that the distance workers have to travel to lockout stations is, in their view, too far and burdensome.

It is a near certainty that sometimes workers will not follow the written standard operating procedures. If the system's design and the purchasing contract merely say "Meet OSHA requirements," the result could be an error-provocative system.

- Saying in purchasing (construction) contracts that confined spaces should be designed to meet OSHA's standard may also result in creating error-provocative situations. An appropriate goal is, first, to try to design out confined spaces, and then consider the safety entry and exit needs in the design process where confined spaces must exist.

CONCLUSION

Too much stress cannot be placed on the significance of the procurement provisions in Z10 and the benefits that will derive from their implementation. It stands to reason that if the purchasing process limits bringing hazards and risks into the workplace, the probability of incidents resulting in injury or illness will be diminished. That's what Z10 is all about: "To reduce the risk of occupational injuries, illnesses, and fatalities."

Since very few organizations have processes in place that comply with Z10's procurement provisions, safety professionals are faced with an enormous task as they attempt to convince managements to adopt those provisions. Surely, undertaking to do so is a worthy and noble task. Although Addendum A to this chapter is lengthy, it is recommended that safety professionals read it to gain an appreciation of how extensive design and purchasing specifications can be and how they can serve well to avoid bringing hazards and risks into the workplace. Also, it must be understood that design engineers may not agree with some of the specifics included in the guidelines and would write different safety-related specifications.

Addendum B is a find. It is brief but is included because of the entity that issued it. The purchasing guidelines developed at the University of Wollongong are an excellent resource and highly recommended.

REFERENCES

ANSI/AIHA Z10-2012. American National Standard, *Occupational Health and Safety Management Systems*. Fairfax, VA: American Industrial Hygiene Association, 2012. ASSE is now the secretariat. Available at https://www.asse.org/cartpage.php?link=z10_2005.

ANSI/ASSE Z590.3-2011. *Prevention Through Design: Guidelines for Addressing Occupational Hazards and Risks in Design and Redesign Processes*. Des Plaines, IL: American Society of Safety Engineers, 2011.

Ergonomic Design Criteria. DaimlerChrysler. http://www.docstoc.com/docs/25321410/150-Ergonomic-Design-Criteria.

OSHA. Summary of OSHA Permit-Required Confined Spaces Rule. The rule citation is 29 CFR 1910.146. Available at http://www.dol.gov/elaws/osha/confined/PRCSGEN.asp.

OSHA. General Environmental Controls. Standard Number 1910.147. Titled "The Control of Hazardous Energy (Lockout/Tagout)." At http://www.osha.gov/pls/oshaweb/owadisp.show_document?p_id=9804&p_table=STANDARDS.

SEMI S2-0706. *Environmental, Health, and Safety Guideline for Semiconductor Manufacturing Equipment*. San Jose, CA: SEMI (Semiconductor Equipment and Materials International), 2006. (*Related Information 1: Equipment/Product Safety Program* is an adjunct to these *Guidelines*.)

WHS Purchasing Guidelines. University of Wollongong, New South Wales, Australia. Available at http://us.yhs4.search.yahoo.com/yhs/search?p=HRD-WHS-GUI-070.9+WHS+Purchasing+Guidelines+2013+&hspart=att&hsimp=yhs-att-001&type=att_lego_portal_home.

ADDENDUM A

GENERAL DESIGN AND PURCHASING GUIDELINES

1. **Purpose**

   1.1. This document provides general technical requirements and guidelines for the design, build, and purchase of equipment and fixtures.

2. **Scope**

   2.1. This document applies to the Arlington manufacturing facility in Campbell, IL.

3. **References**

   3.1. AR-15—General Safety Standards

   3.2. AR-19—Procedure for Processing Purchase Requisitions and Purchase Orders

4. **Definitions**

   4.1. None.

5. **Material and Equipment**

   5.1. Any relevant material or equipment as needed.

6. **Responsibilities**

   6.1. All Arlington associates in the Campbell manufacturing facility must adhere to the guidelines provided in this document when purchasing new equipment or modifying existing equipment.

Advanced Safety Management: Focusing on Z10 and Serious Injury Prevention, Second Edition. Fred A. Manuele.
© 2014 John Wiley & Sons, Inc. Published 2014 by John Wiley & Sons, Inc.

7. **Requirements**

7.1. Purchasing—Follow the guidelines for purchasing as found in AR-19.

7.2. Safety

    7.2.1. Machine guarding—Refer to AR-57 Machine Guarding Procedure for Safety.

    7.2.2. Guarding for construction—Follow AR-62.

        7.2.2.1. Give consideration to the best appropriate finish of all parts, including functionality and aesthetic purposes.

        7.2.2.2. Use stainless steel or ceramics for parts that contact the product. Approval of any parts/material that contact product must be obtained on a case-by-case basis.

        7.2.2.3. Anodize aluminum parts.

        7.2.2.4. Use industrial paint or primer when the painting of steel is necessary.

        7.2.2.5. Finish all surfaces to prevent corrosion.

        7.2.2.6. Consider material compatibility for long component life and the prevention of corrosion.

        7.2.2.7. Allow no direct contact of the product with aluminum, bronze, lead, or oil.

        7.2.2.8. Use steel or aluminum framing for large transparent doors in high stress or vibration areas.

        7.2.2.9. Use only polycarbonate materials such as RAND or a similar material for transparent doors and guards since Plexiglas materials often become brittle and shatter upon impact.

        7.2.2.10. Make all exposed edges (i.e., where contact with a body part may occur during operation of equipment) with a 3/16-inch radius (minimum).

        7.2.2.11. Method of fastening panels to framework:

            1. Use only "TORX" screws.

            2. Use through bolt, flat washers, and anti-vibratory nut, if possible—Tapping into transparent panels is NOT ACCEPTABLE.

            3. Place bolts spaced not more than 6 inches center to center.

            4. Use hinged panels with captive-style fastener for the latch.

    7.2.3. Signage—Clearly identify all controls and devices with appropriate warning signs, labels, and tags.

7.3. Industrial Hygiene

    7.3.1. Noise

        7.3.1.1. Make the maximum noise level of the equipment 80 dBA, 8-Hour Time Weighted Average when measured on the "A" scale of a standard sound level meter or noise dosimeter within 3 feet of the equipment.

        7.3.1.2. Make the maximum peak noise level 115 dBA when measured on the "C" scale of a standard sound level meter, within 3 feet of the equipment.

7.3.2. Ventilation—Use a systematic approach when designing or modifying exhaust ventilation systems.

    7.3.2.1. Refer to "Industrial Ventilation—A Manual of Recommended Practices," published by the American Conference of Governmental Industrial Hygienists (ACGIH), Section 6.0, which presents a systematic approach for designing or modifying exhaust ventilation systems.

    7.3.2.2. As a baseline, design to a minimum of 90 fpm and a maximum of 150 fpm face velocity at point source of exhaust.

    7.3.2.3. Equip each exhaust flow vent system with a continuous air flow monitoring device to detect any loss in air flow: Tie the system into the site emergency power system.

7.3.3. OSHA Hazard Communication

    7.3.3.1. Submit a Material Safety Data Sheet to site safety representative for approval prior to design and build of any new or upgraded equipment or process that requires the use of solvents, chemicals, or other fluids.

    7.3.3.2. An investigation will be conducted to determine if the equipment or process will pose any physical or health hazard.

    7.3.3.3. Measures will then be recommended to minimize the risk.

       • This includes descriptions of features and safeguards protecting the operator from direct or fugitive exposure to chemicals, solvents, or generated waste streams.

7.4. Ergonomics

7.4.1. General Workstation Design—Consider the following ergonomic guidelines for general workstation design as optimal dimensions and are not intended to restrict or limit your ability to design effective workstations. Above all, general workstation design should factor in the amount of risk employees will be subjected to when using workstations.

    7.4.1.1. Typical risk factors associated with industrial designs include but are not limited to:

       1. excessive forces

       2. poor body postures

       3. high repetition

       4. vibration

    7.4.1.2. Design workstations, when possible, to allow operators to work in both sitting and/or standing positions.

    7.4.1.3. Make work surface height for a sit/stand station 38" to 40".

    7.4.1.4. "Seated only" work surface height shall be 28–30" (28" is preferred) when "sit/stand" workstation is not feasible.

    7.4.1.5. Minimize work surface thickness, including underside support members (1" preferred, 2" maximum).

    7.4.1.6. Design work fixtures to allow hands to be positioned no higher than 4" above the work surface. The total dimension from the underside of the workstation to the point where the work is performed should not exceed 6". If visual requirements necessitate higher work positioning, arm rests should be provided.

7.4.1.7. Provide unobstructed leg room for the operator to sit comfortably, (24" wide, 26" from the floor to the underside of work surface, 18" deep from the front edge of work surface).

7.4.1.8. Allow a 4" × 4" toe cut-out at the bottom of the station when Standing Stations are used.

7.4.1.9. Design equipment/fixtures to be adaptable for convenient use by right- or left-handed operators, wherever possible.

7.4.1.10. Round all leading edges which the operator may come into direct contact with whenever possible. Try to recess the external hardware like hinges, door pulls, knobs, etc. as much as possible, to avoid contacting the operator.

7.4.1.11. Minimize equipment/fixture size to limit forward bending or reaching, as much as possible, to allow for tote pans and/or parts trays to be positioned in front of the operator. All operator/equipment interaction issues must be considered.

7.4.1.12. When fixed (non-adjustable) features are included into the design, the following principles are important to remember:
1. Design clearance dimensions for a tall operator (74" in height).
2. Design reaching dimensions for a short operator (60" in height).
3. Design fixed height dimensions designed for an average operator (66" in height).

7.5. Machine and Process Control

7.5.1. Master Control Relay—Provide Master Control Relay for emergency shutdown.

7.5.1.1. Hardwire emergency Stop pushbuttons and all Safety Switches in series to the Master Control Relay.

7.5.1.2. If any of these devices opens, the Master Control Relay should de-energize and remove power from the control circuit.

7.5.2. Emergency Stop—Provide Emergency Stop pushbuttons in an obvious location and within easy reach of either hand of the operator.

7.5.2.1. Design the E-stop such that to restart after Emergency shutdown it is necessary to pull out the E-STOP button and then press the appropriate buttons to initiate normal operation.

7.5.2.2. Design the E-STOP to interrupt power from the outputs, drives and other powered devices.

7.5.3. Interrupt DC Power Supply—Use DC power supply, interrupted on the DC side for faster response.

7.5.4. PLC Power—Wire power to the PLC outputs through a set of master control relay contacts.

7.5.5.   Interlocks—Design machine to not be capable of running in continuous RUN mode with interlocked guards out of position or removed.

7.5.6.   Mechanical Design:
1. No sharp edges or corners.
2. No shear points or pinch points.
3. Design turntable machines to not catch arms, hands, fingers, or clothes and with filled access areas.
4. Consider torque or force limiting devices for any part-moving device such as turntables, carriages and slide assemblies, etc. (for instance, magnetic couplings used for emergency breakaway on linear shuttle assemblies).
5. Use four-way, spring-centered, pneumatic valves for equipment working off of two hand controls.

7.6.   Environmental Impact/Hazard Evaluation

7.6.1.   Have materials used for all new or upgraded equipment and processes that are to be located in the facility evaluated to assure that they comply with Arlington's Policy and Governmental Regulations concerning such materials. This applies particularly to chemicals and generated waste streams.

7.6.1.1.   Environmental Impact—Have any equipment or process which may release any chemicals to the environment evaluated for environmental impact. All information necessary for this impact study must be submitted to the site environmental coordinator early enough in the development stage to avoid costly rework or delays.

7.6.1.2.   Material Safety Data Sheets—Have any chemical proposed for a new process approved by the site safety representative. A Material Safety Data Sheet (MSDS) and any other hazard data information must be submitted for evaluation prior to release of any purchase order for new equipment.

8. Procedure

8.1.   Use this document as a guide whenever purchasing new (or modifying existing) equipment.

8.2.   Codes and Standards

8.2.1.   At a minimum, all equipment shall comply with the latest revisions of the applicable specifications, codes, and standards. When deemed applicable by Engineering, documentation/labeling of said compliance shall be provided. *Note*: In case of conflicting specifications, the more stringent shall apply.

8.3.   Equipment/Fixture Design

8.3.1.   Make sure that all surfaces that may come into contact with the operator's body are free of sharp edges and corners (minimum 3/16" radius).

8.3.2. Tilt or orient fixtures to allow the operator to perform all work with a neutral body posture.

8.3.2.1. A general guideline is to tilt the fixture 15° toward the operator to enhance access and visibility, and to minimize awkward postures.

8.3.2.2. Specific recommendations will be made upon review of the job function in question.

8.3.3. Locate frequently accessed controls in front of and close to the operator to minimize reach distances.

8.3.4. Minimize repetitive reaches in front of the body to never exceed 16".

8.3.5. Repetitive reaches above chest height, below work surface height, or behind the body are not acceptable.

8.3.6. Design repetitively used control buttons (e.g., cycle start buttons) to require nominal activation of 1 pound or less. Where possible, it is also desirable for the pushbutton to be 2–3" in diameter.

8.3.7. Provide control knobs and handles with a nominal diameter of 1¼ inch. Clearance must be provided in equipment to avoid bumping the fingers or hands when parts are being positioned.

8.4. Mechanical—Design and Construction

8.4.1. Design equipment in a manner to prevent operator mistakes. *Note*: This includes preventing incorrect installation of tooling for set-up, incorrect loading of parts, installing wrong parts or incomplete parts. The goal is to make the machine capable of producing zero defects without relying on correct operational procedure.

8.4.2. Materials/Finishes

8.4.2.1. Materials "IN" Contact with the Product

Approval of any parts/material which contact product must be obtained on a case-by-case basis. Listed below are some guidelines to aid in choosing a material.

1. Use nonabrasive and non-marring materials.

2. Use Stainless Steel—300 series is preferred, 400 series in specific applications. Passivation and/or electropolish finish.

3. Use Aluminum—hard anodize only.

4. Use Plastics—fluorocarbons, polycarbonates, acrylics, ABS, polypropylenes, polyethylenes, and nylons, with approval of site project coordinator.

8.4.2.2. Materials "NOT" in Contact with the Product

1. Use Carbon Steels—properly prepared and painted, flash chromed, or electroless nickel plated.

2. Use Stainless Steels—anodize.

3. Use Aluminum—anodize.

8.4.2.3. Miscellaneous Finishes

1. Cover any tabletop surface or equipment that comes in contact with product with either stainless steel or

laminate (see specifications for requirements). Approval of any parts/material which contact product must be obtained on a case-by-case basis.

2. Do not use Wood Products in the manufacturing area.

8.4.3. Equipment Size and Weight Restrictions

    8.4.3.1. Free-Standing Equipment—No restriction

8.4.4. Tooling Requirements

    8.4.4.1. Locating Datum—Dimension all tooling from and locate the part from the primary datum of the component part. Features—Design tooling with the following features.

1. Interchangeability—Use dowels, bushings, and standard dowel patterns for locating to the equipment when practical all like tooling.

2. Quickchange—Use quickchange features on tooling to accomplish the changeover within the time requirement and in a repeatable manner. Quickchange techniques or devices such as single fastener size, quarter-turn fasteners, locking knobs, slotted holes, setup blocks, special tools, etc. may be used.

3. Adjustment/Installation—Design tooling to be fixed and not adjustable to the equipment. Adjustment features should be incorporated into the equipment (locking slide) rather than the tooling (slotted holes) if possible.

4. Design tooling to assure the correct installation of the tooling onto the equipment, using asymmetrical dowel patterns or similar feature.

8.4.5. Hardware Items

    8.4.5.1. Include one (1) resettable stroke counter and one (1) non-resettable stroke counter with each piece of automatic equipment.

8.4.6. Working Environment

    8.4.6.1. Design all equipment to operate in a Class 100,000 working environment unless otherwise specified.

8.4.7. Preferred Mechanical Components

    8.4.7.1. The following Preferred Components are listed in two groups, "A" and "B". "A" components are preferred as "first choice" and "B" components as "second choice."

    8.4.7.2. Use of "B" components shall require the approval of the Equipment Engineer.

    8.4.7.3. Use Preferred Components except when their use is not practical or jeopardizes project delivery.

    8.4.7.4. Obtain approval from Equipment Engineer prior to substitution for Preferred Components.

    8.4.7.5. Use standard commercially available components wherever possible.

8.4.7.6.   General Equipment and Machinery

[In a resource for this Addendum, a chart appears listing equipment items and brand names categorized in the aforementioned groups "A" and "B." It is not duplicated here.]

8.5.   Electrical—Design and Construction
    8.5.1.   Enclosures
        8.5.1.1.   Use enclosures properly rated for the environment for which they will be exposed.
        8.5.1.2.   Equip the main control enclosure with a fusible and lockable disconnect. Alternatively, if appropriate, a fusible lockable disconnect may be provided directly upstream of the main control enclosure. It shall be readily accessible as stated in the NEC.
        8.5.1.3.   **If any live circuit should exist by design in the system after opening the main disconnect, the main disconnect and the enclosure(s) containing that live circuit shall be obviously labeled stating this condition and the source and voltage of the live circuit.**
        8.5.1.4.   Provide spare space in the electrical enclosure for future additions.
            1. Allow for panels less than 500 square inches: a minimum of 40% spare capacity shall be provided.
            2. Allow for panels greater than 500 square inches, a minimum of 15% spare capacity shall be provided.
        8.5.1.5.   Do not mount control equipment on the door or sides of the enclosure except devices such as pushbuttons, pilot lights, selector switches, meters and instruments.
        8.5.1.6.   Mount terminal strips on inside enclosure sides and door, when appropriate. All terminal strips must be permanently, mechanically affixed—not glued or attached with a sticky strip.
        8.5.1.7.   Appropriately protect all wiring against excessive flex and pinching as a result of enclosure door operation.
        8.5.1.8.   During operation, the control enclosure interior temperature shall not experience a rise in temperature greater than 40° Fahrenheit over ambient temperature. Should the installed equipment cause this condition to occur in a static environment, filtered forced air shall be provided to maintain this temperature requirement. Appropriate alarms on this air flow must be provided.
    8.5.2.   Transformers
        8.5.2.1.   Use control transformers, where applicable, sized for 10% to 25% spare capacity (but not less than 100 VA).

Transformers 2 KVA and larger shall be of the dry type and shall be mounted externally.

8.5.2.2. Supply Source for the main control transformer shall be taken from the load side of the main disconnecting device. The isolated secondary of the main control transformer shall provide, nominally 120 VAC, single phase, and shall provide a bonded ground.

8.5.2.3. Fusing—Provide on primary and secondary.

8.5.3 Component Mounting

8.5.3.1. Use meters and operating controls placed so as to conform with Arlington ergonomic and AR-19 methods. Discuss with the electrical engineering representative.

8.5.3.2. Make all panel-mounted components removable without having to remove the control panel.

8.5.3.3. Identify control equipment mounted on the machine by use of engraved Lamicoid-type labels. Labels shall be permanently secured to the machine in order to facilitate their identification and location on wiring diagrams.

8.5.3.4. Identify control equipment mounted internal to the control cabinet by Brady labels or its equivalent. Dymo-type labels shall not be used.

8.5.3.5. Permanently label all components (relays, transformers, fuses, terminal blocks, etc.) mounted on the equipment and in the control enclosures according to a scheme agreed upon by an electrical engineering representative. Labels must be affixed so that replacing parts does not remove or change the intended identification.

8.5.4. Push Buttons

8.5.4.1. Use "Start" of the fully guarded momentary contact type on equipment or operations sensitive or very dangerous in the event of inadvertent start-up. Otherwise, flush momentary type buttons shall be used.

8.5.4.2. Use Normal "Stop" buttons of the extended momentary contact type.

8.5.4.3. Locate the "Stop" button near the start push button.

8.5.4.4. Use Emergency stop push buttons of the illuminated, maintained type when depressed.

1. Arrange the circuit so that depressing an emergency stop pushbutton will inhibit all machine motions and drop out the hardwired start/stop circuit. The E-stop light will be wired independent of other E-stop lights so that any depressed E-stop can be identified regardless of the position of any other E-stop pushbutton.

8.5.4.5. Pushbutton Colors

Color	Typical Function	Example
Red	Stop	Stop motors
	Emergency stop	Master stop
Green	On/start	Start of a cycle or motor
	Autocycle	
Black	Inch, Jog	Inching
	Horn silence	Jogging
Yellow	Return, reset	Return of machine to safe or normal condition
White	By approval only	
Gray	By approval only	
Blue	By approval only	
Orange	By approval only	

8.5.5. Position Sensors—Digital: Solid-state optical or proximity sensors are preferred to electromechanical limit switches. Sensors with built-in status indicating lights are preferred to those without.

8.5.6. Sensors/Signal Conditioners—Analog: Use solid-state sensors or signal conditioners with a 4–20 mA DC output.

8.5.7. Power Distribution

8.5.7.1. Power Supply Protection—Provide suitable power supply protection/conditioning equipment if the equipment is susceptible to power brownouts, voltage surges/spikes, or line conducted electromagnetic interference.

8.5.7.2. Surge Suppression—Include an MOV or diode in parallel located as close to the load as practical on all inductive loads—motor starters, solenoids, relays, transformers, etc.

8.5.7.3. Overloads—Use overloads of the manual reset type. Overload reset buttons shall be installed in the panel enclosures to allow manual reset, unless specifically provided for in separate motor starter control boxes.

8.5.8. Wiring Practices

8.5.8.1. Remote Interlocks—See Section 8.5.13.1 for labeling and Section 8.5.11 for color coding, for installations and equipment containing different sources of power.

8.5.8.2. Failsafe Operation—Utilize sensors, controls, and logic in such a manner that their failure or loss of power would produce the least undesirable consequences. In case of power failure, circuits must be manually re-energized to restart. The overall start and stop function must be hardwired, which eliminates the dependence on an electronic device or controller.

8.5.8.3. Push-To-Test Devices—Use pilot lights of the push-to-test type. However, if they are originating from a programmable logic controller (PLC) output, then a dedicated pushbutton input to the PLC shall be utilized to test all of such PLC output pilot lights. All other devices, such as displays and horns, must be tested by a push-to-test function, and should be wired to the dedicated push-to-test button where possible.

8.5.8.4. Interconnections/Terminals

1. Provide all components such as sensors, temperature sensors, heaters, and small motors which require frequent replacement with an appropriately designed quick disconnecting means. Terminals are one acceptable means.

2. Package and terminate all panel-mounted solid-state devices and/or components (resistors, diodes, etc.) to be easily removable with the use of hand tools. They must be installed to allow easy access for troubleshooting and testing. They shall not be mounted near large heat-producing components such as transformers. They shall not be soldered and/or "butt-spliced."

3. Calibrate critical process control variables on a regular basis. Where applicable and practicable, provide banana jacks/switches or accessible quick disconnects to allow convenient access for calibration. Discuss calibration requirements with the Electrical Engineer or an Instrument Mechanic. A calibration label shall be affixed to the device (or in close proximity), indicating the equipment I.D., calibration date, calibrator, and calibration due date.

4. Terminate devices external to any control enclosure at a terminal block in the control enclosure. Wiring from external devices shall not be connected directly to a device in the control enclosure. However, 440-volt motors may be wired directly to the motor starter. Thermocouples and similar very low voltage signal-carrying conductors may be terminated directly on the device to which it interfaces. If terminals are used, they must be of a special type appropriate to signals being conducted. One terminated and identified wire shall be returned for test purposes from a connection between limit switches, pushbuttons, or other devices connected in series. This test point termination may be in field junction boxes where easily accessible if appropriate.

5. Wire control circuit voltage reference points to a terminal in each terminal enclosure external to the control panel. All terminals shall be marked to correspond with terminal markings as specified on wiring diagrams.

6. Protect sensor devices with cord leads such that the leads cannot be damaged. Open wiring is not permitted.

7. Do not make electrical connection to control devices with soldered connections or wire nuts. Sensors shall be terminated on terminals in a junction box. A field junction box should be mounted to provide terminals if necessary.

8. Use terminal blocks with terminal clamp screws with no more than two wires under one screw.

9. Wire terminal blocks and mount so that internal and external wiring does not cross over the terminals.

10. Provide spare terminals in each terminal enclosure including enclosures external to the control panel. The number shall be at least 10% of the total in use or a minimum of six, whichever is greater. If fewer than 10 terminals are used and space prohibits, then less than 6 spares are adequate. There shall be spares appropriately spaced on each terminal block. A general rule would be to include 2 to 3 spares for each 10 terminals.

11. While maintaining functional appearance, placement of conduit and sealtite runs shall not restrict access for repair or replacement of machine parts. All conduit and sealtite shall be secured to permanent fixtures, walls, or bracing to avoid loose and damageable runs.

8.5.9. Circuit Installation

8.5.9.1. Separate AC wiring from DC wiring on both inside control panels and in wire runs, and signal wiring shall be separated from power wiring. Signal wiring shall be properly shielded. Thermocouple wiring shall be run separately unless shielded in which case it may be run with signal wiring. Where AC and DC wiring must be crossed, the wires or wire bundles shall cross at 90° to each other.

8.5.9.2. Sufficiently loop wiring to components mounted on doors to allow easy opening of the door and protect form excessive flex or pinching.

8.5.9.3. Use stranded copper of type MTW or THHN on all control wiring.

8.5.9.4. Wireways, sealtight, and conduit must meet NEC requirements and provide for adequate spares without overfilling. Conduit or sealtight is required for machine wiring.

8.5.9.5. Use appropriately sized cable/wire anchors that are permanently affixed to the surface. Self-adhesive anchors are not acceptable unless mounted by screws.

8.5.9.6. All cable/wire straps shall be appropriately sized, spaced, and tensioned to avoid cable/wire insulation deformation/damage and provide adequate support.

8.5.9.7–9. Not included.

8.5.9.10. Replace nicked or cracked wire insulation.

8.5.9.11. Use a plug-type removable terminal block at all locations to be separated for shipment where large equipment with interconnecting control panels are required, in order to minimize installation wiring.

8.5.9.12. Protect all wiring running alongside or over sharp edges.

8.5.9.13. Shield and ground all low-voltage DC (signal) wires and cable to prevent noise interference.

8.5.9.14. Utilize a ground fault circuit Interrupter (GFCI) whenever the process is a wet process. (See the NEC.)

8.5.10 Maintenance Considerations

8.5.10.1. **Give consideration to accessing electrical equipment for maintenance troubleshooting**. This may involve maintenance bypass switches for certain interlocks. However, the machine should not be allowed to run production in the maintenance bypass mode. The equipment should be so designed to provide access to sensors, switches, and motors, etc. for preventive maintenance procedures performed on a regular basis.

8.5.10.2. Provide all wiring with a service loop sufficient to re-make the connection at least three times.

8.5.10.3. Label all wire ends as indicated on the wiring diagrams. Where no wiring label is indicated, a to/from designation shall be used. Jacketed power wiring which is color coded can have the labeling on the jacket.

8.5.11. Wire Colors

8.5.11.1. Three-Phase Power and Motor Wiring

Volt	A = L1	B = L2	C = L3	Neutral	G = Ground
480	Brown	Orange	Yellow	Gray	Green
240	Black	Red	Blue	White	Green
208	Black	Black	Black	White	Green

8.5.11.2. One-Phase Power and Motor Wiring

Volt	Phase	N = Neutral	G = Ground
480(277)	App. Phase	Gray	Green
240(120)	App. Phase	White	Green
208(120)	App. Phase	White	Green
120	Black	White	Green

8.5.11.3. AC Control Wiring

Volt	Phase	N = Neutral	G = Ground
5 to 60	Pink	White/pink	Green
61 to 120	Red	White/red	Green

8.5.11.4.   DC Control Wiring and DC Motor Leads

Volt	(+)	(−)
0 to 11	Purple	White/blue
5 to 60	Blue	White/blue
61 to 120	Blue/black	White/blue

8.5.11.5.   Use green or green w/yellow tracer for grounding circuits ONLY.

8.5.11.6.   Cable sheathing and noise suppression conductors do not suffice for fault-carrying conductors. A separate ground shall be used.

8.5.11.7.   For external power or control wiring which is not de-energized when the equipment main disconnect is opened, color code is as follows: use a wire with a yellow tracer—the base color being that which corresponds to the circuit voltage in use.

8.5.11.8.   Make any temporary wiring or jumpers yellow or orange and installed in a manner indicating that this is obviously the purpose.

8.5.11.9.   Select appropriate conductor and jacket insulation for the environment and the service intended. All low voltage conductor insulation shall be 300 volts minimum while high voltage conductor insulation shall be $5 \times$ the nominal conductor voltage or 600 volts minimum. The insulation shall be appropriately selected to withstand the minimum bend radii expected and the effects of the environment (UV, ozone, oil, etc.).

8.5.11.10.   "HI-POT" test all power wiring should be at the appropriate voltage/duration for the nominal conductor voltage carried.

8.5.11.11.   Allowed Exceptions
1. Intrinsically safe wiring may require all conductors to be blue. Under certain conditions, sheathing may be the acceptable fault current conductor in intrinsically safe wiring. (See the NEC.)
2. If AC and AC neutrals or AC and DC neutrals are tied together, then the color of the corresponding AC neutral of the highest voltage will be used on all those common neutrals.
3. Wiring on devices purchased completely wired.
4. Where insulation is used that is not available in the colors specified.
5. Equipment for use outside of the United States when the above color coding is not in agreement with the established local electrical codes.

8.5.12. Methods of Grounding

8.5.12.1. Use a separate grounding conductor for grounding of equipment. A stranded or braided copper conductor shall be used for grounding where subject to vibration.

8.5.12.2. Grounding by attaching the device enclosure to the machine with bolts or other approved means shall be considered satisfactory if all paint and dirt are removed from joint surfaces. Moving machine parts having metal-to-metal bearing surfaces shall not be considered as a grounding conductor.

8.5.12.3. Do not use the grounded neutral conductor of a circuit for the grounding of equipment. No neutral shall be grounded, unless electrically isolated from the plant power distribution system in or immediately adjacent the equipment in question. This is to prevent current imbalances in the power distribution system which would affect ground fault detection.

8.5.12.4. Do not mix signal and power grounds.

8.5.12.5. Terminate signal grounds at one central location whenever possible to eliminate ground loops and noise.

8.5.13. Remote Interlocks

8.5.13.1. Attach caution labels, for installations and equipment containing different sources of power, adjacent to the main disconnects stating:

CAUTION: THIS PANEL CONTAINS MORE THAN ONE SOURCE OF POWER

DISCONNECT THE FOLLOWING SOURCES BEFORE SERVICING:
(Identity and location of all disconnects shall be shown on the label)

8.5.13.2. Use labels of red Lamicoid with white lettering or of equivalent quality, secured in place by screws or rivets. Where practical, the remote sources of power shall be interlocked with the disconnect means. If not, a manual means shall be provided to quickly disconnect these sources.

8.5.13.3. Make all ungrounded wiring which contain remote sources of power yellow throughout the control panel.

8.5.14. Operator Interface—Apply Human Engineering criteria (Ergonomics) to the design of the operator interface. Except on the smallest systems, message board-type status and alarm indicators are preferred to an array of indicator lights and associated Nameplates.

8.5.15. Variable Speed Motor drives—Above 1/4 hp, solid-state variable frequency AC motor drives are preferred to solid-state variable voltage

DC motor drives. Below 1/4 hp stepper motor drives are preferred to variable voltage DC drives. The selection should be discussed with the electrical engineering representative to provide standardization when possible.

8.5.16. Programmable Controllers

8.5.16.1. Discuss the type of PLC, CPU version, and I/O selection with the electrical engineering representative.

8.5.16.2. Ensure that the supplier follows all design criteria established by the PLC manufacturer for installation of PLCs.

8.5.16.3. Ground I/O cards according to function (i.e., DC inputs, AC input, AC outputs, etc.), and spare slots should be left between these function groupings.

8.5.16.4. At least one hardwired Emergency Stop function shall be generated to create an emergency shutdown independent of the PLC and shall function even if a component of the PLC fails. The Emergency Stop function shall be interlocked in to the PLC software. The E-Stop hardwiring shall open appropriate power circuits to PLC outputs.

8.5.16.5. Shield low-voltage DC wiring with the following: Below 15 volts which interfaces to I/O modules, sink or source currents less than 8 mA, or input signal delay times less than 10 mA.

8.5.16.6. Maintain the shield continuity throughout the system when shielded cable is used.

8.5.16.7. Provide a minimum of 20% spare slots for future expansion.

8.5.16.8. Not included.

8.5.16.9. PLC Digital Inputs—Use input modules of 110 VAC or 24 V (first preference is for AC, with DC being the second preference). Discuss with the electrical engineering representative.

8.5.16.10. PLC Digital Outputs—Use 24 V for indicator lights and alarms. AC is first preference, DC is second preference. For pneumatic or hydraulic solenoids and motor starters, 115 VAC modules can be used. Solenoids should be rated for continuous duty.

8.5.16.11. PLC Analog Inputs/Outputs—The preference is 4–20 mA DC. Should 1–5 VDC input/output be required, precision 250-ohm resistors (1%) shall be used and installed at the terminating location.

8.5.16.12. Communication—In order to achieve the goal of integrating manufacturing machine raw data, components and process parameters should be shareable to the higher level of the computer system, and a communication system shall be provided. It includes an interface module, communication software, and communication ports. Coaxial cable and twisted pair wire are common for the communication media; however, the optic fiber cable is preferred for the working environment with high electrical noise.

8.5.17. Instrumentation—All instruments and measuring devices used must be approved by an electrical engineering and/or instrumentation technician prior to use. Special considerations must be given to standardization, precision and accuracy, calibration requirements, and maintenance. All original manuals and specifications shall be provided.

8.5.18. Machine Installation Drawings

8.5.18.1. General—Include an installation drawing showing physical dimensions for mechanical, electrical, and service requirements with respect to the space needed for proper installation for each machine that is to be installed.

8.5.18.2. Service Requirements—Clearly indicate the appropriate electrical service (i.e., 480 V, 3 phase) for operation on the drawing.

8.5.19. Machine Documentation

8.5.19.1. ANSI/USAS Y32.10-1967 drawing symbols should be used. The electrical designation and descriptions on related pneumatic/hydraulic piping diagrams shall match those on the electrical diagrams.

8.5.19.2. Provide a P&ID diagram if appropriate.

8.5.19.3. Provide electrical control drawings in AUTOCAD in all cases. Formats, numbering systems, symbology, details, annotation to be discussed with the electrical engineering representative. Examples will be provided. Standard symbols generally follow standard ABCD.

8.5.19.4. All programs for drives, PLCs, etc. shall be sufficiently annotated and correspond accurately to electrical drawings.

8.5.19.5. Include all original manuals for components with machinery such as the following: motor drive manuals, operator interfaces, special programming devices, message displays, etc.

8.5.19.6. Provide a Bill of Materials with detailed parts description, manufacturer, and part number. The format should be discussed with the electrical engineering representative. Examples can be provided.

8.5.20. Preferred Electrical Components

8.5.20.1. The following Preferred Components are listed in two groups, "A" and "B". "A" components are preferred as "first choice" and "B" components as "second choice" if no "A" components are applicable.

8.5.20.2. Use of "B" components shall require the approval of the equipment engineer.

8.5.20.3. Obtain approval for substitution of Preferred Components from the equipment engineer.

8.5.20.4. Electrical Components

[This section contains a list of electrical components and brand names for "A" and "B" choices. It is not duplicated here.]

8.6. Pneumatics—Design and Construction

8.6.1. Air Supply—Operate equipment from a single incoming air supply drop. Maximum Supply Pressure—110 psig; Design Pressure—60 psig recommended where possible.

8.6.2. Hardware/Circuit

8.6.2.1. Non-Lubricated System—Use pneumatic devices and circuits which do not require lubrication.

8.6.2.2. Disconnect—Design equipment/fixture to operate from a single quick disconnect (with exhausting, locking valve).

8.6.2.3. Cylinders

1. Use ports with lockable flow control valves.
2. Cylinders used in the vertical orientation and which could pose a hazard to the operator during an air dump shall have a pilot-operated check valve in the exhausting port to keep the cylinder from lowering during an air dump.
3. Use cylinders of the permanently lubricated type.
4. Adjustable cushioning at both ends is preferred.
5. Use rods with self-aligning rod end coupling.

8.6.2.4. Exhaust—to be reclassified, filtered, and directed away from the operator and conform to Class 100,000 working environment.

8.6.2.5. Directional Valves—Modular packages or manifold style are preferred.

8.6.2.6. Do not use Air Over Oil or Air Pressure Intensifiers.

8.6.2.7. Fittings—Use plastic, stainless or brass.

8.6.2.8. Ports—Make all ports NPT wherever possible. If NPT is not available, then BSP shall be used. Devices with ports other than NPT shall be labeled to indicate the port type.

8.6.3. Labels—Clearly identify all devices. Labels shall be located on the structure of the equipment next to the device so that the label remains when the device is changed. Bi-lingual labeling considerations are to be addressed at the discretion of Site Project Coordinator.

8.6.4. Preferred Pneumatic Components

8.6.4.1. The following Preferred Components are listed in two groups, "A" and "B". "A" components are preferred as "first choice" and "B" components as "second choice" if no "A" components are applicable.

8.6.4.2.   Use of "B" components shall require the approval of the equipment engineer.

8.6.4.3.   Use Preferred Components except when their use is not practical or jeopardizes project delivery.

8.6.4.4.   Obtain approval for substitution for preferred components of the equipment engineer.

8.6.4.5.   Use standard commercially available components (ISO Standard Preferred) wherever possible.

8.6.4.6.   Pneumatic Components

[This section lists pneumatic components and the names of "A" and "B" supplier companies. It is not duplicated here.]

8.7.   Software

8.7.1.   PLC Software Control Philosophies

8.7.1.1.   Logic Location/Recovery—Locate all start/stop logic and equipment control logic in the PLC. This logic shall be retained in PLC memory while the system is down due to normal or emergency stop, or due to power failure. The goal of the PLC logic will be zero recovery; i.e., after a failure has been corrected, restarting the system shall be accomplished by pushing the "start" button.

8.7.1.2.   Event Driven—Write software to sequence on events (switch closures) rather than time. Counters, Timers, and One-Shots shall be used only if necessary. If they are used, use as many conditions as practical.

8.7.1.3.   Structure—Write PLC software in modules, using tables wherever possible, and arranged in the following order:
1. System start-up
2. Non-motion logic (lights, horns, alarms)
3. Motion logic (solenoids, motors)
4. CIM/HMI interface logic

8.7.1.4.   All alarms shall be latched. Alarms shall be cleared by use of an acknowledge button.

8.7.2.   Testing Objectives: Test software to ensure proper operation in the following areas.

8.7.2.1.   The system performs in compliance with the statement of requirements.

8.7.2.2.   The software is error free and executes correctly as defined by the process specifications.

8.7.2.3.   That operating faults, alarms, interlocks, and error conditions are detected, and recover as specified.

8.7.2.4.   That automatic and manual abort and recovery functions perform as specified.

8.7.2.5.   That operator interfaces are correct as specified.

8.8. Machine Guarding
  8.8.1. Refer to MG2468-18 for guarding requirements and suggested types.

9. Appendices
  9.1. Appendix I – Operation and Maintenance Manual
  9.2. Appendix II – Codes and Standards

Appendix I Operation and Maintenance Manual [To be provided by the equipment supplier.] An Operation Maintenance Manual is required for all Automatic Equipment to the extent appropriate to communicate proper operation and maintenance activities. The scope of the manual may range from one page of instructions for simple equipment to a full comprehensive manual for complex equipment.

Appendix II This appendix lists certain Codes and Standards to which the supplier is to adhere.

ADDENDUM B

UNIVERSITY OF WOLLONGONG

A complete copy of these WHS (Workplace Health and Safety) purchasing guidelines can be found at http://us.yhs4.search.yahoo.com/yhs/search?p=HRD-WHS-GUI-070.9+WHS+Purchasing+Guidelines+2013+&hspart=att&hsimp=yhs-att_001&type=att_lego_portal_home.

A footnote on the first page says: "Hardcopies of this document are considered uncontrolled. Please refer to UOW website or intranet for latest version." This is the latest version, having a 2013 March date.

Contents

Advanced Safety Management: Focusing on Z10 and Serious Injury Prevention,
Second Edition. Fred A. Manuele.
© 2014 John Wiley & Sons, Inc. Published 2014 by John Wiley & Sons, Inc.

1 INTRODUCTION

The purpose of this guideline is to ensure that suitable consideration is given when purchasing equipment, materials, facilities, or substances which may have an adverse impact on health and safety. The most effective method of reducing the risk of a hazard in the workplace is through the process of eliminating the hazard from the workplace. Many hazards can be eliminated before they are introduced into the workplace by conducting a risk assessment prior to purchase and upon receipt of goods.

In essence, there are two questions which need to be asked during the procurement process to ensure potential risk of goods and services are identified and controlled before being introduced into the workplace:

- What WHS risks does the intended purchase pose to health & safety?
- How are the WHS risks managed for the intended item to ensure health & safety? e.g., What is the supplier or University required to implement in order to eliminate or minimize the risks associated with the proposed purchase?

2 SCOPE

This guideline is to be applied in conjunction with the University's Purchasing and Procurement Policy and related procedures and applies to any item purchased by the University via orders, tenders, contracts, petty cash, and credit card transactions. The guidelines apply to any purchase of equipment, materials, or substances, including those which are hired, leased, or donated to the University. The purchase of services and/or labor hire should refer to the WHS Contractor Management Guidelines.

3 RESPONSIBILITIES

Deans, Directors, Heads, and Managers of Units are to ensure that the WHS Purchasing Guidelines are implemented within their area of responsibility. Any person in the University who purchases, leases, or hires goods is responsible for identifying WHS requirements for any item which may pose a reasonably foreseeable injury in the workplace and upon receipt, verify that WHS specifications or control measures are in place to eliminate or reduce the risk.

Persons responsible for purchasing activities and payments are to complete the internal training course "Introduction to eProcurement (Purchasing and Payment) System." Persons undertaking the purchase of goods should have the capability to ensure that the item being received is fit for purpose by identifying WHS requirements prior to purchase and then verifying those requirements upon receipt.

Knowledge of this document shall be the base skill, experience, or qualifications used for the safe purchase of products. Other skills, experience, or qualifications for the identification of WHS requirements of specific goods are outlined in Appendix 1.

4 PROCEDURE

Figure 20.B1 illustrates the procedure for identifying WHS requirements and specifications in the procurement process.

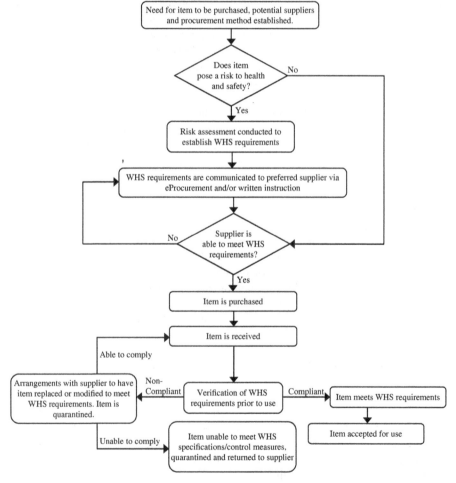

FIGURE 20.B1   Procurement process.

4.1 Determining If an Item Impacts Health and Safety

Any item to be purchased should be evaluated to determine whether the item poses a risk to health and safety. The following questions are an aid in determining whether the item could raise an WHS issue:

- Could a reasonably foreseeable injury or incident occur in the course of normal or unanticipated storage or transport of the item to be purchased?
- Are there any specifications which are required to ensure safe operation or use?
- Does the item need to comply with legislation, codes of practice, or Australian Standards?
- Will a safe work procedure need to be developed to ensure health and safety?

If the answer to any of the above is "yes", then WHS requirements are to be identified to ensure that all safety-related specifications are communicated to the supplier and verified upon receipt.

4.2 Risk Assessment

In some instances the risk assessment is a straightforward process where the WHS specifications or control measures can easily be determined.

Where the item being purchased requires a detailed risk assessment to be undertaken, the appropriate *risk assessment form* shall be used. Examples of when the formal risk assessment is to be used for items which have a risk to health and safety include, but is not limited to:

- Lasers;
- Radiation apparatus;
- Radiation isotopes;
- Biological substances;
- Hazardous substances that are colour coded red on ChemAlert or if the substance is going to be used outside the scope of the MSDS;
- Medical and scientific equipment;
- Mobile vehicles, i.e., forklifts, carts;
- Machinery and plant, i.e., lathes,
- Construction plant and equipment;
- Personal protective equipment, i.e., safety glasses, safety boots, face shields, gloves;
- Ergonomic equipment, i.e., chairs, seating, desks, etc.;
- Custom built equipment;
- Heavy and awkward items which pose a manual handling risk;
- Items which have an "extreme" or "high" risk after completing the UOW Risk Matrix in consultation with users.

The risk assessment should consider the following:

- legal requirements,
- codes of practice or relevant standards,
- potential impact on affected personnel,
- training requirements,
- changes to work procedures,
- personal protective equipment:
- technical data or information.

Once the hazards have been identified, attempts to eliminate the risk from being introduced into the workplace should be attempted prior to supply. Where this is not possible, risk control measures shall be determined to minimise the risk of injury or illness.

Control measures are the requirements which are needed to reduce the risk associated with the hazard to an acceptable level to prevent injury or illness. The method of risk control shall follow the hierarchy of control as outlined in the *WHS Risk Management Guidelines*.

Examples of controls may include substituting the item for something less hazardous which will perform the same function, guarding of moving parts, provision of training and competency assessment, licensing requirements, signage, development of or modification to safe work procedures, or identification of personal protective equipment.

4.3   Risk Control Measures

Requirements and/or risk control measures for WHS are be detailed on the eProcurement purchase order form and provided to the supplier. See Appendix 1 for examples of WHS requirements for common items.

When items are purchased outside the eProcurement system the supplier is to be provided with details of any WHS requirements or risk control measures in writing. This may include WHS requirements listed on either of the following:

- purchase order,
- via a completed risk assessment, or
- a letter outlining specifications.

A list of applicable legislation and WHS specifications is provided in Appendix 1, Examples of WHS Specifications and Control Measures.

4.4   Supplier's Capacity to Comply

It is necessary to ensure that the supplier can meet the requirements as stated in the WHS specifications or control measures. This can be derived through discussions with the supplier on the required WHS specifications in the pre-purchase stage of procurement to ensure the item is fit-for-purpose.

The process of measuring the capacity of a supplier to meet the WHS specifications related to an item shall be documented on the purchase requisition form or other supporting documentation. The process of measuring a supplier's capacity to meet WHS specification includes checking to ensure that the supplier can meet or exceed the WHS specifications or control measures through the risk assessment process or specifications as outlined in Appendix 1. Should the supplier not have the capacity to comply with identified WHS requirements, the purchase is to be made through another supplier capable of meeting the requirements. Alternatively, the selection of another item that is capable of meeting the WHS requirements could be explored.

4.5 Verification of WHS Requirements

The verification of WHS requirements is required upon arrival of goods to ensure that the WHS requirements or control measures have been met as detailed on the purchase order or risk assessment. Verification should be conducted by the person who ordered the item and/or who conducted the risk assessment to determine the WHS requirements. Verification of WHS specifications are to be documented by the person receiving the goods within eProcurement and included in the risk assessment form.

Examples of verification may include:

- checking to ensure that containers are clearly labelled;
- checking that an item is labelled to indicate that it has been made to comply with the relevant Australian Standard;
- checking the compatibility of the item to be stored in compliance with the dangerous goods requirements,
- ensuring that an item is fitted with physical control measures such as guarding of moving parts.

When WHS requirements or control measures cannot be verified, the item must be quarantined and/or tagged out until the verification is complete. Items that are unable to be verified must be returned to the supplier.

If a hazard is not identified prior to purchase but becomes apparent once the item has been received or used, a hazard report shall be lodged using SafetyNet. The hazard report shall detail the corrective actions required to eliminate or minimise the risk of injury to an acceptable level.

4.6 Repeat Purchases

A risk assessment can be re-used for repeated purchases of the same item or where the supplier has previously demonstrated compliance to WHS requirements. However, if the use or quantity of the item differs and has a greater impact on health and safety, the risk assessment should be reviewed and modified accordingly.

4.7 Standing Orders

Where a standing order has been raised with a supplier, the supplier must indicate in writing via a Memorandum of Understanding that all products being supplied to the University will conform to applicable legislation, codes of practice, or Australian Standards not limited to those outlined in Appendix 1 of these Guidelines.

Where WHS requirements are identified outside of the Memorandum of Understanding, these shall be outlined in writing to the supplier using the *risk assessment form* prior to purchase.

Products received by the University from a supplier with a standing order are required to be verified for compliance to WHS requirements prior to use.

4.8   Credit Card/Petty Cash Purchases

To ensure compliance with these guidelines the preferred purchase method for items with WHS considerations is the eProcurement process in preference to credit card or petty cash transactions.

When materials or substances are required to be purchased using a credit card or petty cash, the person purchasing the item shall consider the potential for the equipment, material, facility, or substance to pose a risk to health and safety. In particular, prior to the purchase of:

- hazardous substances, the Material Safety Data sheet (MSDS) should be consulted to ensure that controls can be put in place to minimise the risk;
- Personal Protective equipment (PPE) items must comply with relevant Australian Standards.

This will be documented using the *risk assessment form* prior to purchase.

4.9   Consultation

Any proposed changes to the working environment that could place a risk to health and safety must be communicated to all employees who are likely to be affected. This can be achieved via email notification, or as a standing agenda item in the local area's WHS Consultation Structure.

5   RELATED DOCUMENTATION

Use the links below for the following related documentation:

- *Purchasing and Procurement Policy*
- *Risk Management Guidelines*
- *Risk Assessment Form*

6   PROGRAM EVALUATION

In order to ensure that these guidelines continue to be effective and applicable to the University, these guidelines will be reviewed regularly by the WHS Unit in consultation with the WHS Committee. Conditions which might warrant a review of the guidelines on a more frequent basis would include:

- reported hazards or injuries;
- non-conforming systems;
- WHS Committee concern.

Following the completion of any review, the program will be revised/updated in order to correct any deficiencies. These changes will be communicated via the WHS committee.

7 VERSION CONTROL TABLE

Author's note: Example recordings only appear here.

Version Control	Date Released	Approved by	Amendment
4	February 2008	WHS Manager	Inclusion of training requirements and expansion of Section 4
9	March 2013	WHS Manager	WHS Unit name change and incorporation of Section 4.9

8 APPENDIX 1, EXAMPLES OF WHS SPECIFICATIONS AND CONTROL MEASURES

Author's note: This is a four-page listing of items (Personal Protective Equipment, Hazardous Substances, and Dangerous Goods, Plant, and Equipment) for which Pre-purchase WHS Requirements are recorded that are to be forwarded to the supplier. Also, "Y" or "N" indicators are to be entered in columns indicating, upon receipt, training is needed, inspection and testing is to be done, or there are licensing and registration requirements. Examples follow.

Item	Pre-purchase WHS Requirement (Forward to Supplier)	Upon Receipt		
		Training	Inspecting and Testing	License and Registration
Eye protection	AS 1337: Eye protection for industrial applications	Y	N	N
Radioactive sources	Qualification Required: License to use	Y	N	Y
Machinery	AS 4024.1: Safeguarding machinery—general principles	Y	Y	N

CHAPTER 21

EVALUATION AND CORRECTIVE ACTION: SECTION 6.0 OF Z10

In applying the Plan-Do-Check-Act concept for an occupational health and safety management system, the last two steps are to evaluate performance (6.0A) and take corrective action when nonconformance is found (6.0B). The following is a depiction of the applied PDCA concept and how it relates to the processes required in Section 6.0, "Evaluation and Corrective Action."

Plan: Identify the problem(s) (hazards, risks, management system deficiencies, and opportunities for improvement, as in the planning section, 4.0).

Plan: Analyze the problem(s).

Plan: Develop solutions.

Do: Implement solutions.

Check: Evaluate the results to determine that:

1. The problems were resolved, only partially resolved, or not resolved.

2. The actions taken did or did not create new hazards.

3. Acceptable risk levels were or were not achieved.

Act: Accept the results, or take additional corrective action, as needed.

It says in Section 6.0C that processes are to be in place to "include results of evaluation activities as part of the planning process and management review (6.5)."

Advanced Safety Management: Focusing on Z10 and Serious Injury Prevention,
Second Edition. Fred A. Manuele.
© 2014 John Wiley & Sons, Inc. Published 2014 by John Wiley & Sons, Inc.

The intent is that organizations have feedback processes in place to communicate back to management regarding lessons learned about system shortcomings, so that appropriate activities can be included in the planning process.

SECTION 6.1: MONITORING, MEASUREMENT, AND ASSESSMENT

Methods for monitoring, measurement, and assessment include workplace inspections, exposure assessments, incident tracking, employee input, occupational health assessment, assessment of performance relative to applicable legal and other requirements as determined by the organization, and other methods as required by the employer's occupational health and safety management system. Findings deriving from those processes are to be communicated to interested parties.

Literature is abundant on workplace inspections. That subject is not addressed here further. Measurements of effectiveness with respect to exposure assessments and occupational health assessments are to determine how well the requirements in the assessment and prioritization processes set forth Section 4.2 have been fulfilled. They require that organizations have processes in place to assess risks pertaining to health and safety exposures.

Although establishing performance measures is not one of the subjects listed in the "shall" provisions in Section 6.0, the advisory comments say in E6.1C that "organizations should develop measures of performance that enable them to see how they are doing in preventing injuries and illnesses."

To have statistical validity, the performance measures adopted should consider the extent of the exposures (perhaps hours worked) as well as evaluations of the effectiveness of safety and health management systems. Although the advisory information in E6.1C refers to occupational injury and illness rates as performance measures, a precaution is given indicating that such rates should not be the sole or primary measurement tool. A dissertation that speaks of performance measures suitable for organizations of various sizes may be found in the chapter "Measurement of Safety Performance" in *On the Practice of Safety, 4th edition.*

The effectiveness of the processes outlined in Section 3.0, "Management Leadership and Employee Participation," would be the basis of performance measurements on employee input. The provisions in Section 3.2, "Employee Participation," state: "The organization shall establish and implement processes to ensure effective participation in the occupational health and safety management system by its employees at all levels of the organization, including those working closest to the hazard(s)."

SECTION 6.2: INCIDENT INVESTIGATION

Since I now give greater emphasis to the importance of incident investigation within the spectrum of safety and health management systems, a separate chapter on the subject appears here—Chapter 22. Incident investigations, well made, can be a good source to identify cultural, operational, and technical contributing factors, particularly for incidents that result in serious injury or damage.

SECTION 6.3: AUDITS

Having audits made of safety and health management systems to determine their effectiveness and to identify opportunities for improvement is the subject of Section 6.3. The goal of a safety audit is to provide management with an assessment of the reality of the safety culture in place and to provide recommendations on how the culture can be improved. This important measurement process is also the subject of a separate chapter, Chapter 23.

SECTION 6.4: CORRECTIVE AND PREVENTIVE ACTION

Although the requirements for corrective and preventive action are set forth briefly, the importance of this section should not be minimized. To fulfill its requirements, organizations are to have processes in place so that corrective actions are taken expeditiously: on the deficiencies in occupational safety and health management systems, inadequately controlled hazards, and newly identified hazards that are discovered in the monitoring process.

This section also requires that processes be in place to "review and ensure the effectiveness of corrective and preventive actions taken." Item E6.4, an advisory, says that an effective occupational health and safety management system would "identify system deficiencies and control hazards in any part of the system to an acceptable level of risk." Item E6.4B offers a precaution—zero risk is not attainable and should not be sought. It says that "risk cannot typically be eliminated entirely, although it can be substantially reduced through application of the hierarchy of controls. Residual risk is the remaining risk after controls have been implemented."

SECTION 6.5: FEEDBACK TO THE PLANNING PROCESS

The purpose of this section is to assure that hazards, risks, and safety and health management system deficiencies observed in the monitoring, measurement, audit, incident investigation, and corrective and preventive action activities are communicated to the appropriate parties and considered in the ongoing planning and management review process. As a result of that communication, objectives are to be revised and modifications are to be made in implementation plans to achieve a more effective health and safety management system.

CONCLUSION

When applying the Plan-Do-Check-Act continual improvement process, an important element is to determine whether the management systems put in place achieve what is intended. That is the purpose of Section 6.0, to provide an evaluation mechanism so that system deficiencies can be identified and acted upon. This is an important continual improvement function.

REFERENCE

Manuele, Fred A. *On the Practice of Safety,* 4th ed. Hoboken, NJ: Wiley, 2013.

CHAPTER 22

INCIDENT INVESTIGATION: SECTION 6.2 OF Z10

In Chapter 3, "Innovations in Serious Injury and Fatality Prevention," comments were made about analyses of over 1800 incident investigation reports to assess their quality, with an emphasis on contributing and causal factors. It was also said that the gap between established procedures on incident investigation and what actually takes place can be enormous. Even in the best safety management systems, the quality of incident investigation can be substantially less than adequate. For example, in a very large organization, it was agreed that if the safety staff promoted adoption of a system as uncomplicated as the Five Why Technique to improve incident investigation and the organization achieved a B + grade in two years, a huge step forward would have been taken.

Safety professionals are encouraged to make internal evaluations of the quality of incident investigation to establish a database from which improvements can be proposed. But they should be aware that such evaluations often indicate that culture problems exist. They may find that it had become accepted practice for supervisors, management personnel above the supervisor level, and safety professionals to sign-off on shallow investigation reports, indicating that they were acceptable.

In my studies I observed that safety professionals would better serve their clients' interests if they:

- Viewed incident investigation as a potential source for selecting the improvements that should be made in safety management systems. Because—if incident investigation is done well, the reality of the technical and organizational

Advanced Safety Management: Focusing on Z10 and Serious Injury Prevention,
Second Edition. Fred A. Manuele.
© 2014 John Wiley & Sons, Inc. Published 2014 by John Wiley & Sons, Inc.

methods of operation and cultural causal factors for incidents and exposures that result in serious injuries and illnesses will be revealed.

- Sought to have incident investigation given a much higher place within all of the elements of a safety management system. Because—the quality of incident investigation is a principal marker in evaluating an organization's safety culture.

To provide an information base for safety professionals who choose to promote improvement in the incident investigation process, in this chapter we:

- Discuss the incident investigation provisions in Z10
- Comment on the cultural difficulties facing safety professionals who try to have incident investigations improved if an organization has condoned a low quality of incident investigation
- Promote having compassion for supervisors
- Explain why supervisors who complete incident investigations may be reluctant to record the reality of contributing and causal factors
- Discuss why supervisors may not be adequately qualified to make thorough incident investigations
- Suggest studies of needs and opportunities and courses of action for improvement
- Comment on incident investigation forms
- Promote use of the Five Why System
- Make observations on selected resources

INCIDENT INVESTIGATION PROVISIONS IN Z10

The requirements for incident investigation are set forth concisely in Section 6.2 of Z10. They are contained in one paragraph with no subsections. To fulfill the standard's requirements:

> Organizations shall establish processes to report, investigate and analyze incidents in order to address occupational health and safety management system non-conformances and other factors that may be causing or contributing to the occurrence of incidents. The investigations shall be performed in a timely manner.

That is the whole of it—one brief paragraph on incident investigation sets forth the requirements for this very important subject. It might seem that this safety management process is dealt with too briefly. On the other hand, within an ANSI management system standard, all that needs to be said is said. Advisory comments on incident investigation are more extensive. They say that:

- Incidents should be viewed as possible symptoms of problems in the occupational health and safety management system.

- The goal is to identify and correct hazards and system deficiencies before incidents occur.
- Experience shows that incident investigations should be begun as soon as practical.
- Incident investigations should be used for root-cause analysis to identify system or other deficiencies for developing and implementing corrective action plans.
- Lessons learned from investigations are to be fed back into the planning and corrective action processes.

POSSIBLE EXPLANATIONS FOR INCIDENT INVESTIGATIONS BEING DONE POORLY

In my studies of incident investigation reports, on a scale of 10, with 10 being best, companies were given scores ranging from 2 to 8, with an average of 5.7, and that average could be a stretch. These relatively poor scores were troubling and prompted inquiry into situations that exist in the investigation process that might be barriers to in-depth determinations of contributing factors and why this important safety management function is often done superficially.

Typically, first-line supervisors are given the responsibility to initiate an incident investigation report. It is presumed that they are closest to the work and they know more of the details regarding what has occurred.

I ran a Five Why exercise to examine the incident investigation process to try to determine why there was such a huge gap between procedures adopted regarding incident investigation and what actually takes place. As the Five Why exercise proceeded, it became apparent that our model is flawed on several counts.

THE POSITION IN WHICH SUPERVISORS ARE PLACED

When supervisors complete incident investigations, they are being asked to write performance reviews on themselves and on the people to whom they report—all the way up to the board of directors.

It is understandable that supervisors would tend to avoid expounding on their own shortcomings. The supervisor probably would not write:

This accident occurred in my area of supervision and I take full responsibility for it. I overlooked .... I should have done .... My boss did not forward the work order for repairs I sent him two months ago ....

It is not surprising that supervisors would be reluctant to write about shortcomings in the management systems for which the people to whom they report are responsible. If supervisors do write about such systems shortcomings, adverse personnel relationships could result.

With respect to operators (first-line employees) and incident causation, James Reason wrote in *Human Error* that:

> Rather than being the main instigator of an accident, operators tend to be the inheritors of system defects created by poor design, incorrect installation, faulty maintenance and bad management decisions. Their part is usually that of adding the final garnish to a lethal brew whose ingredients have already been long in the cooking." (p. 173)

Supervisors are one step above line employees. They also work in a "lethal brew whose ingredients have already been long in the cooking." They have little or no influence on the original design of operations and work systems and are hampered in being able to have major changes made in them.

In some organizations, the procedure is to have a team of two or three investigate certain accidents, such as all OSHA recordables. Then, if the team consists of fellow supervisors, the team is expected to write a performance appraisal on the supervisor for the area in which the accident occurred, and that supervisor's boss, and her boss—all the way up to the board of directors.

It is not difficult to understand that supervisors would be averse to criticizing another supervisor and management personnel above the supervisor's level. At every level of management above the line supervisor, it would also be normal to try to avoid being factually critical of themselves. Self-preservation dominates at all levels.

CULTURAL IMPLICATIONS THAT ENCOURAGE GOOD INCIDENT INVESTIGATIONS

In some organizations, senior management insists on being informed of the factors contributing to an accident and that reports be factual. An example follows.

In a company where management is fact based and sincere when they say that they want to know about the contributing factors for accidents, regardless of where the responsibility lies, a special investigation procedure is in place for serious injuries and fatalities.

Management recognized that it was difficult for leaders at all levels to complete factual investigation reports that may be critical of themselves. Thus, an independent facilitator serves as the investigation and discussion team leader. At least five knowledgeable people serve on the team. All members of the team know that a factual report is expected.

It is known that the CEO reads the reports, asks questions to assure that they are complete, and sees that the people who report to him or her resolve to a proper conclusion all of the recommendations made. Thus, the CEO demonstrates by his or her actions that the organization's culture requires fact determination and continual improvement.

In some companies, incident investigation is done well because the safety culture will not tolerate other than superior performance. In the studies made of the quality of investigations, some companies scored 8 out of a possible 10. (More than one safety director accused me of being a hard marker.)

In those companies, the positive safety culture is driven by the senior executives, and in some instances, by the board of directors. At those levels, incident experience is reviewed and personnel are held accountable for results.

An example of the absence of an interest in safety by executives and the board of directors and how that absence was turned into positive and active leadership can be found in the article "Building a Better Safety Vehicle: Leadership-Driven Culture Change at General Motors" published in the January 2005 issue of *Professional Safety*. The authors were Patrick Frazee and Steven Simon. Excerpts follow.

> Safety culture change at GM was driven from the top and realized through the commitment and engagement of the leadership at every level. What follows is the story of how this was accomplished. Paul O'Neill, chair at Alcoa, joined the GM board of directors in 1993. His commitment to worker safety was key to the dramatic turnaround at Alcoa, where he not only improved safety, but also generated quantifiable bottom-line results. So perhaps GM's directors should not have been surprised when, as they prepared to adjourn the first board meeting O'Neill attended, he asked, "Where's the safety report?" There was none. O'Neill's question—and its exposure of the status of safety at the company—would become a watershed in GM's history. The President's Council … decided to meet the challenge and take a close look at GM's safety performance and do whatever was necessary to improve it.

What interpretation can be given to the foregoing? For this important aspect of safety management—incident investigation—senior executive direction and involvement are needed to drive improvement. In every company with which I am familiar, that has achieved stellar safety results, incident experience is reviewed regularly at the chief executive officer level.

To conclude on this subject: It is strongly recommended that if practicable, selected categories of incidents be investigated by teams consisting mostly of personnel not directly related to the area in which the accident occurred.

In my studies of the quality of investigations, reports prepared by well-chosen teams that were encouraged to be factual got the highest scores. That's an idea that safety professionals could promote.

CULTURAL IMPLICATIONS THAT MAY IMPEDE GOOD INCIDENT INVESTIGATIONS

Throughout this book the significance of an organization's culture and how it affects safety-related decision making favorably or unfavorably has been emphasized. There is a relative and all-too-truthful paragraph in *Guidelines for Preventing Human Error in Process Safety* where comments are made on the "Cultural Aspects of Data Collection System Design."

A company's culture can make or break even a well-designed data collection system. Essential requirements are minimal use of blame, freedom from fear of reprisals, and feedback which indicates that the information being generated is being used to make changes that will be beneficial to everybody. All three factors are vital for the success of a data collection system and are all, to a certain extent, under the control of management. (p. 259)

In relation to the foregoing, the title of R. B. Whittingham's book, *The Blame Machine: Why Human Error Causes Accidents*, is particularly appropriate. Whittingham says that his research shows that in some organizations, a "blame culture" has evolved whereby the focus in their investigations is on individual human error and that the corrective action stops at that level. That avoids seeking data on and improving the management systems that may have enabled the human error.

What Whittingham wrote is indicative of an inadequate safety culture. As an example of one aspect of a negative safety culture, consider the following scenario. It represents a culture of fear.

An electrocution occurred. As required in that organization, the corporate safety director visited the location to expand on the investigation. During discussion with the deceased employee's immediate supervisor, it became apparent that the supervisor knew of the design shortcomings in the lockout/tagout system, of which there were many at the location.

When asked why the design shortcomings were not recorded as causal factors in the investigation report, the supervisor's response was: "Are you crazy? I would get fired if I did that. Correcting all these lockout/tagout problems will cost money and my boss doesn't want to hear about things like that."

This culture of fear arose from the system of expected performance that management created. The supervisor completed the investigative report in accord with what he believed management expected. He recorded the causal factor as "employee failed to follow the lockout/tagout procedure," and the investigation stopped there.

Overcoming such a culture of fear in the process of improving incident investigation processes, wherever and to what extent it exists, will require careful analysis and much persuasive diplomacy.

Whittingham wrote: "Organizations, and sometimes whole industries, become unwilling to look too closely at the system faults which caused the error. Instead the attention is focused on the individual who made the error and blame is brought into the equation." (p. xii)

Actions necessary to remove error-provocative system faults can be taken only by management. For an incident investigation system to be effective, management must demonstrate by what it does that it wants to know what the contributing causal factors are.

Assume that the safety culture does not require effective incident investigations. Consider the following examples, limited to ten, of statements that could legitimately be made in investigation reports but may be perceived as self-incriminating. More important, they could be interpreted as being accusatory of management levels above the supervisor—and thus avoided.

1. We did not take the time to train this employee because we were too busy.
2. The injured employee mentioned the hazard to me, but it was the kind of thing that we have tolerated.
3. We have had work orders in maintenance for three months to fix the wiring on this equipment.
4. We did not do pre-job planning, and hazardous situations came up that we did not expect.
5. It was the kind of a rush situation that often happens, and we understand that sometimes the workers take shortcuts and don't follow the SOPs.
6. The work is overly stressful and risky. Hazards were not properly considered in the design process.
7. The equipment is being run beyond its normal life cycle, and the risks in operating it are high.
8. What we are asking our people to do is exhausting, and they make mistakes.
9. We haven't had time to write a Standard Operating Procedure for this job.
10. The stuff that purchasing bought is cheaply made, and it falls apart.

Not doing thorough incident investigations is normal in organizations where management is not fact based and does not promote and require that hazard and risk problems be identified and acted upon. Situations of that sort define safety culture problems.

If safety professionals promote improving the quality of incident investigation, the reality of the safety culture in place must be evaluated and defined accurately as an action plan for improvement is being formulated.

HAVING COMPASSION FOR SUPERVISORS

In a huge number of published investigation procedures, it is said that the first-line supervisor is best qualified to complete incident investigations because he or she is closest to the work and knows the most about the hazards and risks. That premise needs rethinking.

A safety professional should ask: How much training with respect to hazards and risks do supervisors get, does the training make them knowledgeably and technically qualified, and how often is training provided?

I doubt that supervisors become exceptionally knowledgeable about hazards and risks and thus well qualified to make good incident investigations after taking a one- or two-day course on incident investigation.

A question that logically follows pertains to all personnel at all levels who do incident investigations. How often do they complete incident investigations, and do the forms and procedure manuals provide adequate support? It is unusual for a supervisor to complete two or three incident investigations in a year. That may also be the case for members of teams that are given the responsibility to investigate accidents.

Also, consideration needs to be given to the time lapse between when supervisors and others attend a training session and when they complete an incident investigation report. It is generally accepted that knowledge obtained in a training session will not be retained without frequent use.

Supervisors, and others, should be provided with readily available reminder references, the content of which should be comparable to a combination of the two addenda to this chapter.

HOW SERIOUS CAN THE PROBLEM BE?

What follows is an extraction from material sent to me recently by a colleague who has had extensive experience with incident investigation and who tries to influence managements to examine and improve their systems.

This colleague made a presentation on incident investigation to a leadership group for one of the largest manufacturers in the world. The group consisted of about 150 to 175 plant managers, personnel directors, union presidents, and union bargaining chairs.

Participants were asked to choose as many of the subjects in Table 22.1 that they believed could be contributing factors for the occurrence of accidents in the operations in which they were involved.

Subjects from 1 to 7 received positive scores at varying levels, indicating that they could be contributing factors in accidents. Then the group was asked to record—How often these subjects appear in incident investigation reports? Results are shown in Table 22.2.

TABLE 22.1

1. Safety culture
2. Lack of employee participation
3. Inadequate hazard identification
4. Inadequate hazard controls
5. Inadequate management of change
6. Lack of preventive maintenance
7. Inadequate risk sensitivity
8. Don't know

TABLE 22.2

1. Very often	7%
2. Often	22%
3. Seldom	44%
4. Never	20%
5. Don't know	7%

For emphasis: The audience consisted of managers and other upper-level personnel. Forty-four percent indicated that **Seldom** would the contributing factors appear in investigation reports. Twenty percent recorded **Never**. Is this broadly descriptive of reality?

Obviously, the performance level expected by management for incident investigations is very low. That is embedded in the organization's culture. And improvement will not be made until management upgrades its expectations and provides the necessary leadership and accountability to achieve the required culture change.

A few years ago, I had an involvement with that company and studied its incident investigation reports. The culture problem was obvious at that time: Management condoned very shallow incident investigations—and it was determined that the same applies in 2013.

ON THE WAY TO IMPROVEMENT, START WITH A SELF-EVALUATION OF THE CULTURE

Safety professionals who undertake to improve the quality of incident investigation should begin with the first step in the Plan-Do-Check-Act process—define the problem. They should begin with an evaluation of a sampling of completed incident investigation reports. In my studies the identification entries in incident investigation forms—such as name, department, location of the accident, shift, time, occupation, age, time in the job—got relatively high scores for thoroughness of completion.

Thus, it is suggested that the evaluation concentrate on the incident descriptions, causal and contributing factor determination, and the corrective actions taken. Considering efficient time usage, a safety professional may want to have the evaluation include only incidents resulting in serious injury or illness, and incidents and near misses that could have had serious results under slightly different circumstances.

In chapter 3, "Innovations in Serious Injury and Fatality Prevention", an outline for such a study was presented under the heading "Proposing a Study of Serious Injuries." Such a study will not be time consuming since the data to be collected and analyzed should already exist or can easily be obtained. To assist in the study, two addenda to this chapter are provided. Both are taken from the fourth edition of *On the Practice of Safety*. Addendum A describes a sociotechnical causation model for hazards-related incidents, and Addendum B is a reference for causal factors and corrective actions. Another good reference for this evaluation is Chapter 4 "Human Error Avoidance and Reduction" in this book because of its comments on human errors that may be made above the worker level.

A safety professional who undertakes such a study should keep in mind that its outcome is to be an analysis of:

- Activities in which serious injuries occur, for which concentrated prevention efforts will be beneficial
- The quality of causal factor determination and corrective action taking

- The culture that has been established over time with respect to good or not so good causal factor determination and corrective action taking
- Organizational levels that are to be influenced if improvements are to be made

From that analysis, a plan of action would be drafted to favorably influence the system of performance expected. The organization's safety culture with respect to the quality of incident investigation will not be changed without support from senior management.

So, the plan of action must be well crafted to convince management of the value of making the changes proposed: avoiding injuries to employees, good business practice, cost reduction, waste reduction (lean), personnel relations, and fulfilling community responsibility.

It is much, much easier to write all this than it will be for safety professionals to get it done. Culture changes are not accomplished easily. They require considerable time and patience to achieve small steps forward.

AN INTERESTING OBSERVATION

In the previous section it was said, among other things, that studies undertaken should identify activities in which serious injuries occur for which concentrated prevention efforts will be beneficial. An example of such an initiative follows.

In certain studies, safety directors in manufacturing operations were asked to have computer runs made of worker compensation cases valued at $25,000 or more. For those studies, safety directors agreed that employees in their organizations who had the following job titles were ancillary and support personnel. They did not work on the production line producing product.

millwright, welder, carpenter	maintenance, painter, electrician, machinist
service technician, accounting	administrator, salesperson, janitor, driver
service engineer, storekeeper	warehouse, shipping

Percentages of serious injuries that occurred to nonproduction personnel—that is, to ancillary and support personnel—in six companies were 78%, 72%, 72%, 67%, 63%, and 61%. All of those companies are large, have good safety management systems, and the statistical base is large enough to have some credibility.

There were some noted exceptions. In other manufacturing organizations that had OSHA rates higher than industry averages, the percentage of serious injuries that occurred to line employees was much higher. Also, if an organization had a division that was more inherently hazardous than manufacturing, such as mining, that operation skewed the statistics.

Data such as in the foregoing prompt the observation that the type of analyses suggested in the Z10 advisory column for incident investigation could produce meaningful results from which proposals can be made on where activities could profitably be focused.

OTHER SUBJECTS TO BE REVIEWED

As a part of the improvement endeavor, other evaluations should be made, such as what is being taught about incident investigation, what guidance is given in procedure manuals, and whether the content and structure of the incident investigation form assist or hinder thorough investigations.

The following define real-world situations as discovered in the studies made. Consider how such situations would affect the quality of investigations.

- In courses taught on incident investigation, the instructor leads attendees to conclude that 80 to 90% of accidents are caused principally by the unsafe acts of workers.
- Instruction is plainly given that the corrective actions proposed in incident investigation forms should focus on improving worker behavior.
- In the incident investigation procedure manual the same thought is conveyed and little guidance is given on causal factors at levels above the worker.
- After a description of an incident is recorded, the first instruction in an incident investigation form is identify the unsafe act committed by the worker.
- Instructions for a computer-based data-entry system with respect to accidents say: "Enter the unsafe act code." The system allows the entry of only one causal factor code.

When there is a lack of understanding of the fundamentals of incident causation and there is no need to identify contributing causal factors, supervisors, upper levels of management, and safety professionals put their signatures on forms, indicating approval, when the reality is that investigations are shallow and of little value. Making the additional reviews proposed in this chapter will help a safety professional define the extent of the problem and assist in crafting a course of action for improvement.

INCIDENT INVESTIGATION FORMS

Appendix K in Z10 provides a brief dissertation on the value and outcome of an incident investigation—to prevent similar incidents from occurring—and a sample investigation form. This form presents a good basic outline, and its content can serve as a resource from which to craft an incident investigation form particularly suited to an organization's operations. It has several positive characteristics that safety professionals should consider as they draft or revise investigation forms.

1. No entry is required that would lead an investigator to focus on what a worker did or did not do, to the exclusion of other causal factors.
2. Provision is made to enter observations concerning the incident at three levels: supervisor, witnesses, and employees with insight.

3. Assistance is given in determining possible causal factors by listing major categories: Equipment; Tools; Environment; Procedure; and Personnel.

4. Recommended corrective actions are to be listed along with the originator's name. A provision is made to enter "Accepted/Rejected." Actions or rationale and completion dates are to be recorded.

5. A sign-off is required by the "Responsible/Approving Department Manager/ Process Owner."

6. The report ends with the investigator's signature and recordings of the persons to whom the report is sent.

THE FIVE-WHY TECHNIQUE

As incident investigation procedures are improved, the goal is to have contributing causal factors determined and acted upon properly. As a beginning step where the incident investigation system needs much improvement, it is suggested that a problem-solving process be considered for which the training and administrative requirements are not extensive; that is the five-why technique.

Highly skilled incident investigators may say that the five-why process is inadequate because it does not promote the identification of causal factors resulting from decisions made at a senior executive level. That is not necessarily so. Usually, when inquiry gets to the fourth "why," considerations are at the management levels above the supervisor and may consider decisions made by the board of directors.

For many organizations, achieving competence in applying the five-why technique for incident investigations will be a major step forward.

The origin of the five-why process is attributed to Taiichi Ohno while he was at Toyota. He developed and promoted a practice of asking "why" five times to determine what caused a problem so that root causal factors could be identified and effective countermeasures could be implemented. The five-why process is applied in a large number of settings for a huge variety of problems.

Since the premise on which the five-why concept is based is uncomplicated, it can easily be adopted in the incident investigation process, as some safety professionals have done. For the complex incident situation occasionally encountered, starting with the five-why system may lead to the use of event trees or fishbone diagrams or more sophisticated investigation systems.

Given an incident description, the investigator would ask "why" five times to get to the contributing causal factors and outline the necessary corrective actions. A not-overly-complex example follows.

The written incident description says that a tool-carrying wheeled cart tipped over on to an employee while she was trying to move it. She was seriously injured.

1. Why did the cart tip over? The diameter of the casters is too small and the carts are tippy.

2. Why is the diameter of the casters too small? They were made that way in the fabrication shop.

3. Why did the fabrication shop make carts with casters that are too small? They followed the dimensions given to them by engineering.

4. Why did engineering give fabrication dimensions for casters that have been proven to be too small? Engineering did not consider the hazards and risks that would result from using small casters.

5. Why did engineering not consider those hazards and risks? It never occurred to the designer that use of the small casters would create hazardous situations.

Conclusion: I [the department manager] have made engineering aware of the design problem. In that process, an educational discussion took place with respect to the need to focus on hazards and risks in the design process. Also, engineering was asked to study the matter and has given new design parameters to fabrication: The caster diameter is to be tripled. On a high-priority basis, fabrication is to replace all casters on similar carts. A 30-day completion date for that work was set.

I have also alerted supervisors to the problem in areas where carts of that design are used. They have been advised to gather all personnel who use the carts and instruct them that larger casters are being placed on tool carts and that until that is done, moving the carts is to be a two-person effort. I have asked our safety director to alert her associates at other locations of this situation and how we are handling it.

Sometimes, asking "why" as few as three times gets to the root of a problem: on other occasions, may be necessary to ask "why" six times. Having analyzed incident reports in which the five-why system was used, these precautions are offered:

- Management commitment to identifying the reality of causal factors is an absolute necessity.
- Take care that the first "why" is really a "why" and not a "what" or a diversionary symptom.
- Expect that repetition of five-why exercises will be necessary to get the idea across: doing so in group meetings at several levels, but particularly at the management level, is a good idea.
- Be sure that management is prepared to act on the systemic causal factors identified as skill is developed in applying the five-why process, particularly those that arise from human errors made above the worker level.

WHAT THIS CHAPTER IS NOT

Since the literature giving guidance on incident investigation techniques is especially abundant, comments are not being made here on such items as investigation criteria, immediate actions to be taken, fact determination, objectivity, interviewing witnesses, developing incident investigation teams, or action plans. The chapter "Designer Incident Investigation" in the 4th edition of *On the Practice of Safety* gives a detailed review of the methodology. Similar incident investigation procedure outlines appear in the resources described below.

For safety professionals who choose to be educated on more sophisticated incident investigation methods, the following resources provide information on barrier analysis, change analysis, event tree analyses, failure mode and effects analysis, and fishbone (Ishikawa) diagrams.

INCIDENT INVESTIGATION RESOURCES

Since the names of the authors and publishers for each of the resources listed here are shown, as well as the websites for some of them, they are not repeated in the reference list at the end of this chapter. The first two resources listed are highly recommended for their content. Also, they are well worth the price. They are available on the Internet and can be downloaded at no cost.

"Root Cause Analysis Guidance Document, DOE-NE-STD-1004-92." Washington, DC: Us Department of Energy, 1992. Also at http://us.yhs4.search.yahoo.com/yhs/search?p=DOE-NE-STD-1004-92&hspart=att&hsimp=yhs-att_001&type=att_lego_portal_home.

> This is a 69-page highly informative document. It is an instructive read. Various incident investigation techniques are discussed in an "Overview of Occurrence Investigation." Thus, it is a resource on events and causal factor analysis, change analysis, barrier analysis, the management oversight and risk tree (MORT) analytical logic diagram, human performance evaluation, and Kepner–Tregoe problem solving and decision making.

NRI MORT User's Manual, NRI-I (2002), a Generic Edition For Use with the Management Oversight and Risk Tree Analytical Logic Diagram, is published by The Noordwijk Risk Initiative Foundation in The Netherlands.

> In a discussion as to "What is MORT," these comments are made: "By virtue of public domain documentation, MORT has spawned several variants, many of them translations of the *MORT User's Manual* into other languages. The durability of MORT is a testament to its construction and content; it is a highly logical expression of the functions required for an organization to manage risks effectively." They say that this 2002 version of the *MORT User's Manual* aims to:

- Rephrase the questions in British English
- Improve guidance on the investigative application of MORT
- Restore "freshness" to the 1992 MORT question set
- Simplify the system of transfers in the chart

- Remove DOE-specific references
- Help users tailor the question set to their own organizations

I believe they accomplished their purposes—to improve guidance on the investigation application, to restore "freshness," and to simplify the system. What they have done is fascinating. The 69-page document is available on the Internet at http://www.nri.eu.com/NRI1.pdf#search='NRI%20Mort%20User%27s%20Manual'.

I recommend that safety professionals who want to identify the reality of causal factors acquire an understanding of the thinking on which MORT is based.

A review of the aforementioned documents will provide an inexpensive and valuable education. Now, to extend the resource list, three books on incident investigation and causal factor identification and analysis are listed. There are other resources besides these.

- *Guidelines for Preventing Human Error in Process Safety*. New York: Center for Chemical Process Safety of the American Institute of Chemical Engineers, 1994.

  This is a highly recommended text. Chapter 6 deals with data collection and incident analysis methods. Elsewhere, comments are made on types of human error causal factors, their nature, and how to identify and analyze them.

- Hendrick, Kingsley and Ludwig Benner, Jr. *Investigating Accidents With STEP*. New York: Marcel Dekker, 1987.

  Hendrick and Benner have developed an incident investigation system called "Sequentially Timed Events Plotting." Several authors refer to this thought-provoking system. This book is devoted entirely to the STEP system.

- Oakley, Jeffrey. *Accident Investigation Techniques: Basic Theories, Analytical Methods and Applications*. Des Plaines, IL: American Society of Safety Engineers, 2003.

  This is a relatively short and inexpensive book in which comments are made on the incident investigation process generally, and on several investigation and analytical techniques, such as events and causal factors analysis, change analysis, tree analysis, and specialized computerized techniques.

- Although an Internet search will reveal a large number of companies offering consulting services on root causal factor analysis, I am listing two that have published books on the subject and which have a known history with respect to occupational safety and health.

  Gano, Dean L. *Apollo Root Cause Analysis: A New Way of Thinking*. Portland, OR: Apollonian Publications, 1999. Dean Gano has been a consultant in root-cause analysis for many years. His technique and his literature are well regarded.

TapRoot Manual. Knoxville, TN: System Improvements, Inc. This book describes the root-cause identifying and analysis methods developed by the staff at Systems Improvements. This company has also offered consulting services on root-cause analysis for several years.

CONCLUSION

My studies on incident investigation prompt the conclusion that significant risk reduction can be achieved if investigations are done well. If incident investigations are thorough, the reality of the technical, organizational, methods of operation, and cultural causal factors will be revealed.

If safety professionals want to select leading indicators for safety management system improvement, they would have a good data source for that purpose if incident investigation reports identify causal factors realistically. I now believe that the quality of incident investigation is one of the principal markers in evaluating an organization's safety culture.

Assume that a safety professional decides to take action to improve the quality of incident investigation. It is proposed that the following comments about incident investigation as excerpted from the August 2003 "Report of the *Columbia* Accident Investigation Board" be kept in mind as a base for reflection throughout the endeavor. The report pertains to the *Columbia* spaceship disaster. It is accessed at http://www.nasa.gov/columbia/home/CAIB_Vol1.html.

Many accident investigations do not go far enough. They identify the technical cause of the accident, and then connect it to a variant of "operator error." But this is seldom the entire issue.

When the determinations of the causal chain are limited to the technical flaw and individual failure, typically the actions taken to prevent a similar event in the future are also limited: fix the technical problem and replace or retrain the individual responsible. Putting these corrections in place leads to another mistake—the belief that the problem is solved.

Too often, accident investigations blame a failure only on the last step in a complex process, when a more comprehensive understanding of that process could reveal that earlier steps might be equally or even more culpable. In this Board's opinion, unless the technical, organizational, and cultural recommendations made in this report are implemented, little will have been accomplished to lessen the chance that another accident will follow. (Vol. 1, Chap. 7, p. 177)

For emphasis, I paraphrase: If the cultural, technical, organizational, and methods of operation causal factors are not identified, analyzed, and resolved, little will be done to prevent recurrence of similar incidents.

REFERENCES

ANSI/AIHA Z10-2012. American National Standard, *Occupational Health and Safety Management Systems*. Fairfax, VA: American Industrial Hygiene Association, 2012. ASSE is now the secretariat. Available at https://www.asse.org/cartpage.php?link=z10_2005.

Guidelines for Preventing Human Error in Process Safety. New York: Center for Chemical Process Safety of the American Institute of Chemical Engineers, 1994.

Manuele, Fred A. *On the Practice of Safety*, 4th ed. Hoboken, NH: Wiley, 2013.

NASA. *Columbia Accident Investigation Report, Vol. 1, Chap. 7*. Washington, DC: NASA, 2003. http://www.nasa.gov/columbia/home/CAIB_Vol1.html.

Reason, James. *Human Error*. New York: Cambridge University Press, 1990.

Simon, Steven I., and Patrick R. Frazee. "Building a Better Safety Vehicle: Leadership-Driven Culture Change at General Motors." *Professional Safety*, Jan. 2005.

The Five-Why System: http://www.mapwright.com.au/newsletter/fivewhys.pdf and http://www.moresteam.com/toolbox/5-why-analysis.cfm.

Whittingham, R. B. *The Blame Machine: Why Human Error Causes Accidents*. Burlington, MA: Elsevier–Butterworth–Heinemann, 2004.

ADDENDUM A

A DEPICTION OF A SOCIO-TECHNICAL CAUSATION MODEL FOR HAZARDS-RELATED INCIDENTS IS PRESENTED ON THE FOLLOWING PAGE

Advanced Safety Management: Focusing on Z10 and Serious Injury Prevention,
Second Edition. Fred A. Manuele.
© 2014 John Wiley & Sons, Inc. Published 2014 by John Wiley & Sons, Inc.

A socio-technical causation model for hazards-related incidents

An organization's culture
is established by the board of directors and senior management

Management commitment or non-commitment to providing the controls
necessary to achieve and maintain acceptable risk levels is an expression of the culture

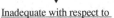

Causal factors may derive from shortcomings in controls when safety policies,
standards, procedures, the accountability system, or their implementation, are

Inadequate with respect to

↓

The design process and operational risk management
and the inadequacies impact negatively on

↓

- Providing resources

- Risk assessment

- Competency and adequacy of staff

- Maintenance for system integrity

- Management of change / pre-job planning

- Procurement – safety specifications

- Risk-related systems

    - Organization of work

    - Training – motivation

    - Employee participation

    - Information — communication

    - Permits

    - Inspections

    - Incident investigation and analysis

    - Providing personal protective equipment

- Third party services

- Emergency planning and management

- Conformance / compliance assurance

- Performance measures

- Management reviews for continual improvement

Multiple causal factors derive from the inadequate controls

The incident process begins with an initiating event.
There are unwanted energy flows or exposures to harmful substances.
Multiple interacting events occur sequentially or in parallel.

Harm or damage results, or could have resulted in slightly different circumstances.

A REFERENCE FOR THE SELECTION OF CAUSAL FACTORS AND CORRECTIVE ACTIONS FOR INCIDENT INVESTIGATION PROCEDURES AND REPORTS

Designers of incident investigation systems should understand that the causal factors and corrective actions included within investigation forms or as separate informational documents must be appropriate for the operations being conducted. The material presented here should not be used without modification to suit needs.

Workplace Design Considerations

1. Hazards derive from basic design of facilities, hardware, equipment, or tooling.
2. Hazardous materials need attention.
3. Layout or position of hardware or equipment presents hazards.
4. Environmental factors (heat, cold, noise, lighting, vibration, ventilation, etc.) presented hazards.
5. Work space for operation, maintenance, or storage is insufficient.
6. Accessibility for maintenance work is hazardous.

Work Method Considerations

1. Work methods are overly stressful.
2. Work methods are error-provocative.
3. Job is overly difficult, unpleasant, or dangerous.
4. Job requires performance beyond what an operator can deliver.
5. Job induces fatigue.

Advanced Safety Management: Focusing on Z10 and Serious Injury Prevention,
Second Edition. Fred A. Manuele.
© 2014 John Wiley & Sons, Inc. Published 2014 by John Wiley & Sons, Inc.

6. Immediate work situation encouraged riskier actions than prescribed work methods.
7. Work flow is hazardous.
8. Positioning of employees in relation to equipment and materials is hazardous.

Job Procedure Particulars

1. No written or known job procedure.
2. Job procedures existed but did not address the hazards.
3. Job procedures existed but employees did not know of them.
4. Employee knew job procedures but deviated from them.
5. Deviation from job procedure not observed by supervision.
6. Employee not capable of doing this job (physically, work habits, or behaviorally).
7. Correct equipment, tools, or materials were not used.
8. Proper equipment, tools, or materials were not available.
9. Employee did not know where to obtain proper equipment, tools, or materials.
10. Employee used substitute equipment, tools, or materials.
11. Defective or worn-out tools were used.

Hazardous Conditions

1. Hazardous condition had not been recognized.
2. Hazardous condition was recognized but not reported.
3. Hazardous condition was reported but not corrected.
4. Hazardous condition was recognized but employees were not informed of the appropriate interim job procedure.

Personal Protective Equipment

1. Proper personal protective equipment (PPE) not specified for job.
2. PPE specified for job but not available.
3. PPE specified for job, but employee did not know requirements.
4. PPE specified for job, but employees did not know how to use or maintain.
5. PPE not used properly.
6. PPE inadequate.

Management and Supervisory Aspects

1. General inspection program is ineffective.
2. Inspection procedure did not detect the hazards.
3. Training as respect identified hazards not provided, inadequate, or didn't take.
4. Maintenance with respect to identified hazards was inadequate.

5. Review not made of hazards and right methods before commencing work for a job done infrequently.

6. This job requires a job hazard/task/ergonomics analysis.

7. Supervisory responsibility and accountability not defined or understood.

8. Supervisors not adequately trained for assigned safety responsibility.

9. Emergency equipment not specified, not readily available, not used, or did not function properly.

Corrective Actions To Be Considered

1. Job study to be recommended: job hazard/task/ergonomics analysis needed.

2. Work methods to be revised to make them more compatible with worker capabilities and limitations.

3. Job procedures to be changed to reduce risk.

4. Changes are to be proposed in work space, equipment location, or workflow.

5. Improvement is to be recommended for environmental conditions.

6. Proper tools to be provided along with information on obtaining them and their use.

7. Instruction to be given on the hazards of using improper or defective tools.

8. Job procedure to be written or amended.

9. Additional training to be given concerning hazard avoidance on this job.

10. Necessary employee counseling will be provided.

11. Disciplinary actions deemed necessary and will be taken.

12. Action is to be recommended with respect to employee who cannot become suited to the work.

13. For infrequently performed jobs, it is to be reinforced that a pre-job review of hazards and procedures is to take place.

14. Particular physical hazards discovered will be eliminated.

15. Improvement in inspection procedures to be initiated or proposed.

16. Maintenance inadequacies are to be addressed.

17. Personal protective equipment shortcomings to be corrected.

CHAPTER 23

AUDIT REQUIREMENTS: SECTION 6.3 OF Z10

Revisions made in the audit provisions in the 2012 version of Z10 have them relate more specifically to the occupational safety and health management systems requirements. They also require that audits be made by competent persons who are independent of the activity being audited. In the following exhibit, words underlined were in the preceding edition and have been removed. Words in bold type are the revisions and additions. Words not underlined or not in bold remain the same.

Section 6.3: Audits

The organization shall establish and implement a process to:

A. **Plan and** conduct periodic audits to determine whether the organization has appropriately applied and effectively implemented OHSMS elements **OHSMS has been established, implemented and maintained in conformance with the requirements of this standard**, including the processes for identifying hazards and controlling risks.

B. **Audits shall be conducted by competent persons who are independent of the activity being audited**.

C. Document and communicate audit results to:

    a. Those responsible for corrective and preventive action;

    b. Area supervision; and

    c. Other affected individuals, including employees and employee representatives.

Advanced Safety Management: Focusing on Z10 and Serious Injury Prevention,
Second Edition. Fred A. Manuele.
© 2014 John Wiley & Sons, Inc. Published 2014 by John Wiley & Sons, Inc.

D. Immediately communicate situations identified in audits that could be expected to cause a fatality, serious injury, or illness in the immediate future, so that prompt corrective action under Section 6.4 is taken.

Having audits made is a part of the Evaluation and Corrective Action processes in Section 6.4. As is the case with every aspect of an organization's endeavors, making a periodic review of progress with respect to stated goals is good business practice. Stated goals, in this instance, would be to have processes in place that meet the requirements of Z10.

Safety audits perform a valuable function in that they determine the effectiveness or ineffectiveness of an organization's safety and health management systems. In accord with the audit requirements in Z10, deficiencies noted during safety audits are to be documented and communicated to those who can take action to eliminate them. The deficiencies are to be prioritized for orderly consideration.

Also, hazardous situations observed during an audit that could be contributing factors for fatalities, serious injuries, or illnesses are to be communicated immediately to the proper decision makers so that actions can be taken on a high-priority basis. That is in concert with one of the principal themes in this book—to improve serious injury prevention.

In the advisory column in Z10 opposite the audit requirements, two particularly important statements are made.

1. The safety audits required are not to be merely "compliance" oriented, meaning that they are not limited to determining compliance with laws, standards, or regulations. Although "compliance" may be considered in the audit process, the intent is to have the audit be "system" oriented so as to evaluate the effectiveness of the standard's management processes.

2. To promote objectivity, audits are to be conducted by persons independent of the activities being audited. But it is made clear that this advisory does not mean that audits must be made by persons "external to the organization."

To assist safety professionals in crafting or re-crafting safety and health audit systems to meet the requirements of Z10, in this chapter we:

- Establishe the purpose of an audit
- Discuss the implications of hazardous situations observed
- Explore management's expectations with respect to audits
- Establish that safety auditors are also being audited during the audit process
- Comment on auditor qualifications
- Discuss the need to have safety and health management system audit guides relate to the hazards and risks in the operations at the location being audited
- Provide information and resources for the development of suitable audit guides

THE PRINCIPAL PURPOSE OF A SAFETY AUDIT: TO IMPROVE THE SAFETY CULTURE

Throughout this book emphasis has been given to the premise that safety is culture driven. Results achieved with respect to safety are a direct reflection of an organization's culture. In their book *Safety Auditing: A Management Tool*, Kase and Wiese stated early in a chapter dealing with successful auditing that:

> Success of a safety auditing program can only be measured in terms of the change it effects on the overall culture of the operation, and enterprise that it audits. (p. 36)

The Kase and Wiese observation can be supported easily. The paramount goal of a safety and health management system audit is to have a beneficial effect on an organization's decision making. Thus, a safety audit report is to serve as a basis for improvement of an organization's safety culture. A safety audit report provides an assessment of the outcomes of the safety-related decisions made by management over the long term. Those outcomes are determined by evaluating the adequacy of what really takes place with respect to the application of existing safety policies, standards, procedures, and operating processes.

SIGNIFICANCE OF OBSERVED HAZARDOUS SITUATIONS

Physical or hazardous situations in operations observed during a safety audit should be viewed principally as indicators of inadequacies in the safety management processes that allowed them to exist. Assume that management takes corrective action to eliminate every hazardous situation noted in an audit report. Still, little will be gained if no change is made in the overall decision making to improve the management systems that allowed the hazardous situations to develop.

REASONABLE MANAGEMENT EXPECTATIONS: THE EXIT INTERVIEW

Safety auditing is an exceptionally valuable process but is time consuming and expensive. Safety professionals should not be surprised if informed managements expect noteworthy results from the audit process that benefit their operations. Safety professionals who make audits should prepare well for the exit interview. That means:

- Having been objective in their evaluations of management systems
- Having good justification for their findings
- Being able to support the prioritizing in management system improvements they propose

In an exit interview with informed management personnel, the auditor or audit team should anticipate and prepare beforehand to respond to questions comparable to the following:

1. What are the most significant risks?
2. What improvements in our management systems do we need to make?
3. In what priority order should I approach what you propose?
4. Are there alternative risk reduction solutions that we can consider?
5. Will you work with me to determine that the actions we take and the money we spend attain sufficient risk reduction?

Audit systems fail if they do not recognize management needs and if they are not looked upon as assisting management in attaining their operational goals. Safety auditors will not be perceived favorably if their work is not considered an asset to managements who seek to improve their safety and health management systems and their safety culture.

Unfortunately, safety auditors cannot absolutely assure managements that every hazard and risk has been identified. Some hazard/risk situations remain obscure, and human beings have not yet developed the perfection necessary to identify all of them. As an example, the negative impact of less-than-adequate decisions affecting design and engineering, purchasing, and maintenance may not be easily observable because their effect may not be felt for several years.

It should be made clear to management that applied safety auditing is based on a sampling technique and that it is patently impossible to identify 100% of the hazard/risk situations and shortcomings in safety management systems.

EVALUATIONS OF AUDITORS BY THOSE AUDITED

Safety professionals should also recognize that the time spent by auditors, the impressions they create, and the time expenditures required of the personnel at the location being audited are also being evaluated. Speculate on the possible comments made upward by location management personnel to executive management for the following situation. It happened.

Four safety and health auditors spent a week making an audit of a 37 employee location. After the second day, employees complained that the auditors were being disruptive because of the amount of their time that the auditors consumed. To make matters worse, during the third day the lead auditor told the location manager that the scorings being given to safety and health management systems by the auditors were higher than usual and that the auditors would have to delve further into operations. Why? Because, the lead auditor said, it was expected that their report would outline management system shortcomings that need attention. Employees at the location became more irritated and complained because of the repetitive, valueless, and duplicatory interferences in their work.

One of the criticisms made by location management pertained to the decision taken by headquarters personnel to send four auditors to their 37 employee location, in light of the fact that maintaining tight cost controls was required by the organization's culture.

AUDITOR COMPETENCY

Throughout Z10 there is an emphasis on identifying, prioritizing, and acting on occupational health and safety management system issues. Those issues are defined in the standard as "hazards, risks, management system deficiencies, and opportunities for improvement." To be able to identify and evaluate those "issues" as they exist in the operation being audited, safety professionals making the audit must have the necessary qualifications and competency.

Managements have often said that auditors had little knowledge of the technical aspects of their hazards and risks and that the audit report was superficial and of little value. If safety audits are to be perceived as having value, the auditors must have the professional qualifications to make them.

Safety professionals who make audits need to consider how well they are prepared for the situation at hand and how best to approach the management personnel in the organization to be audited. Similarly, if persons external to an organization are engaged to make safety audits, the safety professionals who engage them should examine their credentials.

An excerpt from a paper entitled "Auditor Competency for Assessing Occupational Health & Safety Management Systems" will help in that regard. That paper (available on the Internet) was issued jointly by the American Society of Safety Engineers, the American Industrial Hygiene Association, and the American Board of Industrial Hygiene in August 2005. They say that:

> Several studies have raised questions about the value of quality, environmental and occupational health and safety management system certification [audits]. Many of the concerns raised in these studies have focused on the competency of the auditors performing conformity assessment audits.

The foregoing excerpt pertains principally to audits made for management system certifications with respect to quality, environmental, and occupational safety and health management by persons external to an organization. Nevertheless, similar questions have been raised for many years about the value of comparable audits made by in-house personnel.

The paper contains an extensive listing related to specific knowledge and skills of occupational health and safety management system auditors. Only condensed excerpts appear here. They are to serve safety professionals in making a preliminary review of their own capabilities as auditors and in assessing the qualifications of auditors they may engage who are external to their organizations.

Excerpts from the Paper on Auditor Competency

Occupational safety and health management system auditors should have knowledge and skills in occupational health and safety management principles and methods and their application, and related science and technology to enable them to examine occupational health and safety management systems and to generate appropriate audit findings and conclusions. Specific knowledge and skills should include:

- Occupational health and safety management tools (including hazard identification and risk assessment, selection and implementation of appropriate hazard controls, developing proactive and reactive performance measures, understanding techniques to encourage employee participation, and evaluation of work-related accidents and incidents)
- An understanding of the physical, chemical, and biological hazards and other workplace factors affecting human well-being
- The potential interactions of humans, machines, processes, and the work environment
- Methodologies for exposure monitoring and assessment
- Medical surveillance methodologies for monitoring human health and well-being
- Methodologies for accident and incident investigations
- Methodologies used to monitor occupational safety and health performance
- Sector-specific education, experience, and knowledge of operational hazards, risks, processes, products, and services to enable auditors to comprehend and evaluate how the organization's activities, raw materials, production methods and equipment, products, byproducts, and business management systems may impact occupational health and safety performance in the workplace

This list, although abbreviated, provides a good foundation from which a safety and health professional can make a self-evaluation with respect to competency in relation to a particular audit undertaking.

ONE SIZE DOES NOT FIT ALL

Many of the statements made in Dan Petersen's article "What Measures Should We Use, and Why?" concur with my experience. This is what I wrote about his article in *On The Practice of Safety*.

Petersen questions the value of "packaged audits," giving examples of studies that show that audit results did not always correlate to a firm's accident experience. There is a history of that sort of thing with respect to "packaged audits" in which an audit guide is used that may not be sufficiently relative to the actual safety practices and needs in the entity being audited. Petersen concluded that "the self-built audit—one that accurately measures performance of a firm's own safety system— was viewed as the answer." To construct such an audit, Petersen says, a firm must define:

1. Safety system elements
2. The relative importance of each (weighting)
3. Questions to determine what is happening

This is meaty stuff. All elements in a safety management system, while necessary, do not equally impact on those hazards that present the greatest potential for harm, whether measured by incident frequency or severity of injury. Obviously, the safety management elements included, and those emphasized, in an audit system should relate to the hazards that an entity really has to deal with. Keep in mind that hazards include both the characteristics of things and the actions or inactions of people.

To determine what is really happening, an auditor must explore the safety management systems in place, what is expected of them, and which systems are effective or ineffective in controlling an entity's risks. That, in effect, results in a culture appraisal.

In the preceding listing of qualifications for safety and health management system auditors, one of the items pertains to "sector-specific" knowledge of operational hazards, risks, processes, and so on. When drafting an audit guide and in selecting the elements to be emphasized, much should be made of "sector-specific hazards and risks"—meaning those inherent in the operations at the location to be audited.

All hazards are not equal: Neither are the risks deriving from them equal. In a chemical operation where the inherent fire and explosion hazards are significant, the processes in place and their effectiveness with respect to design and engineering, control of fire and explosion potential, occupational health exposures, training, inspection, management of change, and procurement require much greater attention than a warehouse where the only chemicals used are for cleaning purposes. Similarly, provisions to avoid auto accidents are more significant in the operation of a distribution center than if driving by employees is only incidental to operations.

In some organizations the same audit guide is used for all locations and the audit system requires that numerical or alphabetical scorings be recorded for each element being evaluated. The weightings for the elements are the same regardless of their significance at the location being audited. That practice is questionable.

This book emphasizes the prevention of serious injuries. When the audit system requires that identical weightings be given to elements regardless of the nature of the operations being audited, the greater import of a particular management system in the operation may be overlooked. Also, the additional probing into that management system that would be necessary to identify those hazards that may be the causal factors for low-probability/serious-consequence events may be less than adequate.

It seems that greater effectiveness can be achieved if the audit guide is structured so that modifications can be made to suit the hazards and risks at the location being audited. In our communication age, it is appropriate to suggest that safety professionals who craft audit systems consider using a flexible computer-based model in which the descriptive content of the elements to be audited can be abbreviated or expanded and their weightings varied to suit the exposures at individual locations. That would truly be a self-built audit system.

GUIDELINES FOR AN AUDIT SYSTEM

Appendix L gives guidance on how to comply with the standard's audit requirements. The appendix lists all of the sections in Z10 in a table form and includes suggestions on how implementation of them is to be evaluated objectively. Comments concerning adaptation of Appendix L support avoiding the development and implementation of a "packaged" audit system—a one-size-fits-all model. They say that:

> The degree of detail in this table may not be needed for every organization, but may be used as a template that can be modified to match the culture and needs of each organization.

Therefore, modifications are to be made to fit the culture and the inherent hazards and risks in an organization. For example, is it necessary that there be a "documented occupational health and safety policy" as required by Z10 for every location? Or is it appropriate to recognize that for a small operation having as few as ten employees, a verbal and demonstrated commitment by management to achieving superior control of hazards and risks is sufficient?

An example of such a practical adjustment in a safety and health evaluation system can be found in OSHA's "Safety and Health Management Systems eTool—Safety and Health Assessment Worksheet." In OSHA's assessment process, an entry is to be made for this point: *There is a written (or oral, where appropriate) policy.* The implication is that, at times, an orally established safety policy is acceptable.

Appendix L also includes "suggestions for the objective type of evidence that can be used while conducting an OHSMS audit." Suggestions include the types of documents and records to be examined, the titles of the persons to be interviewed, and the activities to be observed. Trying to be objective during a safety and health audit is vital to achieving good results.

The guidance given on objectivity is comparable to the instructions given in the VPP site Worksheet, which is completed to determine whether a location meets the requirements of OSHA's Voluntary Protection Program. OSHA evaluators are to support their conclusions as they evaluate system elements by indicating that they derive from interviews, observations, or documentation.

Since it is proposed that safety professionals not develop a one-size-fits-all audit system, a specifically recommended audit guide to meet Z10 requirements is not being presented in this chapter. Nevertheless, to create or improve an audit guide, it is suggested that the example audit plan in Appendix L combined with the VPP site-based participation evaluation report used by OSHA auditors as they determine whether an organization meets the requirements of its Voluntary Protection Program serve as an excellent reference base. The VPP Site-Based participation evaluation report is available at http://www.osha.gov/dcsp/vpp/vpp_report/site_based.html.

A worksheet is used by OSHA personnel in the evaluation process. Excerpts from the worksheet appear as Addendum A to this chapter. It is an excellent reference for those who want to develop audit guides suited to operations at a particular location. The Worksheet is, in a sense, an audit form.

As shown in Chapter 25, "Comparison: Z10, Other Standards and Guidelines, and VPP Certification", the VPP safety management system requirements are similar to many of the Z10 provisions, but there are also differences. Some of the VPP requirements are not as specific as comparable provisions in Z10. Those provisions are identified below, and references to them in this book are provided.

Subject	Reference
Risk assessment	Chapter 11, "A Primer on Hazard Analysis and Risk Assessment" and Chapter 12, "Provisions for Risk Assessments in Standards and Guidelines"
Design Reviews	Chapter 15, "Safety Design Reviews"
Management of Change	Chapter 19, "Management of Change"
Procurement	Chapter 20, "The Procurement Process"

Similarly, certain VPP requirements are not addressed in Z10 provisions. During a VPP site review, evaluations are made to determine:

- More specifically, the adequacy of the "Occupational Health Care Program and Recordkeeping"

- Whether "Access to experts (for example, Certified Industrial Hygienists, Certified Safety Professionals, Occupational Nurses, or Engineers) is reasonably available to the site, based upon the nature, conditions, complexity, and hazards of the site?"

Safety and health professionals would give appropriate consideration to "the nature, conditions, complexity, and hazards of the site" as stated above to determine whether there was need to include comparable provisions in an audit guide that they draft.

Two other valuable resources that relate closely to the content of the "VPP Site-Based Participation Evaluation Report" and to many of the provisions in Z10 are available on the Internet. Both are OSHA publications. One is the previously mentioned "Safety and Health Management Systems eTool—Safety and Health Assessment Worksheet." Available at http://www.osha.gov/SLTC/etools/safetyhealth/ asmnt_worksheet.html. The other is "The Program Evaluation Profile (PEP)," available at http://www.osha.gov/dsg/topics/safetyhealth/pep.html.

Both of these publications have another feature that will interest some safety and health professionals. They are a resource on scoring systems. If a numerical or alpha scoring system is to be used, the right scoring system is the one with which the auditors and the personnel who review and act on audit reports are comfortable.

CONCLUSION

Auditing performance with respect to established operational goals is good business practice. The audit requirements in Z10 are designed to meet that purpose. Professionally done, safety audits provide valuable information to decision makers who desire to achieve superior safety and health results.

It is suggested that a safety professional who proposes that an organization meet Z10's audit requirements start with a gap analysis. The result would be comparisons between the elements in the safety and health management systems in place and the provisions in Z10. Since the Z10 standard is a state-of-the-art document, it is not surprising that many organizations do not have management systems in place that meet all of its provisions. For a very large percentage of organizations, a gap analysis will reveal shortcomings with respect to design reviews; management of change, risk assessments, a hierarchy of controls, and procurement practices.

After the gap analysis is made, a safety and health professional would assist management in formulating an action plan to fulfill Z10 requirements. As progress is made, the content of audit guides would be adjusted accordingly.

Over time, Z10 will become the benchmark against which the adequacy of occupational safety and health management systems will be measured. Societal expectations of employers with respect to their safety and health management systems will be defined by the standard's provisions. The audit system put in place should assist management in moving closer to compliance with the provisions in Z10.

REFERENCES

ANSI/AIHA Z10-2012. American National Standard, *Occupational Health and Safety Management Systems*. Fairfax, VA: American Industrial Hygiene Association, 2012. ASSE is now the secretariat. Available at https://www.asse.org/cartpage.php?link=z10_2005.

"Auditor Competency for Assessing Occupational Health and Safety Management Systems." Issued jointly by the American Society of Safety Engineers, the American Industrial Hygiene Association, and the American Board of Industrial Hygiene, 2005. Available on the Internet: Enter the title in a search engine.

Kase, Donald W. and Kay J. Wiese. *Safety Auditing: A Management Tool*. New York: Wiley, 1990.

Manuele, Fred A. *On the Practice of Safety*, 4th ed. Hoboken, NJ: Wiley, 2013.

OSHA's VPP Site-Based Participation Evaluation Report. Available at http://www.osha.gov/dcsp/vpp/vpp_report/site_based.html.

Petersen, Dan. "What Measures Should We Use, and Why?" *Professional Safety*, Oct. 1998.

"The Program Evaluation Profile (PEP)." U.S. Department of Labor, OSHA. At. http://www.osha.gov/dsg/topics/safetyhealth/pep.html.

"Safety and Health Management Systems eTool—Safety and Health Assessment Worksheet." U.S. Department of Labor, OSHA, at http://www.osha.gov/SLTC/etools/safetyhealth/form33i.html

ADDENDUM A

OSHA'S VPP SITE-BASED PARTICIPATION SITE WORKSHEET

SECTION I: MANAGEMENT LEADERSHIP AND EMPLOYEE INVOLVEMENT

A. Written Safety and Health Management System

A1. Are all the elements (such as Management Leadership and Employee Involvement, Worksite Analysis, Hazard Prevention and Control, and Safety and Health Training) and sub-elements of a basic safety and health management system part of a signed, written document? (For Federal Agencies, include 29 CFR 1960.) If not, please explain.

A2. Have all VPP elements and sub-elements been in place at least 1 year? If not, please identify those elements that have not been in place for at least 1 year.

A3. Is the written safety and health management system at least minimally effective to address the scope and complexity of worksite hazards? If not, please explain.

A4. Have any VPP documentation requirements been waived (as per FRN, Vol. 74, No. 6, 01/09/09 page 936, IV, and A.4)? If so, please explain.

B. Management Commitment & Leadership

B1. Does management overall demonstrate at least minimally effective, visible leadership with respect to the safety and health management system (as per FRN, Vol. 74, No. 6, 01/09/09 page 936, IV. A.5.a–h)? Provide examples.

Advanced Safety Management: Focusing on Z10 and Serious Injury Prevention,
Second Edition. Fred A. Manuele.
© 2014 John Wiley & Sons, Inc. Published 2014 by John Wiley & Sons, Inc.

B2. How has the site communicated established policies and results-oriented goals and objectives for employee safety to employees?

B3. Do employees understand the goals and objectives for the safety and health management system?

B4. Are the safety and health management system goals and objectives meaningful and attainable? Provide examples supporting the meaningfulness and attainability (or lack thereof if answer is no) of the goal(s). (Attainability can either be unrealistic/realistic goals or poor/good implementation to achieve them.)

B5. How does the site measure its progress towards the safety and health management system goals and objectives? Provide examples.

C. Planning

C1. How does the site integrate planning for safety and health with its overall management planning process (for example, budget development, resource allocation, or training)?

C2. Is safety and health effectively integrated into the site's overall management planning process? If not, please explain.

C3. For site-based construction sites, is safety included in the planning phase of each project?

D. Authority and Line Accountability

D1. Does top management accept ultimate responsibility for safety and health? (Top management acknowledges ultimate responsibility even if some safety and health functions are delegated to others.) If not, please explain.

D2. How is the assignment of authority and responsibility documented and communicated (for example, organization charts, job descriptions, etc.)?

D3. Do the individuals assigned responsibility for safety and health have the authority to ensure that hazards are corrected or necessary changes to the safety and health management system are made? If not, please explain.

D4. How are managers, supervisors, and employees held accountable for meeting their responsibilities for workplace safety and health? (Are annual performance evaluations for managers and supervisors required?)

D5. Are adequate resources (equipment, budget, or experts) dedicated to ensuring workplace safety and health? Provide examples.

D6. Is access to experts (for example, Certified Industrial Hygienists, Certified Safety Professionals, Occupational Nurses, or Engineers), reasonably available, based upon the nature, conditions, complexity, and hazards of the site? If so, under what arrangements and how often are they used?

E. Contract Employees

E1. Does the site utilize contractors? Please explain.

E2. Were there contractors/sub-contractors onsite at the time of the evaluation?

E3.   When selecting onsite contractors/sub-contractors, how does the site evaluate the contractor's safety and health management system and performance (including rates)?

E4.   Are contractors and subcontractors required to maintain an effective safety and health management system and to comply with all applicable OSHA and company safety and health rules and regulations? If not, please explain.

E5.   Does the site's contractor program cover the prompt correction and control of hazards in the event that the contractor/sub-contractor fails to correct or control such hazards? Provide examples.

E6.   How does the site document and communicate oversight, coordination, and enforcement of safety and health expectations to contractors?

E7.   Have the contract provisions specifying penalties for safety and health issues been enforced, when appropriate? If not, please explain.

E8.   How does the site monitor the quality of the safety and health protection of its contract employees?

E9.   Do contract provisions for contractors require the periodic review and analysis of injury and illness data? Provide examples.

E10.   If the contractors' injury and illness rates are above the average for their industries, describe the site's procedures that ensure that all employees are provided effective protection on the worksite? If yes, please explain.

E11.   Based on your answers to the above items, is the contract oversight minimally effective for the nature of the site? (Inadequate oversight is indicated by significant hazards created by the contractor, employees exposed to hazards, or a lack of host audits.) If not, please explain.

F. Employee Involvement

F1.   How were employees selected to be interviewed by the VPP team?

F2.   How many employees were interviewed formally? How many were interviewed informally?

F3.   Do employees support the site's participation in the VPP?

F4.   Do employees feel free to participate in the safety and health management system without fear of discrimination or reprisal? If so, please explain.

F5.   Are employees meaningfully involved in the problem identification and resolution, or evaluation of the safety and health management system (beyond hazard reporting)? (As per FRN page 936 IV, A.6.) For site-based construction sites, does the company encourage strong labor–management communication in the form of supervisor and employee participation in toolbox safety meetings and training, safety audits, incident investigations, etc.?

F6.   Are employees knowledgeable about the site's safety and health management system? If not, please explain.

F7.   Are employees knowledgeable about the VPP? If not, please explain.

F8.   Are the employees knowledgeable about OSHA rights and responsibilities? If not, please explain

F9.   How were employees informed of the safety and health management system, VPP, and OSHA rights and responsibilities? Please explain.

F10.   Did management verify employee's comprehension of the site's safety and health management system, VPP, and OSHA rights and responsibilities?

F11.   Do employees have access to results of self- inspection, accident investigation, appropriate medical records, and personal sampling data upon request? If not, please explain.

G. Safety and Health Management System Evaluation

G1.   Briefly describe the system in place for conducting an annual evaluation.

G2.   Does the annual evaluation cover the aspects of the safety and health management system, including the elements described in the Federal Register? If not, please explain.

G3.   Does the annual evaluation include written recommendations in a narrative format? If not, please explain.

G4.   Is the annual evaluation an effective tool for assessing the success of the site's safety and health management system? Please explain.

G5.   What evidence demonstrates that the site responded adequately to the recommendations made in the annual evaluation?

G6.   Is the annual evaluation conducted by competent site, corporate, or other trained personnel experienced in performing evaluations?

SECTION II: WORKSITE ANALYSIS

A. Baseline Hazard Analysis

A1.   Has the site been at least minimally effective at identifying and documenting the common safety and health hazards associated with the site (such as those found in OSHA regulations, building standards, etc., and for which existing controls are well known)? If not, please explain.

A2.   What methods are used in the baseline hazard analysis to identify health hazards? (Please include examples of instances when initial screening and full-shift sampling were used. See FRN page 937, B.2.b.)

A3.   Does the company rely on historical data to evaluate health hazards on the worksite? If so, did the company identify any operations that differed significantly from past experience and conduct additional analysis such as sampling or monitoring to ensure employee protection? If so, please describe.

A4.   Does the site have a documented sampling strategy used to identify health hazards and assess employees' exposure (including duration, route, and frequency of exposure), and the number of exposed employees? If not, please explain.

A5.   Do sampling, testing, and analysis follow nationally recognized procedures? If not, please explain.

A6. Does the site compare sampling results to the minimum exposure limits or are more restrictive exposure limits (PELs, TLVs, etc.) used? Please explain.

A7. Does the site identify hazards (including health) that need further analysis? If not, please explain. For site-based construction sites, does the hazard analysis include studies to identify potential employee hazards, phase analyses, task analyses, etc.?

A8. Does industrial hygiene sampling data, such as initial screening or full shift sampling data, indicate that records are being kept in logical order and include all sampling information (for example, sampling time, date, employee, job title, concentrated measures, and calculations)? If not, please explain the deficiencies and how they are being addressed.

A9. For site-based construction sites, are hazard analyses conducted to address safety and health for each phase of work?

B. Hazard Analysis of Significant Changes

B1. When purchasing new materials or equipment, or implementing new processes, what types of analyses are performed to determine impact on safety and health, and are these analyses adequate?

B2. When implementing/introducing non-routine tasks, materials or equipment, or modifying processes, what types of analyses are performed to determine impact on safety and health, and are these analyses adequate?

C. Hazard Analysis of Routine Activities

C1. Is there at least a minimally effective hazard analysis system in place for routine operations and activities?

C2. Does hazard identification and analysis address both safety and health hazards, if appropriate? If not, please explain.

C3. What hazard analysis technique(s) are employed for routine operations and activities (e.g., job hazard analysis, HAZ-OPS, fault trees)? Please explain.

C4. Are the results of the hazard analysis of routine activities adequately documented? If not, please explain.

C5. For site-based construction sites, are hazard analyses conducted to address safety and health hazards for specialty trade contractors during each phase of work?

D. Routine Inspections

D1. Does the site have a minimally effective system for performing safety and health inspections (i.e., a minimally effective system identifies hazards associated with normal operations)? If not, please explain.

D2. Are routine safety and health inspections conducted monthly, with the entire site covered at least quarterly (construction sites: entire site weekly.

D3.   For site-based construction sites, are employees required to conduct inspections as often as necessary, but not less than weekly, of their workplace/area and of equipment?

D4.   Does the site incorporate hazards identified through baseline hazard analysis, accident investigations, annual evaluations, etc., into routine inspections to prevent reoccurrence?

D5.   Are employees conducting inspections adequately trained in hazard identification? If not, please explain.

D6.   Is the routine inspection system written, including documentation of results indicating what needs to be corrected, by whom, and by when? If not, please explain.

D7.   Did the VPP team find hazards that were not found/noted on the site's routine inspections? If so, please explain.

E. Hazard Reporting

E1.   Is there a minimally effective means for employees to report hazards and have them addressed? If not, please explain.

E2.   Does the hazard reporting system have an anonymous component?

E3.   Does the site have a reliable system for employees to notify appropriate management personnel in writing about safety and health concerns? Please describe.

E4.   Do the employees agree that they have an effective system for reporting safety and health concerns? If not, please explain.

F. Hazard Tracking

F1.   Does a minimally effective hazard tracking system exist that result in hazards being controlled? If not, please explain.

F2.   Does the hazard tracking system result in hazards being corrected and provide feedback to employees for hazards they have reported? If not, please explain.

F3.   Does the hazard tracking system result in timely correction of hazards with interim protection established when needed? Please describe.

F4.   Does the hazard tracking system address hazards found by employees, hazard analysis of routine and non-routine activities, inspections, and accident or incident investigations? If not, please explain.

G. Accident/Incident Investigations

G1.   Is there a minimally effective system for conducting accident/incident investigations, including near-misses? If not, please explain.

G2.   Is the accident/incident investigation policy and procedures documented and understood by all? If not, please explain.

G3.   Is there a reporting system for near-misses that include tracking, etc.? If not, please explain.

G4.　Are those conducting the investigations trained in accident/incident investigation techniques? Please explain what techniques are used, e.g., Fault-Tree, Root Cause, etc.

G5.　Describe how investigators discover and document all the contributing factors that led to an accident/incident or a near-miss.

G6.　Were any uncontrolled hazards discovered during the investigation previously addressed in any prior hazard analyses (e.g., baseline, self-inspection)? If yes, please explain.

H. Trend Analysis

H1.　Does the site have a minimally effective means for identifying and assessing trends?

H2.　Have there been any injury and/or illness trends over the last three years? If so, please explain.

H3.　Did the team identify trends that should have been identified by the site? If so, please describe.

H4.　If there have been injury and/or illness trends, what adequate courses of action have been taken? Please explain.

H5.　Does the site assess trends utilizing data from hazard reports and/or accident/incident investigations to determine the potential for injuries and illnesses? If not, please explain.

H6.　Are the results of trend analyses shared with employees and management and utilized to direct resources, prioritize hazard controls and modify goals to address trends? If not, please explain.

SECTION III: HAZARD PREVENTION AND CONTROL

A. Hazard Prevention and Control

A1.　Does the site select at least minimally effective controls to prevent exposing employees to hazards?

A2.　When the site selects hazard controls, does it follow the preferred hierarchy (engineering controls, administrative controls, work practice controls [e.g., lockout/tagout, bloodborne pathogens, and confined space programs], and personal protective equipment) to eliminate or control hazards? Please provide examples, such as how exposures to health hazards were controlled.

A3.　Describe any administrative controls used at the site to limit employee exposure to hazards (for example, job rotation).

A4.　Do the work practice controls and administrative controls adequately address those hazards not covered by engineering controls? If not, please explain.

A5.　Are the work practice controls (e.g., lockout/tagout, bloodborne pathogens, and confined space programs) recommended by hazard analyses and implemented at the site? If not, explain.

A6.　Are follow-up studies (where appropriate) conducted to ensure that hazard controls were adequate? If not, please explain.

A7.   Are hazard controls documented and addressed in appropriate procedures, safety and health rules, inspections, training, etc.? Provide examples.

Disciplinary System

A8.   Are there written employee safety procedures including a disciplinary system? Describe the disciplinary system?

A9.   Has the disciplinary system been clearly communicated and enforced equally for both management and employees, when appropriate? If not, please explain

Emergency Procedures

A10.   Does the site have minimally effective written procedures for emergencies?

A11.   Did the site explain the frequency and types of emergency drills held (including at least an evacuation drill annually)?

A12.   Is the emergency response plan updated as changes occur in the work areas e.g., evacuation routes or auditory systems?

A13.   Did the site describe the system used to verify all employees' participation in at least one evacuation drill each year?

Preventive/Predictive Maintenance

A14.   Does the site have a written preventive/predictive maintenance system? If not, please explain.

A15.   Did the hazard identification and analysis (including manufacturers' recommendations) identify hazards that could result if equipment is not maintained properly? If not, please explain.

A16.   Does the preventive maintenance system detect hazardous failures before they occur? If not, please explain. Is the preventive maintenance system adequate?

Personal Protective Equipment (PPE)

A17.   How does the site select Personal Protective Equipment (PPE)?

A18.   Did the site describe the PPE used at the site?

A19.   Where PPE is required, do employees understand that it is required, why it is required, its limitations, how to use it, and how to maintain it? If not, please explain.

A.20.   Did the team observe employees using, storing, and maintaining PPE properly? If not, explain.

Process Safety Management (PSM)

A21.   Is the site covered by the Process Safety Management standard (29 CFR 1910.119)? If yes, please answer questions A22–A25 below. Additionally, please complete either the onsite evaluation supplement A or B, and the onsite evaluation supplement C. If not, skip to Section B.

A22.   Which chemicals that trigger the Process Safety Management (PSM) standard are present?

A23.   Which process(es) were followed from beginning to end and used to verify answers to the questions asked in the PSM application supplement, the PSM Questionnaire, and/or the Dynamic Inspection Priority Lists?

A24.   Verify that contractor employees who perform maintenance, repair, turn-around, major renovation, or specialty work on or adjacent to a covered process have received adequate training and demonstrate appropriate knowledge of hazards associated with PSM, such as non-routine tasks, process hazards, hot work, emergency evacuation procedures, etc.? Please explain.

A25.   Is the PSM program adequate in that it addresses the elements of the PSM standard and the PSM directive? Please explain.

B. Occupational Health Care Program

B1.   Describe the occupational health care program (including availability of physician services, first aid, and CPR/AED) and special programs such as audiograms or other medical tests used.

B2.   How are licensed occupational health professionals used in the site's hazard identification and analysis, early recognition and treatment of illness and injury, and the system for limiting the severity of harm that might result from workplace illness or injury? Is this use appropriate?

B3.   Is the occupational health program adequate for the size and location of the site, as well as the nature of hazards found here? If not, please explain.

C. Recordkeeping

C1.   Are OSHA required recordkeeping forms being maintained properly in terms of accuracy, form completion, etc.? If not, please explain.

C2.   Is the record keeper knowledgeable of 29 CFR 1904, OSHA's recordkeeping standard?

C3.   What records were reviewed to determine compliance with the recordkeeping standard?

C4.   Do the injury and illness rates accurately reflect work performed by contractors/sub-contractors at the site evaluated? Please explain.

C5.   Was there any evidence of recordable injuries/illnesses not being reported due to management pressure, production concerns, incentive programs, etc.? If yes, please explain.

SECTION IV: SAFETY AND HEALTH TRAINING

A. Safety and Health Training

A1.   What are the safety and health training requirements for managers, supervisors, employees, and contractors? Please explain.

A2.   Is the training delivered by qualified instructors?

A3.   Does the training provided to managers, supervisors, and non-supervisory employees (including contract employees) adequately address safety and health hazards?

A4.   Does the company/site operate an effective safety and health orientation program for all employees including new hires? Please explain.

A5.   How are the safety and health training needs for employees determined? Please explain.

A6.   Does the site provide minimally effective training to educate supervisors and employees (including contract employees) regarding the known hazards of the site and their controls? If not, please explain.

A7.   Are managers, supervisors, and non-supervisory employees (including contract employees) taught the safe work procedures to follow in order to protect themselves from hazards during initial job training and subsequent reinforcement training?

A8.   Who is trained in hazard identification and analysis?

A9.   Is training in hazard identification and analysis adequate for the conditions and hazards of the site? If not, please explain.

A10. Does management have a thorough understanding of the hazards of the site? Provide examples that demonstrate their understanding.

A11. Do managers, supervisors, and non-supervisory employees (including contract employees) and visitors on the site understand what to do in emergency situations? Please explain.

CHAPTER 24

MANAGEMENT REVIEW: SECTION 7.0 OF Z10

Section 7.0 opens with this statement: "This section defines the requirements for the periodic review of the occupational health and safety management system." The importance of the management review requirements in Z10 is inverse to the length of this chapter.

In Chapter 1, *"An Overview of ANSI/AIHA Z10-2005—the American National Standard for Occupational Health and Safety Management Systems"* we stated that Section 3.0, Management Leadership and Employee Participation, is the most important section in Z10. Top management leadership is vital because it sets the organization's safety culture and because continual improvement processes cannot be successful without effective top management direction. To achieve superior results, top management must repeatedly "walk the talk".

We also said that Section 7.0, the management review section, was a close second in importance. Maintaining superior management leadership requires that evaluations be made of the effectiveness of safety processes so that improvements can be made where necessary.

Section 7.1 requires that "the organization shall establish process for top management to review the occupational health and safety management system at least annually and to recommend improvements to ensure its continued suitability, adequacy, and effectiveness." This section lists inputs to the management review process to be considered, among which are:

Advanced Safety Management: Focusing on Z10 and Serious Injury Prevention,
Second Edition. Fred A. Manuele.
© 2014 John Wiley & Sons, Inc. Published 2014 by John Wiley & Sons, Inc.

A. Progress in the reduction of risk

G. The objectives to which objectives have been met

H. The performance of the occupational health and safety management systems relative to expectations, taking into consideration changing circumstances, resource needs, alignment with the business plan, and consistency with the occupational health and safety policy.

At the conclusion of the review (Section 7.2), top management shall determine the:

A. Future direction of the OHSMS based on business strategies and conditions; and

B. Need for changes to the organization's policy, priorities, objectives, resources, or other OHSMS elements.

Action items for improvement are to be drafted as the performance assessment is made and "Results and action items from the management reviews shall be documented, communicated to affected individuals, and tracked to completion."

As shown below, the management review process begins with the "check" step in the Plan-Do-Check Act model and provides input to senior management so that processes put in place previously can be accepted as satisfactory or revised, as in the "act" step.

Plan:	Identify the problem(s) (hazards, risks, management system deficiencies, and opportunities for improvement, as in the planning section, 4.0).
Plan:	Analyze the problem(s).
Plan:	Develop solutions.
Do:	Implement solutions.
Check:	Evaluate the results to determine that:
	1. The problems were resolved, only partially resolved, or not resolved.
	2. The actions taken did or did not create new hazards.
	3. Acceptable risk levels were or were not achieved.
Act:	Accept the results, or take additional corrective action, as needed.

The writers of the Z10 standard may appreciate the recognition given to the significance of the standard, particularly this management review section. On March 23, 2005, a serious workplace disaster occurred at the BP Texas City refinery. It resulted in 15 deaths and more than 170 injuries. A blue ribbon panel, populated mostly by known experts, was created with financial support from the U.S. Chemical Safety Review Board to "make a thorough, independent, and credible assessment of the effectiveness of BP's corporate oversight of safety management systems at its five U.S. refineries and its corporate culture."

The Report of the BP U.S. Refineries Independent Safety Review Panel was issued in January 2007 as a document available to the public. It has become known in occupational safety circles as The Baker Report. The panel's chair was James A. Baker III, who served in senior government positions under three U.S. presidents. Several references are made in the report to sections in Z10 as recommended practices. Those references, and by inference the Z10 standard, are testimony that they represent the state of the art in safety and health management systems.

Much of the following, taken from The Baker Report, relates to the content of Section 3.0 in Z10, "Management Leadership and Employee Participation."

In 2005, the American National Standards Institute (ANSI) approved "a voluntary consensus standard on occupational health and safety management systems." While the standard is oriented toward occupational rather than process safety, the Panel believes that the standard provides a useful tool in analyzing safety management systems generally. The ANSI Z10 standard emphasizes "continual improvement and systematically eliminating the underlying or root causes of deficiencies." The standard indicates that an organization's management should provide leadership and assume overall responsibility for

- implementing, maintaining, and monitoring performance of the safety system;
- providing appropriate financial, human, and organizational resources to plan, implement, operate, check, correct, and review the system;
- defining roles, assigning responsibilities, establishing accountability, and delegating authority to implement an effective system for continual improvement
- integrating the system into the organization's other business systems and processes. (p. 131)

Elsewhere in The Baker Report, observations are made on management reviews that are close to a verbatim copy of what appears in the advisory column for Section 7. Whereas *OHSMS* is the term used in Section 7, it has been replaced by the *safety management system* in the Baker Report.

The related commentary to the ANSI Z10 standard provides a useful description of the role of and purpose for management reviews:

- Management reviews are a critical part of the continual improvement of the [safety management system].
- The purpose of reviews is for top management, with the participation of [safety management system] leaders and process owners, to do a strategic and critical evaluation of the performance of the [safety management system], and to recommend improvements.

- This review is not just a presentation or a non-critical review of the system, but should focus on results and opportunities for continual improvement. It is up to the organization to determine appropriate measures of [safety management system] effectiveness. They should also evaluate how well the [safety management system] is integrated with other business management systems, so it supports both health and safety goals and business needs and strategies.

Reviews by top management are required because they have the authority to make the necessary decisions about actions and resources, although it may also be appropriate to include other employee and management levels in the process. "To be effective, the review process should ensure that the necessary information is available for top management to evaluate the continuing suitability, adequacy, and effectiveness of the [safety management system]....Reviews should present results (for example, a scorecard) to focus top management on the [safety management system] elements most in need [of] their attention."

At the conclusions of the reviews, top management should make decisions, give direction, and commit resources to implement the decisions. The management review should include an assessment of the current [safety management system] to address if the system is encompassing all of the risks to which the organization is exposed. This portion of the review should include a review of major risk exposures and ask the question, "Are there any holes" in the current [safety management system] that could allow a risk that might not be considered within the [safety management system]. (p. 225)

Section 3.0, "Management Leadership and Employee Participation," and Section 7.0, "Management Review," are vital and integrated parts of a whole. An organization cannot achieve superior results if the performance in these two sections is not stellar.

A management review is to result in a documentation of the action items necessary to achieve continual improvement in occupational health and safety management systems, the assignment of responsibility for the actions to be taken, completion dates, and requirements for periodic reporting on progress made. One test of the organization's safety culture is whether resources are made available to achieve the improvements decided upon.

Appendix M is to help managements fulfill the Z10 management review requirements. It consists principally of a scorecard, whose purpose is to focus "top management's attention on the parts of the occupational health and safety management system that need their attention and direction mostly." The scorecard lists the major items in Z10 on one page and provides for entering indicators of the implementation status of each of the provisions. When using the scorecard example given, colors are to be entered to show the following performance levels.

- Blue: world-class OHS performance
- Green: strong; conforming/complete, may have minor gaps with action plans

- Yellow: moderate; scattered nonconformances need to be addressed, positive trends/major elements in place
- Violet: significant nonconformances exist, still needs focus
- Red: major effort required; major systematic nonconformances exist

The foregoing color scheme is recorded here as an example of how performance gradations may be expressed. The writers of Appendix M properly recognized that a variety of evaluation systems may be used—qualitative or quantitative. They also make an important statement when they said that management review reports should suit the organization's "size, operations, services, or culture." A summary report will be accepted more readily if it fits the organization's style and culture.

This management review section gives safety and health professionals a meaningful opportunity to assist in providing objective summary reports on the status of occupational health and safety management systems and to present managements with proposals to overcome shortcomings. Such reports will have greater value if a section addresses serious injury potential and risk reduction measures.

In accord with the PDCA concept, the overriding theme of the management review is to achieve continual improvement. Thus, having action items for improvement in the review process and follow-through are vital.

In many companies, a major management review process is conducted annually and a summary progress report carrying the signature of the chief executive officer is published. Such reports may be made available broadly, such as on the Internet. Publication of the reports serves the purposes of good community relations as well as good employee relations.

REFERENCES

ANSI/AIHA Z10-2012. American National Standard, *Occupational Health and Safety Management Systems*. Fairfax, VA: American Industrial Hygiene Association, 2012. ASSE is now the secretariat. Available at https://www.asse.org/cartpage.php?link=z10_2005.

The Report of the BP U.S. Refineries Independent Safety Review Panel, at http://us.yhs4.search.yahoo.com/yhs/search?p=The+Report+of+the+BP+U.S.+Refineries+Independent+Safety+Review+Panel&hspart=att&hsimp=yhs-att_001&type=att_lego_portal_home, Click on the Acrobat indicator.

CHAPTER 25

COMPARISON: Z10, OTHER SAFETY GUIDELINES AND STANDARDS, AND VPP CERTIFICATION

How Z10 compares to other standards and guidelines on safety and health management systems and whether compliance with Z10 will meet "certification" requirements are questions often asked. Writers of the 2012 version of Z10 responded to the question by providing a matrix in Appendix N, "Management System Standard Comparison."

This appears in the introduction to Appendix N: "The matrix is intended to demonstrate the significant similarity in essential elements of management systems." In four pages of fine print, the matrix compares Z10 with:

- ISO 14001:2004. *Environmental management systems—Requirements with guidance for use*
- OHSAS 18001:2007. *Occupational health and safety management systems— Specifications*
- ILO OHSMS:2001. *Guidelines on Occupational Health and Safety Management Systems*
- VPP (2008). *OSHA's Voluntary Protection Program*
- ISO 9001:2008. *Quality management systems—Requirements*

I attempted to make a similar but detailed comparison and was not satisfied with the accuracy of the results. It was found, as stated in the introduction to Annex N: "Element to element comparison is difficult and check marks simply indicate that the element is present in a standard or document." An element may stand alone in one

Advanced Safety Management: Focusing on Z10 and Serious Injury Prevention, Second Edition. Fred A. Manuele.
© 2014 John Wiley & Sons, Inc. Published 2014 by John Wiley & Sons, Inc.

standard with specifics as to requirements: In another standard, the subject may be encompassed within other elements and not have comparable requirements. Nevertheless, Annex N is an important work and deserves recognition.

In this chapter we:

- Provide my response to the questions on comparison
- Comment on certification
- Discuss the "certification" available through OSHA's Voluntary Protection Program (VPP)
- Explore the similarities and differences in Z10 and VPP
- Encourage organizations to seek VPP certification
- Duplicate the VPP requirements for the Star Designation

COMPARISONS

I made an attempt to compare the provisions in Z10 to the following selected and well-known safety guidelines and standards.

- OHSAS 18001:2007. A guideline issued by the British Standards Institute, commonly known as 18001
- ILO-OSH 2001. An International Labor Organization publication
- BS 8800–2004. *Occupational health and safety management systems—Guide*. A comprehensive British standard issued by the British Standards Institute
- ISO 31000:2009. *Risk Management and Practices*, a standard published by the International Organization for Standardization that has been adopted in the United States as ANSI/ASSE Z690.2-2011.
- *OSHA's Voluntary Protection Program (VPP)*. As described in the *Policies and Procedures Manual*, placed on the Internet in 2003

A MESS OF FALSE POSITIVES

It was found that the comparison exercise, intended to provide detailed and accurate relationships between elements was not fruitful. Subjective judgments were necessary on the intent of the provisions in the standards and guidelines reviewed that used similar words but which were determined not to be substantially the same. I placed the provisions of Z10 in columnar form and observed that attempting to record meaningful conclusions for questionably comparable sections in other standards and guidelines resulted in an inaccurate, bewildering, and cumbersome output.

Safety professionals are obligated to be cognizant of the state of the art as the practice of safety evolves. The source of an innovation in safety and health management is irrelevant. If applying an innovation serves to reduce risk, safety and health professionals are compelled to promote its adoption, regardless of the standard or guideline in which it first appears. In the following comments, broad assumptions are made.

Z10 and 18001 are comparable. Some sections in 18001 could be stated more precisely as separate items.

ILO-OSH 2001 was a foundational guideline when it was published in 2001. Recognition is given to the value of this guideline in the introduction to Z10 as follows: "The standard also draws from the approaches used by the International Labor Organization's (ILO) guidelines on Occupational Health and Safety Management Systems." But provisions in ILO-OSH 2001 are not as extensive as those in Z10.

BS 8800–2004 is a good, well-written, and comprehensive standard. BS 8800 and Z10 are close matches.

The ISO 31000:2009 standard, also known as Z690.2, covers all forms of risk that an organization encounters. The following appears in the introduction.

The generic approach described in this standard provides the principles and guidelines for managing any form of risk in a systematic, transparent and credible manner and within any scope and context.

This standard has provisions for both the "external and the internal" context of an organization's risk management needs. An example of the extensiveness of the standard's external coverage is found in Section 4.3.1, "Understanding of the Organization and Its Context." It reads as follows, partially.

Before starting the design and implementation of the framework for managing risk, it is important to evaluate and understand both the external and internal context of the organization, since these can significantly influence the design of the framework. Evaluating the organization's external context may include, but is not limited to: a) the social and cultural, political, legal, regularity, financial, technological, economic, natural and competitive environment, whether international, national, regional or local.

Obviously, the intent of ISO 31000:2009—Z690.2—is much broader than that of Z10. The internal context is also notably broad. But in so far as the internal context pertains to occupational hazards and the risks deriving from those hazards and their avoidance, elimination, reduction, and control, Z690.2-2011 and Z10 are a good match.

OSHA's Voluntary Protection Program (VPP) and Z10 have remarkable similarities and differences. Since I promote obtaining certification under the VPP as a step toward improving the effectiveness of a safety management system, considerable space in this chapter is given to the VPP.

Major differences relate to elements in a safety management system that are now given greater significance in Z10 because of what has been learned in recent years and elements in the VPP program that are not addressed in Z10. Although there are substantial similarities, only the differences between Z10 and the VPP are highlighted here.

RISK ASSESSMENT

Comments are made in Chapters 11, "A Primer on Hazard Analysis and Risk Assessment" and 12 "Provisions for Risk assessments in Standards and Guidelines" on the significance now given throughout the world to having risk assessment processes as an integral part of a safety and health management system. That is a fairly recent development.

A 1989 OSHA publication entitled "Safety and Health Program Management Guidelines" is based on four basic elements:

- Management leadership and employee involvement
- Worksite analysis
- Hazard prevention and control
- Safety and health training

From the beginning of the VPP in 1982 and through the years following the 1989 publication, those four elements were considered sound and largely sufficient. I distributed a good number of excerpts from the guidelines.

The emphasis in VPP is on hazard identification and analysis. The term *hazard* is not defined. But the wording with respect to hazards in the VPP requirements puts a heavy emphasis on conditions, and it is limiting. The only place in the VPP requirements where risk is mentioned is in the opening paragraph in the section "Worksite Analysis," as in the following.

A hazard identification and analysis system must be implemented to systematically identify basic and unforeseen safety and health hazards, evaluate their risks, and prioritize and recommend methods to eliminate or control hazards to an acceptable level of risk. Through this system, management must gain a thorough knowledge of the safety and health hazards and employee risks.

Z10 now has a specific requirement for Risk Assessment in Section 5.1.3. Hazard identification and analysis are vital processes. The outcome establishes the severity of harm that can result from an incident. But event probability must also be considered to properly assess, prioritize, and act on risks. When making risk assessments, a mindset must be adopted that focuses on both the probability of incident occurrence and the severity of harm or damage that may result.

It is now commonly accepted that safety professionals are obligated to consider both aspects of risk, probability and severity, as they give counsel to prioritize actions to be taken and to attain acceptable risk levels.

Z10 says that risk assessments "shall" be made and gives comments in an advisory column on what is to be considered in the process. Further guidance is given in Appendix F, a six-page dissertation on risk assessment. Z10 makes a great deal of prioritizing risks—an important point in these economic times where risk reduction resources are limited.

In VPP, the section on Tracking of Hazard Correction states: "A documented system must be in place to ensure that hazards identified…are assigned to a responsible party and corrected in a timely fashion. This system must include methods for: Recording and prioritizing hazards; and Assigning responsibilities, time-frames for correction, interim protection, and follow-up to ensure abatement." Note that it is hazards that are to be prioritized. That cannot be done properly if the probability of an incident occurring because a hazard's potential is realized is not considered.

DESIGN REVIEWS AND PROCUREMENT REQUIREMENTS

There are two provisions in Z10 for which only briefly mention is made in the VPP Star requirements. Those provisions serve to avoid bringing hazards into the workplace, a concept that is gaining recognition as an important element in safety management systems. Some companies have applied those provisions with good success. Z10 says that processes "shall" be in place to have:

- Safety design reviews made
- Safety specifications included in purchasing orders, agreements, and contracts

The idea is that if a safety management system includes processes that reduce the probability of bringing new hazards into the workplace, the risks deriving from those hazards are also reduced, and resources do not have to be allocated in retrofitting to achieve acceptable risk levels. Retrofitting can be costly and is often less effective than eliminating or controlling risks in the design process and including safety specifications in procurement documents.

The VPP requirements are not quite as precise. They do say, in "Pre-use analysis," that the safety and health impact of new equipment must be assessed and that the "practice *should* [emphasis added] be integrated in the procurement/design phase to maximize the opportunity for proactive hazard controls."

MANAGEMENT OF CHANGE

I believe that having an effective management of change system is vital for safety management system effectiveness, particularly with respect to serious injury prevention. Since only a few companies have good management of change systems in place, OSHA will serve its purposes well if this provision is given greater emphasis. The VPP provision in "Hazard Analysis of Significant Changes" does say very briefly that hazard analyses must be made of significant changes.

REQUIREMENTS IN VPP NOT ADDRESSED SIMILARLY IN Z10

There are four subjects in the VPP requirements for which there are no comparable sections in Z10. They are on the list of subjects to be reviewed for the next version of Z10.

- Industrial hygiene requirements for the VPP are extensive. "A written IH program is required. The program must establish procedures and methods for identification, analysis, and control of health hazards for prevention of occupational disease." As will be seen in Addendum A to this chapter, VPP requirements for the prevention of occupational disease are broad and detailed. Z10 does not give separate treatment to occupational health exposures. Hazards and risks pertaining to injuries and illnesses are treated as parts of a whole.

- VPP requirements also include an "Occupational Health Care Program" which requires that licensed health care professionals be available, arrangements for needed health care services are made, employees trained in first aid and CPR must be available for all shifts, and emergency procedures and services, including provisions for ambulance and emergency technicians, are available and explained to employees.

- The VPP says, in a section entitled "Certified Professional Resources," that "access to certified safety and health professionals and other licensed health care professionals is required. They may be provided by offsite sources such as corporate headquarters, insurance companies, or private contractors. OSHA will accept certification from any recognized accrediting organization." Although there is no provision in Z10 requiring access to certified professional resources, applying many of the provisions in Z10 will require the counsel of highly qualified professionals.

- Z10 is quiet on Preventive Maintenance of Equipment. VPP states: "A written preventive and predictive maintenance system must be in place for monitoring and maintaining workplace equipment. Equipment must be replaced or repaired on a schedule, following manufacturers' recommendations, to prevent it from failing and creating a hazard. Documented records of maintenance and repairs must be kept. The system must include maintenance of hazard controls such as machine guards, exhaust ventilation, mufflers, etcetera."

ON CERTIFICATION

The British Standards Institute (BSI) is a major player in certification and seems to have made a financial success of it. Both 18001 and BS 8800–2004 are published by BSI. There is a bit of a mystery here. BSI is a standards developer in the UK. Yet it runs a profitable fee-based business providing certification of health and safety management systems based on the 18001 guideline. It should be understood that 18001 is a guideline, not a standard. It is not equivalent to BS 8800, a standard that was updated and reissued in 2004.

Requiring certification by organizations for quality management (ISO 9000 series) and environmental management (ISO 14000 series) is common. Sometimes, particularly in a competitive business situation in Asia, providers of goods or services are required to present documents indicating that their occupational safety and health management systems are also certified. The certifying entities help meet that requirement. Comments made by colleagues reflecting their experiences with certifying agencies are at the extremes:

- "They did a thorough and valuable job. They identified improvements in our safety processes that needed to be made."
- "The certification process was a paper exercise. A review of our safety and health manual was made in our office. It did us no operational good at all. But since we needed the certificate in a competitive situation, we didn't complain."

Nevertheless, it seems that companies that have attained superior safety results more often want their achievements recognized by having some sort of certification. That will be demonstrated in the statistics shown later with respect to the growth in the number of companies attaining VPP recognition. For superior performers in the United States, VPP certification has become a mark of distinction.

To emphasize: The focus of this book is on injury and illness prevention, with an emphasis on avoiding serious injuries and illnesses. That purpose will be moved forward if, as a company seeks certification and meets the certification requirements, its safety and health management system is improved. Such a form of certification is available in the United States through the Occupational Safety and Health Administration, which administers the Voluntary Protection Program.

Payment of a fee for the process to attain VPP status is not required. Organizations that certify for the 18001 guideline are mostly fee-based entities.

THE VPP PROGRAM

On July 2, 1982, OSHA announced the establishment of the Voluntary Protection Program (VPP) to recognize and promote effective work-site-based safety and health management systems. Granting the VPP designation is OSHA's "official recognition of the outstanding efforts of employers and employees who have created exemplary worksite safety and health management systems."

In practice, the VPP sets performance-based criteria for a managed safety and health system, invites sites to apply, and then assesses applicants against these criteria. OSHA's verification includes an application review and a rigorous on-site evaluation by a team of OSHA safety and health experts. VPPs are designed to recognize and promote effective safety and health management. VPP participants are a select group of facilities that have designed and implemented outstanding safety and health programs.

Statistical evidence for VPP's success is impressive. The average VPP work site has a Days Away Restricted or Transferred (DART) case rate 52% below the average for its industry. These sites do not necessarily start out with such low rates. For some entities, reductions in injuries and illnesses begin when the site commits to the VPP approach to safety and health management and the challenging VPP application process.

The VPP concept recognizes that compliance with OSHA standards alone can never fully achieve the objectives of the Occupational Safety and Health Act. Good safety and health management programs that go beyond OSHA standards can protect workers more effectively. Features of the VPP program include the following, among others:

- Management agrees to lead an effective program that meets established criteria.
- Employees agree to participate in the program and work with management to assure a safe and healthful workplace. (*Note*: Unions must provide written assurances.)
- OSHA initially verifies that the program meets the VPP criteria through an on-site review. OSHA then publicly recognizes the site's exemplary program and removes the site from targeted inspection lists.
- Accidents, employee complaints, and chemical spills are handled according to established enforcement procedures and policies.
- OSHA periodically reassesses a site's program to confirm that the site continues to meet VPP criteria.

WHAT ARE THE BENEFITS?

- Fewer injuries. VPP participants generally experience from 60 to 80% fewer lost workday injuries than would be expected of an "average" site of the same size in their industry classifications.
- Improved employee motivation to work safely, leading to better quality and productivity.
- Reduced workers' compensation costs.
- Recognition in the community.
- Improvement of programs that are already good, through the internal and external review that's part of the VPP application process.

INCIDENCE EXPERIENCE REQUIREMENT

To qualify for the Star designation, an entity must show that its three-year illness and injury Total Case Incidence Rate (TCIR) and its three-year Days Away from Work, Restricted Work Activity, and Job Transfer (DART rate) are below the entity's industry average. That suggests exclusivity and deserved recognition for those companies that have superior safety and health management systems and stellar performance.

CERTIFICATION

An approval ceremony follows attaining Star recognition. Usually, an OSHA representative visits the site to recognize its achievements, presents the VPP certificate or plaque, and gives the site a VPP flag. Therefore, the desired certification is achieved.

THE VPP CORPORATE PILOT PROGRAM

In April 2004, OSHA offered corporations that have made a commitment to achieving VPP status a streamlined application and on-site evaluation process called the VPP Corporate Pilot Program. One purpose is to reduce duplication for employers that implement standardized safety management systems throughout their organizations.

Being admitted to the program fosters cooperation with OSHA and forms sort of a partnership, which should result in more locations within a company achieving Star status.

The following is taken verbatim from the OSHA paper on the corporate program at http://www.osha.gov/dcsp/vpp/corporate.html.

VPP corporate is designed for corporate applicants, who demonstrate a strong commitment to employee safety and health and VPP. These applicants, typically large corporations or Federal Agencies, have adopted VPP on a large scale for protecting the safety and health of its employees.

VPP corporate applicants must have established, standardized corporate-level safety and health management systems, effectively implemented organization-wide as well as internal audit/screening processes that evaluate their facilities for safety and health performance. Under VPP corporate, streamlined processes have been established to eliminate these redundancies and expand VPP participation for corporate applicants in a more efficient manner.

INTEREST IN ACHIEVING VPP RECOGNITION

It is common at safety conferences and meetings that safety directors whose companies have obtained VPP recognition to let others know about their achievements. Attaining VPP Star status has become a mark of distinction. It provides entities that have better than average performance with a certification that they are in a superior class.

Many of those companies have certifications for their quality management and environmental management systems, so obtaining certification for their safety and health management systems adds to their prestige.

OSHA says that granting the VPP designation is its "official recognition of the outstanding efforts of employers and employees who have created exemplary work-site safety and health management systems."

The upward trend with respect to attaining that recognition is notable, as shown in the statistics in Table 25.1. The source of these data is slide 9 in a presentation placed on the Internet by OSHA. It is at http://www.osha.gov/dcsp/vpp/charts/slide9.html.

VPP REQUIREMENTS FOR THE STAR DESIGNATION

OSHA's *VPP Policies and Procedures Manual* covers all aspects of the VPPs. Chapter III in the *Manual* sets forth the Requirements for Star, Merit, Resident Contractor, Construction Industry, and Federal Agency Worksites. Safety and health

TABLE 25.1   Growth of Federal and State VPP Recognitions as of April 15, 2012

Year	VPP Sites
1986	46
1996	285
2006	1,665
2008	2,161
2010	2436
2012	2370

management system requirements for Star status are contained in Section C of Chapter III. They are duplicated here in Addendum A in their entirety, to provide a base for review.

CONCLUSION

The focus of this book is on preventing incidents that result in injuries and illnesses, with an emphasis on serious injury and illness avoidance. That purpose will be moved forward if companies analyze their safety and health management systems and develop action plans to achieve VPP recognition. The Z10 standard and its annexes, and the VPP requirements shown in this chapter, serve as excellent resources in making the analysis proposed.

I believe strongly that having safety and health management systems in place that meet Z10 requirements will result in significant reductions in injuries and illnesses. Since Z10 represents best practices for a standard, its existence places special responsibilities on safety professionals to be familiar with and promote adoption of its provisions. Regardless of where an innovation appears in a safety standard or guideline, safety professionals are obligated to be cognizant of the state of the art as the practice of safety evolves and give counsel on best practices and how to apply available resources so as to achieve acceptable risk levels.

REFERENCES

ANSI/AIHA Z10-2012. American National Standard, *Occupational Health and Safety Management Systems*. Fairfax, VA: American Industrial Hygiene Association, 2012. ASSE is now the secretariat. Available at https://www.asse.org/cartpage.php?link=z10_2005.

BS 8800–2004. *Occupational health and safety management systems—Guide*: London: British Standards Institute, 2004.

ILO-OSH 2001. *Guidelines on occupational safety and health management systems*: Geneva, Switzerland: International Labor Organization, 2001.

ISO 14001:2004. *Environmental management systems—Requirements with guidance for use*. Geneva, Switzerland, International Organization for Standardization, 2004.

ISO 31000:2009. *Risk Management Practices*. Geneva, Switzerland, International Organization for Standardization, 2004. Also as ANSI/ASSE Z690.2-2001. Des Plaines, IL: American Society of Safety Engineers, 2011.

ISO 9001:2008. *Quality management systems—Requirements*. Geneva, Switzerland, International Organization for Standardization, 2008.

OSHA's Voluntary Protection Programs (VPP): Policies and Procedures Manual, CSP 03-01-002–TED 8.4. Chapter III, Requirements for Star, Merit, Resident Contractor, Construction Industry, and Federal Agency Worksites at http://www.osha.gov/pls/oshaweb/owadisp.show_document?p_table=DIRECTIVES&p_id=2976.

OSHA: VPP program—Internet entries at http://www.osha.gov/dcsp/vpp/all_about_vpp.html and http://www.osha.gov/dcsp/vpp/vppflyer.html.

OSHA's VPP Corporate Pilot Program. Enter "OSHA Fact Sheet, VPP Corporate Pilot" into a search engine.

OHSAS 18001: 1999. *Occupational health and safety management systems—Specification*: London: British Standards Institute, 1999.

Safety and Health Program Management Guidelines. Occupational Safety and Health Administration, *Federal Register*, Jan. 26, 1989.

U.S. Department of Labor. *All About VPP*. At http://www.osha.gov/dcsp/vpp/all_about_vpp.html.

U.S. Department of Labor. *Voluntary Protection Programs for Federal Worksites*. At http://www.osha.gov/dcsp/vpp/vppflyer.html.

U.S. Department of Labor. Occupational Safety and Health Administration. *We Can Help*. Slide 9 at www.osha.gov/dcsp/vpp/charts/html. Directly to slide 9 at http://www.osha.gov/dcsp/vpp/charts/slide9.html.

ADDENDUM A

This presentation is an adaptation from CSP 03-01-002–TED 8.4 Voluntary Protection Programs (VPP): Policies and Procedures Manual. Chapter III, Requirements for Star, Merit, Resident Contractor, Construction Industry, and Federal Agency Worksites.

C. COMPREHENSIVE SAFETY AND HEALTH MANAGEMENT SYSTEM REQUIREMENTS

The following safety and health management system elements and sub-elements must be implemented [for Star recognition]. For small sites, at the discretion of the onsite team, some of the requirements may be implemented and documented less formally.

1. Management Leadership and Employee Involvement

a. **Management Commitment**   Management demonstrates its commitment by:

- Establishing, documenting, and communicating to employees and contractors clear goals that are attainable and measurable, objectives that are relevant to workplace hazards and trends of injury and illness, and policies and procedures that indicate how to accomplish the objectives and meet the goals.

Advanced Safety Management: Focusing on Z10 and Serious Injury Prevention,
Second Edition. Fred A. Manuele.
© 2014 John Wiley & Sons, Inc. Published 2014 by John Wiley & Sons, Inc.

- Signing a statement of commitment to safety and health.
- Meeting and maintaining VPP requirements.
- Maintaining a written safety and health management system that documents the elements and sub-elements, procedures for implementing the elements, and other safety and health programs, including those required by OSHA standards.
- Identifying persons whose responsibilities for safety and health includes carrying out safety and health goals and objectives, and clearly defining and communicating their responsibilities in their written job descriptions.
- Assigning adequate authority to those persons who are responsible for safety and health, so they are able to carry out their responsibilities.
- Providing and directing adequate resources (including time, funding, training, personnel, etc.) to those responsible for safety and health, so they are able to carry out their responsibilities.
- Holding those assigned responsibility for safety and health accountable for meeting their responsibilities through a documented performance standards and appraisal system.
- Planning for typical as well as unusual/emergency safety and health expenditures in the budget, including funding for prompt correction of uncontrolled hazards.
- Integrating safety and health into other aspects of planning, such as planning for new equipment, processes, buildings, etc.
- Establishing lines of communication with employees and allowing for reasonable employee access to top management at the site.
- Setting an example by following the rules, wearing any required personal protective equipment, reporting hazards, reporting injuries and illnesses, and basically doing anything that they expect employees to do.
- Ensuring that all workers (including contract workers) are provided equal, high-quality safety and health protection.
- Conducting an annual evaluation of the safety and health management system in order to:
  - Maintain knowledge of the hazards of the site.
  - Maintain knowledge of the effectiveness of system elements.
  - Ensure completion of the previous years' recommendations.
  - Modify goals, policies, and procedures.

b. Employee Involvement Employees must be involved in the safety and health management system in at least three meaningful, constructive ways in addition to their right to report a hazard. Avenues for employees to have input into safety and health decisions include participation in audits, accident/incident investigations, self-inspections, suggestion programs, planning, training, job hazard analyses, and appropriate safety and health committees and teams. Employees do not meet this requirement by participating in incentive programs or simply working in a safe manner.

- Employees must be trained for the task(s) they will perform. For example, they must be trained in hazard recognition to participate in self-inspections.
- Employees must receive feedback on any suggestions, ideas, reports of hazards, etc. that they bring to management's attention. A site must provide documented evidence that employees' suggestions were followed up and implemented when appropriate and feasible.
- All employees, including new hires, must be notified about the site's participation in VPP and employees' rights (such as the right to file a complaint) under the OSH Act. Orientation training curriculum must include this information.
- Employees and contractors must demonstrate an understanding of and be able to describe the fundamental principles of VPP.

c. Contract Worker Coverage   Contract workers must be provided with safety and health protection equal in quality to that provided to employees.

- All contractors, whether regularly involved in routine site operations or engaged in temporary projects such as construction or repair, must follow the safety and health rules of the host site.
- VPP participants must have in place a documented oversight and management system covering applicable contractors. Such a system must:
    - Ensure that safety and health considerations are addressed during the process of selecting contractors and when contractors are onsite.
    - Encourage contractors to develop and operate effective safety and health management systems.
    - Include provisions for timely identification, correction, and tracking of uncontrolled hazards in contractor work areas.
    - Include a provision for removing a contractor or contractor's employees from the site for safety or health violations. *Note*: A site may have been operating effectively for 1 year without actually invoking this provision if just cause to remove a contractor or contractor's employee did not occur.
- **Injury and Illness Data Requirements**
    - Nested contractors (such as contracted maintenance workers) and temporary employees who are supervised by host site management are governed by the site's safety and health management system and are therefore included in the host site's rates.
    - Site management must maintain copies of the TCIR (total case incident rate) and DART (three year day away, restricted, and/or transfer case incident) rate data for all applicable contractors based on hours worked at the site. (See Appendix A.)

- Sites must report all applicable contractors' TCIR and DART rate data to OSHA annually.
- **Training**. Managers, supervisors, and non-supervisory employees of contract employers must be made aware of:
  - The hazards they may encounter while on the site.
  - How to recognize hazardous conditions and the signs and symptoms of workplace-related illnesses and injuries.
  - The implemented hazard controls, including safe work procedures.
  - Emergency procedures.

d. Safety and Health Management System Annual Evaluation   There must be a system and written procedures in place to annually evaluate the safety and health management system. The annual evaluation must be a critical review and assessment of the effectiveness of all elements and sub-elements of a comprehensive safety and health management system. An annual evaluation that is merely a workplace inspection with a brief report pointing out hazards or a general statement of the sufficiency of the system is inadequate for purposes of VPP qualification.

- The written annual evaluation must identify the strengths and weaknesses of the safety and health management system and must contain specific recommendations, time lines, and assignment of responsibility for making improvements. It must also document actions taken to satisfy the recommendations.
- The annual evaluation may be conducted by site employees with managers, qualified corporate staff, or outside sources who are trained in conducting such evaluations.
- At least one annual evaluation and demonstrated corrective action must be completed before VPP approval.
- The annual evaluation must be included with the participant's annual submission to OSHA. Appendix D provides a suggested format.

2. Worksite Analysis

A hazard identification and analysis system must be implemented to systematically identify basic and unforeseen safety and health hazards, evaluate their risks, and prioritize and recommend methods to eliminate or control hazards to an acceptable level of risk.

Through this system, management must gain a thorough knowledge of the safety and health hazards and employee risks. The required methods of hazard identification and analysis are described below.

a. Baseline Safety and Industrial Hygiene Hazard Analysis   A baseline survey and analysis is a first attempt at understanding the hazards at a worksite. It establishes initial levels of exposure (baselines) for comparison to future levels, so that changes can be recognized. Systems for identifying safety and industrial hygiene hazards, while often integrated, may be evaluated separately. Baseline surveys must:

- Identify and document common safety hazards associated with the site (such as those found in OSHA regulations or building standards, for which existing controls are well known), and how they are controlled.
- Identify and document common health hazards (usually by initial screening using direct-reading instruments) and determine if further sampling (such as full-shift dosimetry) is needed.
- Identify and document safety and health hazards that need further study.
- Cover the entire work site, indicate who conducted the survey, and when it was completed.

The original baseline hazard analysis need not be repeated subsequently unless warranted by changes in processes, equipment, hazard controls, etc.

b. Hazard Analysis of Routine Jobs, Tasks, and Processes   Task-based or system/process hazard analyses must be performed to identify hazards of routine jobs, tasks, and processes in order to recommend adequate hazard controls. Acceptable techniques include, but are not limited to: Job Hazard Analysis (JHA), and Process Hazard Analysis (PrHA).

- Hazard analyses should be conducted on routine jobs, tasks and processes that:
  - Have written procedures.
  - Have had injuries/illnesses associated with them or have experienced significant incidents or near-misses.
  - Are perceived as high-hazard tasks, i.e., they could result in a catastrophic explosion, electrocution, or chemical over-exposure.
  - Have been recommended by other studies and analyses for more in-depth analysis.
  - Are required by a regulation or standard.
  - Any other instance when the VPP applicant or participant determines that hazard analysis is warranted.

c. Hazard Analysis of Significant Changes   Hazard analysis of significant changes, including but not limited to non-routine tasks (such as those performed less than once a year), new processes, materials, equipment and facilities, must be conducted to identify uncontrolled hazards prior to the activity or use, and must lead to hazard elimination or control.

   If a non-routine or new task is eventually to be done on a routine basis, then a hazard analysis of this routine task should subsequently be developed.

d. Pre-use Analysis   When a site is considering new equipment, chemicals, facilities, or significantly different operations or procedures, the safety and health impact to the employees must be reviewed. The level of detail of the analysis should be commensurate with the perceived risk and number of employees affected.

This practice should be integrated in the procurement/ design phase to maximize the opportunity for proactive hazard controls.

e. Documentation and Use of Hazard Analyses

Hazard analyses performed to meet the requirements of **c.** or **d.** above must be documented and must:

- Consider both health and safety hazards.
- Identify the steps of the task or procedure being analyzed, hazard controls currently in place, recommendations for needed additional or more effective hazard controls, dates conducted, and responsible parties.
- Be used in training in safe job procedures, in modifying workstations, equipment or materials, and in future planning efforts.
- Be easily understood.
- Be updated as the environment, procedures, or equipment change, or errors are found that invalidate the most recent hazard analyses.

f. Routine Self-Inspections

A system is required to ensure routinely scheduled self-inspections of the workplace. It must include written procedures that determine the frequency of inspection and areas covered, those responsible for conducting the inspections, recording of findings, responsibility for abatement, and tracking of identified hazards for timely correction. Findings and corrections must be documented.

- Inspections must be made at least monthly, with the actual inspection schedule being determined by the types and severity of hazards.
- The entire worksite must be covered at least once each quarter.
- Top management and others, including employees who have knowledge of the written procedures and hazard recognition, may participate in the inspection process.
- Personnel qualified to recognize workplace hazards, particularly hazards peculiar to their industry, must conduct inspections.
- Documentation of inspections must evidence thoroughness beyond the perfunctory use of checklists.

g. Hazard Reporting System for Employees

The site must operate a reliable system that enables employees to notify appropriate management personnel in writing—without fear of reprisal—about conditions that appear hazardous, and to receive timely and appropriate responses. The system can be anonymous and must include timely responses to employees and tracking of hazard elimination or control to completion.

h. Industrial Hygiene (IH) Program   A written IH program is required. The program must establish procedures and methods for identification, analysis, and control of health hazards for prevention of occupational disease.

- **IH Surveys**. Additional expertise, time, technical equipment, and analysis beyond the baseline survey may be required to determine which environmental contaminants (whether physical, biological, or chemical) are present in the workplace, and to quantify exposure so that proper controls can be implemented.
- **Sampling Strategy**. The written program must address sampling protocols and methods implemented to accurately assess employees' exposure to health hazards. Sampling should be conducted when:
  - Performing baseline hazard analysis, such as initial screening and grab sampling.
  - Baseline hazard analysis suggests that more in-depth exposure analysis, such as full-shift sampling, is needed.
  - Particularly hazardous substances (as indicated by an OSHA standard, chemical inventory, material safety data sheet, etc.) are being used or could be generated by the work process.
  - Employees have complained of signs of illness.
  - Exposure incidents or near-misses have occurred.
  - It is required by a standard or other legal requirement.
  - Changes have occurred in such things as the processes, equipment, or chemicals used.
  - Controls have been implemented and their effectiveness needs to be determined.
  - Any other instance when the VPP applicant or participant determines that sampling warranted.
- **Sampling Results**. Sampling results must be analyzed and compared to at least OSHA permissible exposure limits (PELs) to determine employees' exposure and possible overexposure. Comparison to more restrictive levels, such as action levels, threshold limit values (TLVs), or self-imposed standards is encouraged to reduce exposures to the lowest feasible level.
  - **Documentation**. The results of sampling must be documented and must include a description of the work process, controls in place, sampling time, exposure calculations, duration, route, and frequency of exposure, and number of exposed employees.
  - **Communication**. Sampling results must be communicated to employees and management.
  - **Use of Results**. Sampling results must be used to identify areas for additional, more in-depth study, to select hazard controls, and to determine if existing controls are adequate.
- **IH Expertise**. IH sampling should be performed by an industrial hygienist, but initial sampling, full-shift sampling, or both may be performed by safety staff

members with special training in the specific procedures for the suspected or identified health hazards in the workplace.

- **Procedures**. Standard, nationally recognized procedures must be used for surveying and sampling as well as for testing and analysis.
- **Use of Contractors**. If an outside contractor conducts industrial hygiene surveys, the contractor's report must include all sampling information listed above and must be effectively communicated to site management. Any recommendations contained in the report should be considered and implemented where appropriate. Use of contractors does not remove responsibility for the IH program, including identification and control of health hazards, from the VPP applicant or participant.

i. Investigation of Accidents and Near-Misses   The site must investigate all accidents and near-misses and must maintain written reports of the investigations. Accident and near-miss investigations must:

- Be conducted by personnel trained in accident investigation techniques. Personnel who were not involved in the accident or who do not supervise the injured employee(s) should conduct the investigation to minimize potential conflicts of interest.
- Document the entire sequence of relevant events.
- Identify all contributing factors, emphasizing failure or lack of hazard controls.
- Determine whether the safety and health management system was effective, and where it was not, provide recommendations to prevent recurrence.
- Not place undue blame or reprisal on employees, although human error can be a contributing factor.
- Assign priority, time frames, and responsibility for implementing recommended controls.
- The results of investigations (to include, at a minimum, a description of the incident and the corrections made to avoid recurrence) must be made available to employees on request, although the actual investigation records need not be provided.

j. Trend Analysis   The process must include analysis of information such as injury/illness history, hazards identified during inspections, employee reports of hazards, and accident and near-miss investigations for the purpose of detecting trends. The results of trend analysis must be shared with employees and management and utilized to direct resources; prioritize hazard controls; and determine or modify goals, objectives, and training to address the trends.

3. Hazard Prevention and Control

Management must ensure the effective implementation of systems for hazard prevention and control and ensure that necessary resources are available, including the following:

a. Certified Professional Resources Access to certified safety and health professionals and other licensed health care professionals is required. They may be provided by offsite sources such as corporate headquarters, insurance companies, or private contractors. OSHA will accept certification from any recognized accrediting organization.

b. Hazard Elimination and Control Methods The types of hazards employees are exposed to, the severity of the hazards, and the risk the hazards pose to employees should all be considered in determining methods of hazard prevention, elimination, and control. In general, the following hierarchy should be followed in determining hazard elimination and control methods.

When engineering controls have been studied, investigated, and implemented, yet still do not bring employees' exposure levels to below OSHA permissible exposure limits; or when engineering controls are determined to be infeasible, then a combination of controls may be used. Whichever controls a site chooses to employ, the controls must be understood and followed by all affected parties; appropriate to the site's hazards; equitably enforced through the disciplinary system; written, implemented, and updated by management as needed; used by employees; and incorporated in training, positive reinforcement, and correction programs.

- **Engineering**. Engineering controls directly eliminate a hazard by such means as substituting a less hazardous substance, by isolating the hazard, or by ventilating the workspace. These are the most reliable and effective controls.
- **Protective Safety Devices**. Although not as reliable as true engineering controls, such methods include interlocks, redundancy, failsafe design, system protection, fire suppression, and warning and caution notes.
- **Administrative**. Administrative controls significantly limit daily exposure to hazards by control or manipulation of the work schedule or work habits. Job rotation is a type of administrative control.
- **Work Practices**. These controls include workplace rules, safe and healthful work practices, personal hygiene, housekeeping and maintenance, and procedures for specific operations.
- **Personal Protective Equipment** (PPE). PPE to be used are determined by hazards identified in hazard analysis. PPE should only be used when all other hazard controls have been exhausted or more significant hazard controls are not feasible.

c. Hazard Control Programs Applicants and participants must be in compliance with any hazard control program required by an OSHA standard, such as PPE, Respiratory Protection, Lockout/Tagout, Confined Space Entry, Process Safety Management, or Bloodborne Pathogens. VPP applicants and participants must periodically review these programs (most OSHA standards require an annual review) to ensure they are up to date.

d. Occupational Health Care Program

- Licensed health care professionals must be available to assess employee health status for prevention, early recognition, and treatment of illness and injury.
- Arrangements for needed health services such as pre-placement physicals, audiograms, and lung function tests must be included.
- Employees trained in first aid, CPR providers, physician care, and emergency medical care must be available for all shifts within a reasonable time and distance. The applicant or participant may consider, based on site conditions, providing Automated External Defibrillators (AEDs) and training in their use.
- Emergency procedures and services including provisions for ambulances, emergency medical technicians, emergency clinics or hospital emergency rooms should be available and explained to employees on all shifts. Also see paragraph h below.

e. Preventive Maintenance of Equipment

A written preventive and predictive maintenance system must be in place for monitoring and maintaining workplace equipment. Equipment must be replaced or repaired on a schedule, following manufacturers' recommendations, to prevent it from failing and creating a hazard. Documented records of maintenance and repairs must be kept. The system must include maintenance of hazard controls such as machine guards, exhaust ventilation, mufflers, etc.

f. Tracking of Hazard Correction

A documented system must be in place to ensure that hazards identified by any means (self-inspections, accident investigations, employee hazard reports, preventive maintenance, injury/illness trends, etc.) are assigned to a responsible party and corrected in a timely fashion. This system must include methods for:

- Recording and prioritizing hazards, and
- Assigning responsibility, time-frames for correction, interim protection, and follow-up to ensure abatement.

g. Disciplinary System

A documented disciplinary system must be in place. The system must include enforcement of appropriate action for violations of the safety and health policies, procedures, and rules. The disciplinary policy must be clearly communicated and equitably enforced to employees and management. The disciplinary system for safety and health can be a sub-part of an all-encompassing disciplinary system.

h. Emergency Preparedness and Response

Written procedures for response to all types of emergencies (fire, chemical spill, accident, terrorist threat, natural disaster, etc.) on all shifts must be established, must follow OSHA standards, must be communicated to all employees, and must be practiced at least annually. These procedures must list requirements or provisions for:

- Assessment of the emergency.
- Assignment of responsibilities (such as incident commander).
- First aid.
- Medical care.
- Routine and emergency exits.
- Emergency telephone numbers.
- Emergency meeting places.
- Training drills, minimally including annual evacuation drills. Drills must be conducted at times appropriate to the performance of work so as not to create additional hazards. Coverage of critical operations must be provided so that all employees have an opportunity to participate in evacuation drills.
- Documentation and critique of evacuation drills and recommendations for improvement.
- Personal protective equipment where needed.

4.   Safety and Health Training

a. Training must be provided so that managers, supervisors, non-supervisory employees, and contractors are knowledgeable of the hazards in the workplace, how to recognize hazardous conditions, signs and symptoms of workplace-related illnesses, and safe work procedures.

b. Training required by OSHA standards must be provided in accordance with the particular standard.

c. Managers and supervisors must understand their safety and health responsibilities and how to carry them out effectively.

d. New employee orientation/training must include, at a minimum, discussion of hazards at the site, protective measures, emergency evacuation, employee rights under the OSH Act, and VPP.

e. Training should be provided for all employees regarding their responsibilities for each type of emergency. Managers, supervisors, and non-supervisory employees, including contractors and visitors, must understand what to do in emergency situations.

f. Persons responsible for conducting hazard analysis, including self-inspections, accident/incident investigations, job hazard analysis, etc., must receive training to carry out these responsibilities, e.g., hazard recognition training, accident investigation techniques, etc.

g. Training attendance must be documented. Training frequency must meet OSHA standards, or for non-OSHA required training, be provided at adequate intervals. Additional training must be provided when s in work processes, new equipment, new procedures, etc. occur.

h. Training curricula must be up-to-date, specific to worksite operations, and modified when needed to reflect changes and/or new workplace procedures, trends, hazards and controls identified by hazard analysis. Training curricula must be understandable for all employees.

i. Persons who have specific knowledge or expertise in the subject area must conduct training.

j. Where personal protective equipment (PPE) is required, employees must understand that it is required, why it is required, its limitations, how to use it, and maintenance.

INDEX

Advanced Safety Management: Focusing on Z10 and Serious Injury Prevention,
Second Edition. Fred A. Manuele.
© 2014 John Wiley & Sons, Inc. Published 2014 by John Wiley & Sons, Inc.